VECTOR IDENTITIES

$\mathbf{u} = u_1\mathbf{i} + u_2\mathbf{j} + u_3\mathbf{k}$ then   (dot product) $\mathbf{u} \bullet \mathbf{v} = u_1v_1 + u_2v_2 + u_3v_3$

$\mathbf{v} = v_1\mathbf{i} + v_2\mathbf{j} + v_3\mathbf{k}$

$\mathbf{w} = w_1\mathbf{i} + w_2\mathbf{j} + w_3\mathbf{k}$ (cross product) $\mathbf{u} \times \mathbf{v} = \begin{vmatrix} \mathbf{i} & \mathbf{j} & \mathbf{k} \\ u_1 & u_2 & u_3 \\ v_1 & v_2 & v_3 \end{vmatrix} = (u_2v_3 - u_3v_2)\mathbf{i} + (u_3v_1 - \cdots$

length of $\mathbf{u} = |\mathbf{u}| = \sqrt{\mathbf{u} \bullet \mathbf{u}} = \sqrt{u_1^2 + u_2^2 + u_3^2}$ angle between $\mathbf{u}$ and $\mathbf{v} = \cos^{-1}\left(\dfrac{\cdots}{|\mathbf{u}||\mathbf{v}|}\right)$

triple product identities $\mathbf{u} \bullet (\mathbf{v} \times \mathbf{w}) = \mathbf{v} \bullet (\mathbf{w} \times \mathbf{u}) = \mathbf{w} \bullet (\mathbf{u} \times \mathbf{v})$ $\mathbf{u} \times (\mathbf{v} \times \mathbf{w}) = (\mathbf{u} \bullet \mathbf{w})\mathbf{v} - (\mathbf{u} \bullet \mathbf{v})\mathbf{w}$

IDENTITIES INVOLVING GRADIENT, DIVERGENCE, CURL, AND LAPLACIAN

$\nabla = \mathbf{i}\dfrac{\partial}{\partial x} + \mathbf{j}\dfrac{\partial}{\partial y} + \mathbf{k}\dfrac{\partial}{\partial z}$ ("del" or "nabla" operator)

$\nabla\phi(x, y, z) = \mathbf{grad}\,\phi(x, y, z) = \dfrac{\partial\phi}{\partial x}\mathbf{i} + \dfrac{\partial\phi}{\partial y}\mathbf{j} + \dfrac{\partial\phi}{\partial z}\mathbf{k}$

$\mathbf{F}(x, y, z) = F_1(x, y, z)\mathbf{i} + F_2(x, y, z)\mathbf{j} + F_3(x, y, z)\mathbf{k}$

$\nabla \bullet \mathbf{F}(x, y, z) = div\,\mathbf{F}(x, y, z) = \dfrac{\partial F_1}{\partial x} + \dfrac{\partial F_2}{\partial y} + \dfrac{\partial F_3}{\partial z}$

$\nabla \times \mathbf{F}(x, y, z) = curl\,\mathbf{F}(x, y, z) = \begin{vmatrix} \mathbf{i} & \mathbf{j} & \mathbf{k} \\ \dfrac{\partial}{\partial x} & \dfrac{\partial}{\partial y} & \dfrac{\partial}{\partial z} \\ F_1 & F_2 & F_3 \end{vmatrix}$

$= \left(\dfrac{\partial F_3}{\partial y} - \dfrac{\partial F_2}{\partial z}\right)\mathbf{i} + \left(\dfrac{\partial F_1}{\partial z} - \dfrac{\partial F_3}{\partial x}\right)\mathbf{j} + \left(\dfrac{\partial F_2}{\partial x} - \dfrac{\partial F_1}{\partial y}\right)\mathbf{k}$

$\nabla(\phi\psi) = \phi\nabla\psi + \psi\nabla\phi$

$\nabla \bullet (\phi\mathbf{F}) = (\nabla\phi) \bullet \mathbf{F} + \phi(\nabla \bullet \mathbf{F})$

$\nabla \times (\phi\mathbf{F}) = (\nabla\phi) \times \mathbf{F} + \phi(\nabla \times \mathbf{F})$

$\nabla \times (\nabla\phi) = \mathbf{0}$ (curl grad $= 0$)

$\nabla^2\phi(x, y, z) = \nabla \bullet \nabla\phi(x, y, z) = \mathbf{div\,grad}\,\phi = \dfrac{\partial^2\phi}{\partial x^2} + \dfrac{\partial^2\phi}{\partial y^2} + \dfrac{\partial^2\phi}{\partial z^2}$

$\nabla \bullet (\mathbf{F} \times \mathbf{G}) = (\nabla \times \mathbf{F}) \bullet \mathbf{G} - \mathbf{F} \bullet (\nabla \times \mathbf{G})$

$\nabla \times (\mathbf{F} \times \mathbf{G}) = \mathbf{F}(\nabla \bullet \mathbf{G}) - \mathbf{G}(\nabla \bullet \mathbf{F}) - (\mathbf{F} \bullet \nabla)\mathbf{G} + (\mathbf{G} \bullet \nabla)\mathbf{F}$

$\nabla(\mathbf{F} \bullet \mathbf{G}) = \mathbf{F} \times (\nabla \times \mathbf{G}) + \mathbf{G} \times (\nabla \times \mathbf{F}) + (\mathbf{F} \bullet \nabla)\mathbf{G} + (\mathbf{G} \bullet \nabla)\mathbf{F}$

$\nabla \bullet (\nabla \times \mathbf{F}) = 0$ (div curl $= 0$)

$\nabla \times (\nabla \times \mathbf{F}) = \nabla(\nabla \bullet \mathbf{F}) - \nabla^2\mathbf{F}$ (curl curl $=$ grad div $-$ laplacian)

VERSIONS OF THE FUNDAMENTAL THEOREM OF CALCULUS

$\displaystyle\int_a^b f'(t)\,dt = f(b) - f(a)$ (the one-dimensional **Fundamental Theorem**)

$\displaystyle\int_C \mathbf{grad}\,\phi \bullet d\mathbf{r} = \phi\big(\mathbf{r}(b)\big) - \phi\big(\mathbf{r}(a)\big)$ if C is the curve $\mathbf{r} = \mathbf{r}(t)$, $(a \le t \le b)$.

$\displaystyle\iint_R \left(\dfrac{\partial F_2}{\partial x} - \dfrac{\partial F_1}{\partial y}\right) dA = \oint_C \mathbf{F} \bullet d\mathbf{r} = \oint_C F_1(x, y)\,dx + F_2(x, y)\,dy$ where C is the positively oriented boundary of R (**Green's Theorem**)

$\displaystyle\iint_S \mathbf{curl}\,\mathbf{F} \bullet \hat{\mathbf{N}}\,dS = \oint_C \mathbf{F} \bullet d\mathbf{r} = \oint_C F_1(x, y, z)\,dx + F_2(x, y, z)\,dy + F_3(x, y, z)\,dz$ where C is the oriented boundary of S.   (**Stokes's Theorem**)

three-dimensional versions: S is the closed boundary of D, with outward normal $\hat{\mathbf{N}}$

$\displaystyle\iiint_D \mathbf{div}\,\mathbf{F}\,dV = \oiint_S \mathbf{F} \bullet \hat{\mathbf{N}}\,dS$ **Divergence Theorem** $\displaystyle\iiint_D \mathbf{curl}\,\mathbf{F}\,dV = -\oiint_S \mathbf{F} \times \hat{\mathbf{N}}\,dS$

$\displaystyle\iiint_D \mathbf{grad}\,\phi\,dV = \oiint_S \phi\hat{\mathbf{N}}\,dS$

FORMULAS RELATING TO CURVES IN 3-SPACE

curve: $\mathbf{r} = \mathbf{r}(t) = x(t)\mathbf{i} + y(t)\mathbf{j} + z(t)\mathbf{k}$ Velocity: $\mathbf{v} = \dfrac{d\mathbf{r}}{dt} = v\hat{\mathbf{T}}$ Speed: $v = |\mathbf{v}| = \dfrac{ds}{dt}$

arc length: $s = \displaystyle\int_{t_0}^t v\,dt$ Acceleration: $\mathbf{a} = \dfrac{d\mathbf{v}}{dt} = \dfrac{d^2\mathbf{r}}{dt^2}$ Tangential and normal components: $\mathbf{a} = \dfrac{dv}{dt}\hat{\mathbf{T}} + v^2\kappa\hat{\mathbf{N}}$

unit tangent: $\hat{\mathbf{T}} = \dfrac{\mathbf{v}}{v}$ Binormal: $\hat{\mathbf{B}} = \dfrac{\mathbf{v} \times \mathbf{a}}{|\mathbf{v} \times \mathbf{a}|}$ Normal: $\hat{\mathbf{N}} = \hat{\mathbf{B}} \times \hat{\mathbf{T}} = \dfrac{d\hat{\mathbf{T}}/dt}{|d\hat{\mathbf{T}}/dt|}$

curvature: $\kappa = \dfrac{|\mathbf{v} \times \mathbf{a}|}{v^3}$ Radius of curvature: $\rho = \dfrac{1}{\kappa}$ Torsion: $\tau = \dfrac{(\mathbf{v} \times \mathbf{a}) \bullet (d\mathbf{a}/dt)}{|\mathbf{v} \times \mathbf{a}|^2}$

the Frenet-Serret formulas: $\dfrac{d\hat{\mathbf{T}}}{ds} = \kappa\hat{\mathbf{N}}$, $\dfrac{d\hat{\mathbf{N}}}{ds} = -\kappa\hat{\mathbf{T}} + \tau\hat{\mathbf{B}}$, $\dfrac{d\hat{\mathbf{B}}}{ds} = -\tau\hat{\mathbf{N}}$

ORTHOGONAL CURVILINEAR COORDINATES

transformation: $x = x(u, v, w)$, $\quad y = y(u, v, w)$, $\quad z = z(u, v, w)$

scale factors: $h_u = \left|\dfrac{\partial \mathbf{r}}{\partial u}\right|$, $\quad h_v = \left|\dfrac{\partial \mathbf{r}}{\partial v}\right|$, $\quad h_w = \left|\dfrac{\partial \mathbf{r}}{\partial w}\right|$

volume element: $dV = h_u h_v h_w \, du \, dv \, dw$

scalar field: $f(u, v, w)$

gradient: $\nabla f = \dfrac{1}{h_u}\dfrac{\partial f}{\partial u}\hat{\mathbf{u}} + \dfrac{1}{h_v}\dfrac{\partial f}{\partial v}\hat{\mathbf{v}} + \dfrac{1}{h_w}\dfrac{\partial f}{\partial w}\hat{\mathbf{w}}$

$$\nabla^2 f = \dfrac{1}{h_u h_v h_w}\left[\dfrac{\partial}{\partial u}\left(\dfrac{h_v h_w}{h_u}\dfrac{\partial f}{\partial u}\right) + \dfrac{\partial}{\partial v}\left(\dfrac{h_u h_w}{h_v}\dfrac{\partial f}{\partial v}\right) + \dfrac{\partial}{\partial w}\left(\dfrac{h_u h_v}{h_w}\dfrac{\partial f}{\partial w}\right)\right]$$

position vector: $\mathbf{r} = x(u, v, w)\mathbf{i} + y(u, v, w)\mathbf{j} + z(u, v, w)\mathbf{k}$

local basis: $\hat{\mathbf{u}} = \dfrac{1}{h_u}\dfrac{\partial \mathbf{r}}{\partial u}$, $\quad \hat{\mathbf{v}} = \dfrac{1}{h_v}\dfrac{\partial \mathbf{r}}{\partial v}$, $\quad \hat{\mathbf{w}} = \dfrac{1}{h_w}\dfrac{\partial \mathbf{r}}{\partial w}$

vector field: $\mathbf{F}(u, v, w) = F_u(u, v, w)\hat{\mathbf{u}} + F_v(u, v, w)\hat{\mathbf{v}} + F_w(u, v, w)\hat{\mathbf{w}}$

divergence: $\nabla\bullet\mathbf{F} = \dfrac{1}{h_u h_v h_w}\left[\dfrac{\partial}{\partial u}\left(h_v h_w F_u\right) + \dfrac{\partial}{\partial v}\left(h_u h_w F_v\right) + \dfrac{\partial}{\partial w}\left(h_u h_v F_w\right)\right]$

curl: $\nabla \times \mathbf{F} = \dfrac{1}{h_u h_v h_w}\begin{vmatrix} h_u\hat{\mathbf{u}} & h_v\hat{\mathbf{v}} & h_w\hat{\mathbf{w}} \\ \dfrac{\partial}{\partial u} & \dfrac{\partial}{\partial v} & \dfrac{\partial}{\partial w} \\ F_u h_u & F_v h_v & F_w h_w \end{vmatrix}$

PLANE POLAR COORDINATES

transformation: $x = r\cos\theta$, $\quad y = r\sin\theta$

scale factors: $h_r = \left|\dfrac{\partial \mathbf{r}}{\partial r}\right| = 1$, $\quad h_\theta = \left|\dfrac{\partial \mathbf{r}}{\partial \theta}\right| = r$

area element: $dA = r\, dr\, d\theta$

scalar field: $f(r, \theta)$

gradient: $\nabla f = \dfrac{\partial f}{\partial r}\hat{\mathbf{r}} + \dfrac{1}{r}\dfrac{\partial f}{\partial \theta}\hat{\boldsymbol{\theta}}$

laplacian: $\nabla^2 f = \dfrac{\partial^2 f}{\partial r^2} + \dfrac{1}{r}\dfrac{\partial f}{\partial r} + \dfrac{1}{r^2}\dfrac{\partial^2 f}{\partial \theta^2}$

position vector: $\mathbf{r} = r\cos\theta\, \mathbf{i} + r\sin\theta\, \mathbf{j}$

local basis: $\hat{\mathbf{r}} = \cos\theta\mathbf{i} + \sin\theta\mathbf{j}$, $\quad \hat{\boldsymbol{\theta}} = -\sin\theta\mathbf{i} + \cos\theta\mathbf{j}$

vector field: $\mathbf{F}(r, \theta) = F_r(r, \theta)\hat{\mathbf{r}} + F_\theta(r, \theta)\hat{\boldsymbol{\theta}}$

divergence: $\nabla \bullet \mathbf{F} = \dfrac{\partial F_r}{\partial r} + \dfrac{1}{r}F_r + \dfrac{1}{r}\dfrac{\partial F_\theta}{\partial \theta}$

curl: $\nabla \times \mathbf{F} = \left[\dfrac{\partial F_\theta}{\partial r} + \dfrac{F_\theta}{r} - \dfrac{1}{r}\dfrac{\partial F_r}{\partial \theta}\right]\mathbf{k}$

CYLINDRICAL COORDINATES

transformation: $x = r\cos\theta$, $\quad y = r\sin\theta$, $\quad z = z$

scale factors: $h_r = \left|\dfrac{\partial \mathbf{r}}{\partial r}\right| = 1$, $\quad h_\theta = \left|\dfrac{\partial \mathbf{r}}{\partial \theta}\right| = r$, $\quad h_z = \left|\dfrac{\partial \mathbf{r}}{\partial z}\right| = 1$

volume element: $dV = r\, dr\, d\theta\, dz$

scalar field: $f(r, \theta, z)$

gradient: $\nabla f = \dfrac{\partial f}{\partial r}\hat{\mathbf{r}} + \dfrac{1}{r}\dfrac{\partial f}{\partial \theta}\hat{\boldsymbol{\theta}} + \dfrac{\partial f}{\partial z}\mathbf{k}$

laplacian: $\nabla^2 f = \dfrac{\partial^2 f}{\partial r^2} + \dfrac{1}{r}\dfrac{\partial f}{\partial r} + \dfrac{1}{r^2}\dfrac{\partial^2 f}{\partial \theta^2} + \dfrac{\partial^2 f}{\partial z^2}$

position vector: $\mathbf{r} = r\cos\theta\, \mathbf{i} + r\sin\theta\, \mathbf{j} + z\mathbf{k}$

local basis: $\hat{\mathbf{r}} = \cos\theta\mathbf{i} + \sin\theta\mathbf{j}$, $\quad \hat{\boldsymbol{\theta}} = -\sin\theta\mathbf{i} + \cos\theta\mathbf{j}$, $\quad \hat{\mathbf{z}} = \mathbf{k}$

surface area element (on $r = a$): $dS = a\, d\theta\, dz$

vector field: $\mathbf{F}(r, \theta, z) = F_r(r, \theta, z)\hat{\mathbf{r}} + F_\theta(r, \theta, z)\hat{\boldsymbol{\theta}} + F_z(r, \theta, z)\mathbf{k}$

divergence: $\nabla \bullet \mathbf{F} = \dfrac{\partial F_r}{\partial r} + \dfrac{1}{r}F_r + \dfrac{1}{r}\dfrac{\partial F_\theta}{\partial \theta} + \dfrac{\partial F_z}{\partial z}$

curl: $\nabla \times \mathbf{F} = \dfrac{1}{r}\begin{vmatrix} \hat{\mathbf{r}} & r\hat{\boldsymbol{\theta}} & \mathbf{k} \\ \dfrac{\partial}{\partial r} & \dfrac{\partial}{\partial \theta} & \dfrac{\partial}{\partial z} \\ F_r & rF_\theta & F_z \end{vmatrix}$

SPHERICAL COORDINATES

transformation: $x = R\sin\phi\cos\theta$, $\quad y = R\sin\phi\sin\theta$, $\quad z = R\cos\phi$

scale factors: $h_R = \left|\dfrac{\partial \mathbf{r}}{\partial R}\right| = 1$, $\quad h_\phi = \left|\dfrac{\partial \mathbf{r}}{\partial \phi}\right| = R$, $\quad h_\theta = \left|\dfrac{\partial \mathbf{r}}{\partial \theta}\right| = R\sin\phi$

local basis: $\hat{\mathbf{R}} = \sin\phi\cos\theta\, \mathbf{i} + \sin\phi\sin\theta\, \mathbf{j} + \cos\phi\, \mathbf{k}$, $\quad \hat{\boldsymbol{\phi}} = \cos\phi\cos\theta\, \mathbf{i} + \cos\phi\sin\theta\, \mathbf{j} - \sin\phi\, \mathbf{k}$, $\quad \hat{\boldsymbol{\theta}} = -\sin\theta\, \mathbf{i} + \cos\theta\, \mathbf{j}$

volume element: $dV = R^2\sin\phi\, dR\, d\phi\, d\theta$

scalar field: $f(R, \phi, \theta)$

gradient: $\nabla f = \dfrac{\partial f}{\partial R}\hat{\mathbf{R}} + \dfrac{1}{R}\dfrac{\partial f}{\partial \phi}\hat{\boldsymbol{\phi}} + \dfrac{1}{R\sin\phi}\dfrac{\partial f}{\partial \theta}\hat{\boldsymbol{\theta}}$

laplacian: $\nabla^2 f = \dfrac{\partial^2 f}{\partial R^2} + \dfrac{2}{R}\dfrac{\partial f}{\partial R} + \dfrac{1}{R^2}\dfrac{\partial^2 f}{\partial \phi^2} + \dfrac{\cot\phi}{R^2}\dfrac{\partial f}{\partial \phi} + \dfrac{1}{R^2\sin^2\phi}\dfrac{\partial^2 f}{\partial \theta^2}$

position vector: $\mathbf{r} = R\sin\phi\cos\theta\, \mathbf{i} + R\sin\phi\sin\theta\, \mathbf{j} + R\cos\phi\mathbf{k}$

surface area element (on $R = a$): $dS = a^2\sin\phi\, d\theta\, d\phi$

vector field: $\mathbf{F}(R, \phi, \theta) = F_R(R, \phi, \theta)\hat{\mathbf{R}} + F_\phi(R, \phi, \theta)\hat{\boldsymbol{\phi}} + F_\theta(R, \phi, \theta)\hat{\boldsymbol{\theta}}$

divergence: $\nabla \bullet \mathbf{F} = \dfrac{\partial F_R}{\partial R} + \dfrac{2}{R}F_R + \dfrac{1}{R}\dfrac{\partial F_\phi}{\partial \phi} + \dfrac{\cot\phi}{R}F_\phi + \dfrac{1}{R\sin\phi}\dfrac{\partial F_\theta}{\partial \theta}$

curl: $\nabla \times \mathbf{F} = \dfrac{1}{R^2\sin\phi}\begin{vmatrix} \hat{\mathbf{R}} & R\hat{\boldsymbol{\phi}} & R\sin\phi\,\hat{\boldsymbol{\theta}} \\ \dfrac{\partial}{\partial R} & \dfrac{\partial}{\partial \phi} & \dfrac{\partial}{\partial \theta} \\ F_R & RF_\phi & R\sin\phi\, F_\theta \end{vmatrix}$

Calculus

A Complete Course

NINTH EDITION

ROBERT A. ADAMS

University of British Columbia

CHRISTOPHER ESSEX

University of Western Ontario

Calculus

A Complete Course

NINTH EDITION

Editorial Director: Claudine O'Donnell

Acquisitions Editor: Claudine O'Donnell

Marketing Manager: Euan White

Program Manager: Kamilah Reid-Burrell

Project Manager: Susan Johnson

Production Editor: Leanne Rancourt

Manager of Content Development: Suzanne Schaan

Developmental Editor: Charlotte Morrison-Reed

Media Editor: Charlotte Morrison-Reed

Media Developer: Kelli Cadet

Compositor: Robert Adams

Preflight Services: Cenveo® Publisher Services

Permissions Project Manager: Joanne Tang

Interior Designer: Anthony Leung

Cover Designer: Anthony Leung

Cover Image: © Hiroshi Watanabe / Getty Images

Vice-President, Cross Media and Publishing Services: Gary Bennett

Pearson Canada Inc., 26 Prince Andrew Place, Don Mills, Ontario M3C 2T8.

ISBN 978-0-13-415436-7

1   17

Library and Archives Canada Cataloguing in Publication

Adams, Robert A. (Robert Alexander), 1940-, author
    Calculus : a complete course / Robert A. Adams, Christopher Essex. -- Ninth edition.

Includes index.
ISBN 978-0-13-415436-7 (hardback)

    1. Calculus--Textbooks. I. Essex, Christopher, author II. Title.

QA303.2.A33 2017   515   C2016-904267-7

To Noreen and Sheran

Contents

Appendices — A-1

Preface

A fashionable curriculum proposition is that students should be given what they need and no more. It often comes bundled with language like "efficient" and "lean." Followers are quick to enumerate a number of topics they learned as students, which remained unused in their subsequent lives. What could they have accomplished, they muse, if they could have back the time lost studying such retrospectively unused topics? But many go further—they conflate unused with useless and then advocate that students should therefore have lean and efficient curricula, teaching only what students need. It has a convincing ring to it. Who wants to spend time on courses in "useless studies?"

When confronted with this compelling position, an even more compelling reply is to look the protagonist in the eye and ask, "How do you know what students need?" That's the trick, isn't it? If you could answer questions like that, you could become rich by making only those lean and efficient investments and bets that make money. It's more than that though. Knowledge of the fundamentals, unlike old lottery tickets, retains value. Few forms of human knowledge can beat mathematics in terms of enduring value and raw utility. Mathematics learned that you have not yet used retains value into an uncertain future.

It is thus ironic that the mathematics curriculum is one of the first topics that terms like *lean* and *efficient* get applied to. While there is much to discuss about this paradox, it is safe to say that it has little to do with what students actually need. If anything, people need more mathematics than ever as the arcane abstractions of yesteryear become the consumer products of today. Can one understand how web search engines work without knowing what an eigenvector is? Can one understand how banks try to keep your accounts safe on the web without understanding polynomials, or grasping how GPS works without understanding differentials?

All of this knowledge, seemingly remote from our everyday lives, is actually at the core of the modern world. Without mathematics you are estranged from it, and everything descends into rumour, superstition, and magic. The best lesson one can teach students about what to apply themselves to is that the future is uncertain, and it is a gamble how one chooses to spend one's efforts. But a sound grounding in mathematics is always a good first option. One of the most common educational regrets of many adults is that they did not spend enough time on mathematics in school, which is quite the opposite of the efficiency regrets of spending too much time on things unused.

A good mathematics textbook cannot be about a contrived minimal necessity. It has to be more than crib notes for a lean and diminished course in what students are deemed to need, only to be tossed away after the final exam. It must be more than a website or a blog. It should be something that stays with you, giving help in a familiar voice when you need to remember mathematics you will have forgotten over the years. Moreover, it should be something that one can grow into. People mature mathematically. As one does, concepts that seemed incomprehensible eventually become obvious. When that happens, new questions emerge that were previously inconceivable. This text has answers to many of those questions too.

Such a textbook must not only take into account the nature of the current audience, it must also be open to how well it bridges to other fields and introduces ideas new to the conventional curriculum. In this regard, this textbook is like no other. Topics not available in any other text are bravely introduced through the thematic concept of *gateway applications*. Applications of calculus have always been an important feature of earlier editions of this book. But the agenda of introducing gateway applications was introduced in the 8th edition. Rather than shrinking to what is merely needed, this 9th edition is still more comprehensive than the 8th edition. Of course, it remains possible to do a light and minimal treatment of the subject with this book, but the decision as to what that might mean precisely becomes the responsibility of a skilled instructor, and not the result of the limitations of some text. Correspondingly, a richer treatment is also an option. Flexibility in terms of emphasis, exercises, and projects is made easily possible with a larger span of subject material.

Some of the unique topics naturally addressed in the gateway applications, which may be added or omitted, include Liapunov functions, and Legendre transformations, not to mention exterior calculus. Exterior calculus is a powerful refinement of the calculus of a century ago, which is often overlooked. This text has a complete chapter on it, written accessibly in classical textbook style rather than as an advanced monograph. Other gateway applications are easy to cover in passing, but they are too often overlooked in terms of their importance to modern science. Liapunov functions are often squeezed into advanced books because they are left out of classical curricula, even though they are an easy addition to the discussion of vector fields, where their importance to stability theory and modern biomathematics can be usefully noted. Legendre transformations, which are so important to modern physics and thermodynamics, are a natural and easy topic to add to the discussion of differentials in more than one variable.

There are rich opportunities that this textbook captures. For example, it is the only mainstream textbook that covers sufficient conditions for maxima and minima in higher dimensions, providing answers to questions that most books gloss over. None of these are inaccessible. They are rich opportunities missed because many instructors are simply unfamiliar with their importance to other fields. The 9th edition continues in this tradition. For example, in the existing section on probability there is a new gateway application added that treats heavy-tailed distributions and their consequences for real-world applications.

The 9th edition, in addition to various corrections and refinements, fills in gaps in the treatment of differential equations from the 8th edition, with entirely new material. A linear operator approach to understanding differential equations is added. Also added is a refinement of the existing material on the Dirac delta function, and a full treatment of Laplace transforms. In addition, there is an entirely new section on phase plane analysis. The new phase plane section covers the classical treatment, if that is all one wants, but it goes much further for those who want more, now or later. It can set the reader up for dynamical systems in higher dimensions in a unique, lucid, and compact exposition. With existing treatments of various aspects of differential equations throughout the existing text, the 9th edition becomes suitable for a semester course in differential equations, in addition to the existing standard material suitable for four semesters of calculus.

Not only can the 9th edition be used to deliver five standard courses of conventional material, it can do much more through some of the unique topics and approaches mentioned above, which can be added or overlooked by the instructor without penalty. There is no other calculus book that deals better with computers and mathematics through Maple, in addition to unique but important applications from information theory to Lévy distributions, and does all of these things fearlessly. This 9th edition is the first one to be produced in full colour, and it continues to aspire to its subtitle: "A Complete Course." It is like no other.

About the Cover

The fall of rainwater droplets in a forest is frozen in an instant of time. For any small droplet of water, surface tension causes minimum energy to correspond to minimum surface area. Thus, small amounts of falling water are enveloped by nearly perfect minimal spheres, which act like lenses that image the forest background. The forest image is inverted because of the geometry of ray paths of light through a sphere. Close examination reveals that other droplets are also imaged, appearing almost like bubbles in glass. Still closer examination shows that the forest is right side up in the droplet images of the other droplets—transformation and inverse in one picture. If the droplets were much smaller, simple geometry of ray paths through a sphere would fail, because the wave nature of light would dominate. Interactions with the spherical droplets are then governed by Maxwell's equations instead of simple geometry. Tiny spheres exhibit Mie scattering of light instead, making a large collection of minute droplets, as in a cloud, seem brilliant white on a sunny day. The story of clouds, waves, rays, inverses, and minima are all contained in this instant of time in a forest.

To the Student

You are holding what has become known as a "high-end" calculus text in the book trade. You are lucky. Think of it as having a high-end touring car instead of a compact economy car. But, even though this is the first edition to be published in full colour, it is not high end in the material sense. It does not have scratch-and-sniff pages, sparkling radioactive ink, or anything else like that. It's the content that sets it apart. Unlike the car business, "high-end" book content is not priced any higher than that of any other book. It is one of the few consumer items where anyone can afford to buy into the high end. But there is a catch. Unlike cars, you have to do the work to achieve the promise of the book. So in that sense "high end" is more like a form of "secret" martial arts for your mind that the economy version cannot deliver. If you practise, your mind will become stronger. You will become more confident and disciplined. Secrets of the ages will become open to you. You will become fearless, as your mind longs to tackle any new mathematical challenge.

But hard work is the watchword. Practise, practise, practise. It is exhilarating when you finally get a new idea that you did not understand before. There are few experiences as great as figuring things out. Doing exercises and checking your answers against those in the back of the book are how you practise mathematics with a text. You can do essentially the same thing on a computer; you still do the problems and check the answers. However you do it, more exercises mean more practice and better performance.

There are numerous exercises in this text—too many for you to try them all perhaps, but be ambitious. Some are "drill" exercises to help you develop your skills in calculation. More important, however, are the problems that develop reasoning skills and your ability to apply the techniques you have learned to concrete situations. In some cases, you will have to plan your way through a problem that requires several different "steps" before you can get to the answer. Other exercises are designed to extend the theory developed in the text and therefore enhance your understanding of the concepts of calculus. Think of the problems as a tool to help you correctly wire your mind. You may have a lot of great components in your head, but if you don't wire the components together properly, your "home theatre" won't work.

The exercises vary greatly in difficulty. Usually, the more difficult ones occur toward the end of exercise sets, but these sets are not strictly graded in this way because exercises on a specific topic tend to be grouped together. Also, "difficulty" can be subjective. For some students, exercises designated difficult may seem easy, while exercises designated easy may seem difficult. Nonetheless, some exercises in the regular sets are marked with the symbols ▌, which indicates that the exercise is somewhat more difficult than most, or ❷, which indicates a more theoretical exercise. The theoretical ones need not be difficult; sometimes they are quite easy. Most of the problems in the *Challenging Problems* section forming part of the *Chapter Review* at the end of most chapters are also on the difficult side.

It is not a bad idea to review the background material in Chapter P (Preliminaries), even if your instructor does not refer to it in class.

If you find some of the concepts in the book difficult to understand, *re-read* the material slowly, if necessary several times; *think about it*; formulate questions to ask fellow students, your TA, or your instructor. Don't delay. It is important to resolve your problems as soon as possible. If you don't understand today's topic, you may not understand how it applies to tomorrow's either. Mathematics builds from one idea to the next. Testing your understanding of the later topics also tests your understanding of the earlier ones. Do not be discouraged if you can't do *all* the exercises. Some are very difficult indeed. The range of exercises ensures that nearly all students can find a comfortable level to practise at, while allowing for greater challenges as skill grows.

Answers for most of the odd-numbered exercises are provided at the back of the book. Exceptions are exercises that don't have short answers: for example, "Prove that . . ." or "Show that . . ." problems where the answer is the whole solution. A *Student Solutions Manual* that contains detailed solutions to even-numbered exercises is available.

Besides ▌ and ❷ used to mark more difficult and theoretical problems, the following symbols are used to mark exercises of special types:

✴ Exercises pertaining to differential equations and initial-value problems. (It is not used in sections that are wholly concerned with DEs.)

▦ Problems requiring the use of a calculator. Often a scientific calculator is needed. Some such problems may require a programmable calculator.

▨ Problems requiring the use of either a graphing calculator or mathematical graphing software on a personal computer.

🖱 Problems requiring the use of a computer. Typically, these will require either computer algebra software (e.g., Maple, Mathematica) or a spreadsheet program such as Microsoft Excel.

To the Instructor

Calculus: a Complete Course, 9th Edition contains 19 chapters, P and 1–18, plus 5 Appendices. It covers the material usually encountered in a three- to five-semester real-variable calculus program, involving real-valued functions of a single real variable (differential calculus in Chapters 1–4 and integral calculus in Chapters 5–8), as well as vector-valued functions of a single real variable (covered in Chapter 11), real-valued functions of several real variables (in Chapters 12–14), and vector-valued functions of several real variables (in Chapters 15–17). Chapter 9 concerns sequences and series, and its position is rather arbitrary.

Most of the material requires only a reasonable background in high school algebra and analytic geometry. (See Chapter P—Preliminaries for a review of this material.) However, some optional material is more subtle and/or theoretical and is intended for stronger students, special topics, and reference purposes. It also allows instructors considerable flexibility in making points, answering questions, and selective enrichment of a course.

Chapter 10 contains necessary background on vectors and geometry in 3-dimensional space as well as some linear algebra that is useful, although not absolutely essential, for the understanding of subsequent multivariable material. Material on differential equations is scattered throughout the book, but Chapter 18 provides a compact treatment of ordinary differential equations (ODEs), which may provide enough material for a one-semester course on the subject.

There are two split versions of the complete book. *Single-Variable Calculus, 9th Edition* covers Chapters P, 1–9, 18 and all five appendices. *Calculus of Several Variables, 9th Edition* covers Chapters 9–18 and all five appendices. It also begins with a brief review of Single-Variable Calculus.

Besides numerous improvements and clarifications throughout the book and tweakings of existing material such as consideration of probability densities with heavy tails in Section 7.8, and a less restrictive definition of the Dirac delta function in Section 16.1, there are two new sections in Chapter 18, one on Laplace Transforms (Section 18.7) and one on Phase Plane Analysis of Dynamical Systems (Section 18.9).

There is a wealth of material here—too much to include in any one course. It was never intended to be otherwise. You must select what material to include and what to omit, taking into account the background and needs of your students. At the University of British Columbia, where one author taught for 34 years, and at the University of Western Ontario, where the other author continues to teach, calculus is divided into four semesters, the first two covering single-variable calculus, the third covering functions of several variables, and the fourth covering vector calculus. In none of these courses was there enough time to cover all the material in the appropriate chapters; some sections are always omitted. The text

is designed to allow students and instructors to conveniently find their own level while enhancing any course from general calculus to courses focused on science and engineering students.

Several supplements are available for use with *Calculus: A Complete Course, 9th Edition.* Available to students is the **Student Solutions Manual** (ISBN: 9780134491073): This manual contains detailed solutions to all the even-numbered exercises, prepared by the authors. There are also such Manuals for the split volumes, for *Single Variable Calculus* (ISBN: 9780134579863), and for *Calculus of Several Variables* (ISBN: 9780134579856).

Available to instructors are the following resources:

- **Instructor's Solutions Manual**
- **Computerized Test Bank** Pearson's computerized test bank allows instructors to filter and select questions to create quizzes, tests, or homework (over 1,500 test questions)
- **Image Library**, which contains all of the figures in the text provided as individual enlarged .pdf files suitable for printing to transparencies.

These supplements are available for download from a password-protected section of Pearson Canada's online catalogue (catalogue.pearsoned.ca). Navigate to this book's catalogue page to view a list of those supplements that are available. Speak to your local Pearson sales representative for details and access.

Also available to qualified instructors are **MyMathLab®** and **MathXL®** Online Courses for which access codes are required.

MyMathLab helps improve individual students' performance. It has a consistently positive impact on the quality of learning in higher-education math instruction. MyMathLab's comprehensive online gradebook automatically tracks your students' results on tests, quizzes, homework, and in the study plan. MyMathLab provides engaging experiences that personalize, stimulate, and measure learning for each student. The homework and practice exercises in MyMathLab are correlated to the exercises in the textbook. The software offers immediate, helpful feedback when students enter incorrect answers. Exercises include guided solutions, sample problems, animations, and eText clips for extra help. MyMathLab comes from an experienced partner with educational expertise and an eye on the future. Knowing that you are using a Pearson product means knowing that you are using quality content. That means that our eTexts are accurate and our assessment tools work. To learn more about how MyMathLab combines proven learning applications with powerful assessment, visit www.mymathlab.com or contact your Pearson representative.

MathXL is the homework and assessment engine that runs MyMathLab. (MyMathLab is MathXL plus a learning management system.) MathXL is available to qualified adopters. For more information, visit our website at www.mathxl.com, or contact your Pearson representative.

In addition, there is an **eText** available. Pearson eText gives students access to the text whenever and wherever they have online access to the Internet. eText pages look exactly like the printed text, offering powerful new functionality for students and instructors. Users can create notes, highlight text in different colours, create bookmarks, zoom, click hyperlinked words and phrases to view definitions, and view in single-page or two-page view.

Learning Solutions Managers. Pearson's Learning Solutions Managers work with faculty and campus course designers to ensure that Pearson technology products, assessment tools, and online course materials are tailored to meet your specific needs. This highly qualified team is dedicated to helping schools take full advantage of a wide range of educational resources by assisting in the integration of a variety of instructional materials and media formats. Your local Pearson Canada sales representative can provide you with more details on this service program.

Acknowledgments

The authors are grateful to many colleagues and students at the University of British Columbia and Western University, and at many other institutions worldwide where previous editions of these books have been used, for their encouragement and useful comments and suggestions.

We also wish to thank the sales and marketing staff of all Addison-Wesley (now Pearson) divisions around the world for making the previous editions so successful, and the editorial and production staff in Toronto, in particular,

Acquisitions Editor:	Jennifer Sutton
Program Manager:	Emily Dill
Developmental Editor:	Charlotte Morrison-Reed
Production Manager:	Susan Johnson
Copy Editor:	Valerie Adams
Production Editor/Proofreader:	Leanne Rancourt
Designer:	Anthony Leung

for their assistance and encouragement.

This volume was typeset by Robert Adams using TEX on an iMac computer running OSX version 10.10. Most of the figures were generated using the mathematical graphics software package **MG** developed by Robert Israel and Robert Adams. Some were produced with Maple 10.

The expunging of errors and obscurities in a text is an ongoing and asymptotic process; hopefully each edition is better than the previous one. Nevertheless, some such imperfections always remain, and we will be grateful to any readers who call them to our attention, or give us other suggestions for future improvements.

May 2016

R.A.A.
Vancouver, Canada
adms@math.ubc.ca

C.E.
London, Canada
essex@uwo.ca

What Is Calculus?

Early in the seventeenth century, the German mathematician Johannes Kepler analyzed a vast number of astronomical observations made by Danish astronomer Tycho Brahe and concluded that the planets must move around the sun in elliptical orbits. He didn't know why. Fifty years later, the English mathematician and physicist Isaac Newton answered that question.

Why do the planets move in elliptical orbits around the sun? Why do hurricane winds spiral counterclockwise in the northern hemisphere? How can one predict the effects of interest rate changes on economies and stock markets? When will radioactive material be sufficiently decayed to enable safe handling? How do warm ocean currents in the equatorial Pacific affect the climate of eastern North America? How long will the concentration of a drug in the bloodstream remain at effective levels? How do radio waves propagate through space? Why does an epidemic spread faster and faster and then slow down? How can I be sure the bridge I designed won't be destroyed in a windstorm?

These and many other questions of interest and importance in our world relate directly to our ability to analyze motion and how quantities change with respect to time or each other. Algebra and geometry are useful tools for describing relationships between *static* quantities, but they do not involve concepts appropriate for describing how a quantity *changes*. For this we need new mathematical operations that go beyond the algebraic operations of addition, subtraction, multiplication, division, and the taking of powers and roots. We require operations that measure the way related quantities change.

Calculus provides the tools for describing motion quantitatively. It introduces two new operations called *differentiation* and *integration*, which, like addition and subtraction, are opposites of one another; what differentiation does, integration undoes.

For example, consider the motion of a falling rock. The height (in metres) of the rock t seconds after it is dropped from a height of h_0 m is a function $h(t)$ given by

$$h(t) = h_0 - 4.9t^2.$$

The graph of $y = h(t)$ is shown in the figure below:

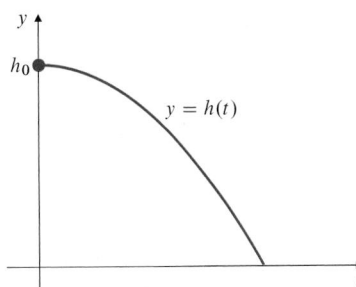

The process of differentiation enables us to find a new function, which we denote $h'(t)$ and call the *derivative* of h with respect to t, which represents the *rate of change* of the height of the rock, that is, its *velocity* in metres/second:

$$h'(t) = -9.8t.$$

Conversely, if we know the velocity of the falling rock as a function of time, integration enables us to find the height function $h(t)$.

Calculus was invented independently and in somewhat different ways by two seventeenth-century mathematicians: Isaac Newton and Gottfried Wilhelm Leibniz. Newton's motivation was a desire to analyze the motion of moving objects. Using his calculus, he was able to formulate his laws of motion and gravitation and *conclude from them* that the planets must move around the sun in elliptical orbits.

Many of the most fundamental and important "laws of nature" are conveniently expressed as equations involving rates of change of quantities. Such equations are called *differential equations*, and techniques for their study and solution are at the heart of calculus. In the falling rock example, the appropriate law is **Newton's Second Law of Motion:**

$$\text{force} \ = \ \text{mass} \ \times \ \text{acceleration.}$$

The *acceleration*, -9.8 m/s^2, is the rate of change (the *derivative*) of the velocity, which is in turn the rate of change (the *derivative*) of the height function.

Much of mathematics is related indirectly to the study of motion. We regard *lines*, or *curves*, as geometric objects, but the ancient Greeks thought of them as paths traced out by moving points. Nevertheless, the study of curves also involves geometric concepts such as tangency and area. The process of differentiation is closely tied to the geometric problem of finding tangent lines; similarly, integration is related to the geometric problem of finding areas of regions with curved boundaries.

Both differentiation and integration are defined in terms of a new mathematical operation called a **limit**. The concept of the limit of a function will be developed in Chapter 1. That will be the real beginning of our study of calculus. In the chapter called "Preliminaries" we will review some of the background from algebra and geometry needed for the development of calculus.

CHAPTER P

Preliminaries

" 'Reeling and Writhing, of course, to begin with,'
the Mock Turtle replied, 'and the different branches
of Arithmetic—Ambition, Distraction, Uglification,
and Derision.' **"**

Lewis Carroll (Charles Lutwidge Dodgson) 1832–1898
from *Alice's Adventures in Wonderland*

Introduction

This preliminary chapter reviews the most important things you should know before beginning calculus. Topics include the real number system; Cartesian coordinates in the plane; equations representing straight lines, circles, and parabolas; functions and their graphs; and, in particular, polynomials and trigonometric functions.

Depending on your precalculus background, you may or may not be familiar with these topics. If you are, you may want to skim over this material to refresh your understanding of the terms used; if not, you should study this chapter in detail.

P.1 Real Numbers and the Real Line

Calculus depends on properties of the real number system. **Real numbers** are numbers that can be expressed as decimals, for example,

$$5 = 5.00000\ldots$$
$$-\tfrac{3}{4} = -0.750000\ldots$$
$$\tfrac{1}{3} = 0.3333\ldots$$
$$\sqrt{2} = 1.4142\ldots$$
$$\pi = 3.14159\ldots$$

In each case the three dots (...) indicate that the sequence of decimal digits goes on forever. For the first three numbers above, the patterns of the digits are obvious; we know what all the subsequent digits are. For $\sqrt{2}$ and π there are no obvious patterns.

The real numbers can be represented geometrically as points on a number line, which we call the **real line**, shown in Figure P.1. The symbol $\mathbb{R}$ is used to denote either the real number system or, equivalently, the real line.

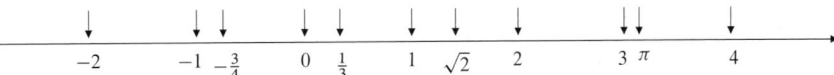

Figure P.1 The real line

The properties of the real number system fall into three categories: algebraic properties, order properties, and completeness. You are already familiar with the *algebraic properties*; roughly speaking, they assert that real numbers can be added, subtracted, multiplied, and divided (except by zero) to produce more real numbers and that the usual rules of arithmetic are valid.

The *order properties* of the real numbers refer to the order in which the numbers appear on the real line. If x lies to the left of y, then we say that "x is less than y" or "y is greater than x." These statements are written symbolically as $x < y$ and $y > x$, respectively. The inequality $x \le y$ means that either $x < y$ or $x = y$. The order properties of the real numbers are summarized in the following *rules for inequalities*:

<table>
<tr><td colspan="2">

Rules for inequalities

If a, b, and c are real numbers, then:

1. $a < b$ $\implies$ $a + c < b + c$
2. $a < b$ $\implies$ $a - c < b - c$
3. $a < b$ and $c > 0$ $\implies$ $ac < bc$
4. $a < b$ and $c < 0$ $\implies$ $ac > bc$; in particular, $-a > -b$
5. $a > 0$ $\implies$ $\dfrac{1}{a} > 0$
6. $0 < a < b$ $\implies$ $\dfrac{1}{b} < \dfrac{1}{a}$

Rules 1–4 and 6 (for $a > 0$) also hold if $<$ and $>$ are replaced by $\le$ and $\ge$.
</td></tr>
</table>

Note especially the rules for multiplying (or dividing) an inequality by a number. If the number is positive, the inequality is preserved; if the number is negative, the inequality is reversed.

The *completeness* property of the real number system is more subtle and difficult to understand. One way to state it is as follows: if A is any set of real numbers having at least one number in it, and if there exists a real number y with the property that $x \le y$ for every x in A (such a number y is called an **upper bound** for A), then there exists a *smallest* such number, called the **least upper bound** or **supremum** of A, and denoted $\sup(A)$. Roughly speaking, this says that there can be no holes or gaps on the real line—every point corresponds to a real number. We will not need to deal much with completeness in our study of calculus. It is typically used to prove certain important results—in particular, Theorems 8 and 9 in Chapter 1. (These proofs are given in Appendix III but are not usually included in elementary calculus courses; they are studied in more advanced courses in mathematical analysis.) However, when we study infinite sequences and series in Chapter 9, we will make direct use of completeness.

The set of real numbers has some important special subsets:

(i) the **natural numbers** or **positive integers**, namely, the numbers 1, 2, 3, 4, ...

(ii) the **integers**, namely, the numbers 0, ± 1, ± 2, ± 3, ...

(iii) the **rational numbers**, that is, numbers that can be expressed in the form of a fraction m/n, where m and n are integers, and $n \ne 0$.

The rational numbers are precisely those real numbers with decimal expansions that are either:

(a) terminating, that is, ending with an infinite string of zeros, for example, $3/4 = 0.750000\ldots$, or

(b) repeating, that is, ending with a string of digits that repeats over and over, for example, $23/11 = 2.090909\ldots = 2.\overline{09}$. (The bar indicates the pattern of repeating digits.)

Real numbers that are not rational are called *irrational numbers*.

The symbol $\implies$ means "implies."

EXAMPLE 1   Show that each of the numbers   (a) $1.323232\cdots = 1.\overline{32}$ and
(b) $0.3405405405\ldots = 0.3\overline{405}$ is a rational number by expressing it as a quotient of two integers.

Solution

(a) Let $x = 1.323232\ldots$ Then $x - 1 = 0.323232\ldots$ and

$$100x = 132.323232\ldots = 132 + 0.323232\ldots = 132 + x - 1.$$

Therefore, $99x = 131$ and $x = 131/99$.

(b) Let $y = 0.3405405405\ldots$ Then $10y = 3.405405405\ldots$ and
$10y - 3 = 0.405405405\ldots$ Also,

$$10,000y = 3,405.405405405\ldots = 3,405 + 10y - 3.$$

Therefore, $9,990y = 3,402$ and $y = 3,402/9,990 = 63/185$.

The set of rational numbers possesses all the algebraic and order properties of the real numbers but not the completeness property. There is, for example, no rational number whose square is 2. Hence, there is a "hole" on the "rational line" where $\sqrt{2}$ should be.[1] Because the real line has no such "holes," it is the appropriate setting for studying limits and therefore calculus.

Intervals

A subset of the real line is called an **interval** if it contains at least two numbers and also contains all real numbers between any two of its elements. For example, the set of real numbers x such that $x > 6$ is an interval, but the set of real numbers y such that $y \neq 0$ is not an interval. (Why?) It consists of two intervals.

If a and b are real numbers and $a < b$, we often refer to

(i) the **open interval** from a to b, denoted by (a, b), consisting of all real numbers x satisfying $a < x < b$.

(ii) the **closed interval** from a to b, denoted by $[a, b]$, consisting of all real numbers x satisfying $a \leq x \leq b$.

(iii) the **half-open interval** $[a, b)$, consisting of all real numbers x satisfying the inequalities $a \leq x < b$.

(iv) the **half-open interval** $(a, b]$, consisting of all real numbers x satisfying the inequalities $a < x \leq b$.

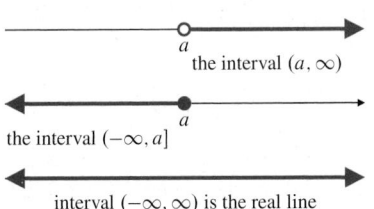

open interval (a, b)

closed interval $[a, b]$

half-open interval $[a, b)$

half-open interval $(a, b]$

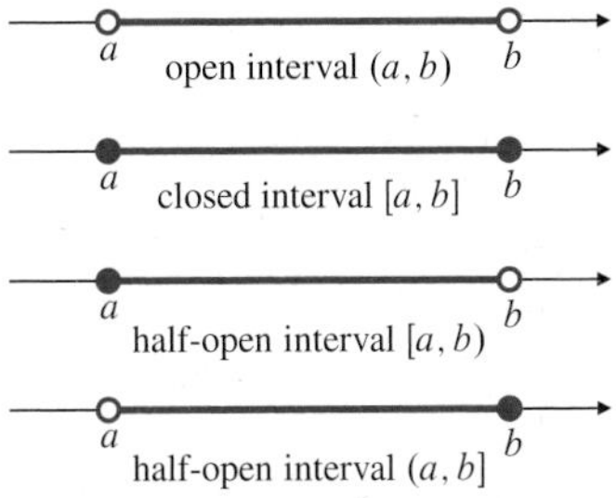

Figure P.2   Finite intervals

These are illustrated in Figure P.2. Note the use of hollow dots to indicate endpoints of intervals that are not included in the intervals, and solid dots to indicate endpoints that are included. The endpoints of an interval are also called **boundary points**.

The intervals in Figure P.2 are **finite intervals**; each of them has finite length $b - a$. Intervals can also have infinite length, in which case they are called **infinite intervals**. Figure P.3 shows some examples of infinite intervals. Note that the whole real line $\mathbb{R}$ is an interval, denoted by $(-\infty, \infty)$. The symbol ∞ ("infinity") does *not* denote a real number, so we never allow ∞ to belong to an interval.

the interval (a, ∞)

the interval $(-\infty, a]$

interval $(-\infty, \infty)$ is the real line

Figure P.3   Infinite intervals

---

[1]   How do we know that $\sqrt{2}$ is an irrational number? Suppose, to the contrary, that $\sqrt{2}$ is rational. Then $\sqrt{2} = m/n$, where m and n are integers and $n \neq 0$. We can assume that the fraction m/n has been "reduced to lowest terms"; any common factors have been cancelled out. Now $m^2/n^2 = 2$, so $m^2 = 2n^2$, which is an even integer. Hence, m must also be even. (The square of an odd integer is always odd.) Since m is even, we can write $m = 2k$, where k is an integer. Thus $4k^2 = 2n^2$ and $n^2 = 2k^2$, which is even. Thus n is also even. This contradicts the assumption that $\sqrt{2}$ could be written as a fraction m/n in lowest terms; m and n cannot both be even. Accordingly, there can be no rational number whose square is 2.

EXAMPLE 2 Solve the following inequalities. Express the solution sets in terms of intervals and graph them.

(a) $2x - 1 > x + 3$ (b) $-\dfrac{x}{3} \geq 2x - 1$ (c) $\dfrac{2}{x - 1} \geq 5$

Solution

(a) $2x - 1 > x + 3$ Add 1 to both sides.

$\qquad 2x > x + 4$ Subtract x from both sides.

$\qquad x > 4$ The solution set is the interval $(4, \infty)$.

(b) $-\dfrac{x}{3} \geq 2x - 1$ Multiply both sides by -3.

$\qquad x \leq -6x + 3$ Add $6x$ to both sides.

$\qquad 7x \leq 3$ Divide both sides by 7.

$\qquad x \leq \dfrac{3}{7}$ The solution set is the interval $(-\infty, 3/7]$.

(c) We transpose the 5 to the left side and simplify to rewrite the given inequality in an equivalent form:

$$\frac{2}{x - 1} - 5 \geq 0 \quad \Longleftrightarrow \quad \frac{2 - 5(x - 1)}{x - 1} \geq 0 \quad \Longleftrightarrow \quad \frac{7 - 5x}{x - 1} \geq 0.$$

The fraction $\dfrac{7 - 5x}{x - 1}$ is undefined at $x = 1$ and is 0 at $x = 7/5$. Between these numbers it is positive if the numerator and denominator have the same sign, and negative if they have opposite sign. It is easiest to organize this sign information in a chart:

The symbol $\Longleftrightarrow$ means "if and only if" or "is equivalent to." If A and B are two statements, then $A \Longleftrightarrow B$ means that the truth of either statement implies the truth of the other, so either both must be true or both must be false.

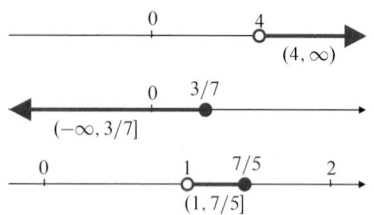

$(4, \infty)$

$(-\infty, 3/7]$

$(1, 7/5]$

Figure P.4   The intervals for Example 2

x				1		7/5	
$7 - 5x$		$+$	$+$	$+$	0	$-$	
$x - 1$		$-$	0	$+$	$+$	$+$	
$(7 - 5x)/(x - 1)$		$-$	undef	$+$	0	$-$	

Thus the solution set of the given inequality is the interval $(1, 7/5]$.

See Figure P.4 for graphs of the solutions.

Sometimes we will need to solve systems of two or more inequalities that must be satisfied simultaneously. We still solve the inequalities individually and look for numbers in the intersection of the solution sets.

EXAMPLE 3 Solve the systems of inequalities:

(a) $3 \leq 2x + 1 \leq 5$ (b) $3x - 1 < 5x + 3 \leq 2x + 15$.

Solution

(a) Using the technique of Example 2, we can solve the inequality $3 \leq 2x + 1$ to get $2 \leq 2x$, so $x \geq 1$. Similarly, the inequality $2x + 1 \leq 5$ leads to $2x \leq 4$, so $x \leq 2$. The solution set of system (a) is therefore the closed interval $[1, 2]$.

(b) We solve both inequalities as follows:

$$\left.\begin{array}{r} 3x - 1 < 5x + 3 \\ -1 - 3 < 5x - 3x \\ -4 < 2x \\ -2 < x \end{array}\right\} \quad \text{and} \quad \left\{\begin{array}{l} 5x + 3 \leq 2x + 15 \\ 5x - 2x \leq 15 - 3 \\ 3x \leq 12 \\ x \leq 4 \end{array}\right.$$

The solution set is the interval $(-2, 4]$.

Solving quadratic inequalities depends on solving the corresponding quadratic equations.

EXAMPLE 4 **Quadratic inequalities**
Solve: (a) $x^2 - 5x + 6 < 0$ (b) $2x^2 + 1 > 4x$.

Solution

(a) The trinomial $x^2 - 5x + 6$ factors into the product $(x - 2)(x - 3)$, which is negative if and only if exactly one of the factors is negative. Since $x - 3 < x - 2$, this happens when $x - 3 < 0$ and $x - 2 > 0$. Thus we need $x < 3$ and $x > 2$; the solution set is the open interval $(2, 3)$.

(b) The inequality $2x^2 + 1 > 4x$ is equivalent to $2x^2 - 4x + 1 > 0$. The corresponding quadratic equation $2x^2 - 4x + 1 = 0$, which is of the form $Ax^2 + Bx + C = 0$, can be solved by the quadratic formula (see Section P.6):

$$x = \frac{-B \pm \sqrt{B^2 - 4AC}}{2A} = \frac{4 \pm \sqrt{16 - 8}}{4} = 1 \pm \frac{\sqrt{2}}{2},$$

so the given inequality can be expressed in the form

$$\left(x - 1 + \tfrac{1}{2}\sqrt{2}\right)\left(x - 1 - \tfrac{1}{2}\sqrt{2}\right) > 0.$$

This is satisfied if both factors on the left side are positive or if both are negative. Therefore, we require that either $x < 1 - \tfrac{1}{2}\sqrt{2}$ or $x > 1 + \tfrac{1}{2}\sqrt{2}$. The solution set is the *union* of intervals $\left(-\infty, 1 - \tfrac{1}{2}\sqrt{2}\right) \cup \left(1 + \tfrac{1}{2}\sqrt{2}, \infty\right)$.

Note the use of the symbol $\cup$ to denote the **union** of intervals. A real number is in the union of intervals if it is in at least one of the intervals. We will also need to consider the **intersection** of intervals from time to time. A real number belongs to the intersection of intervals if it belongs to *every one* of the intervals. We will use $\cap$ to denote intersection. For example,

$$[1, 3) \cap [2, 4] = [2, 3) \quad \text{while} \quad [1, 3) \cup [2, 4] = [1, 4].$$

EXAMPLE 5 Solve the inequality $\dfrac{3}{x - 1} < -\dfrac{2}{x}$ and graph the solution set.

Solution We would like to multiply by $x(x - 1)$ to clear the inequality of fractions, but this would require considering three cases separately. (What are they?) Instead, we will transpose and combine the two fractions into a single one:

$$\frac{3}{x - 1} < -\frac{2}{x} \quad \Longleftrightarrow \quad \frac{3}{x - 1} + \frac{2}{x} < 0 \quad \Longleftrightarrow \quad \frac{5x - 2}{x(x - 1)} < 0.$$

We examine the signs of the three factors in the left fraction to determine where that fraction is negative:

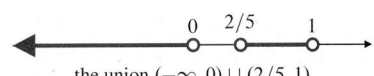

the union $(-\infty, 0) \cup (2/5, 1)$

Figure P.5 The solution set for Example 5

x		0		2/5		1	
$5x - 2$	$-$	$-$	$-$	0	$+$	$+$	$+$
x	$-$	0	$+$	$+$	$+$	$+$	$+$
$x - 1$	$-$	$-$	$-$	$-$	$-$	0	$+$
$\dfrac{5x - 2}{x(x - 1)}$	$-$	undef	$+$	0	$-$	undef	$+$

The solution set of the given inequality is the union of these two intervals, namely, $(-\infty, 0) \cup (2/5, 1)$. See Figure P.5.

The Absolute Value

The **absolute value**, or **magnitude**, of a number x, denoted $|x|$ (read "the absolute value of x"), is defined by the formula

$$|x| = \begin{cases} x & \text{if } x \geq 0 \\ -x & \text{if } x < 0 \end{cases}$$

The vertical lines in the symbol $|x|$ are called **absolute value bars**.

EXAMPLE 6 $|3| = 3, \quad |0| = 0, \quad |-5| = 5.$

Note that $|x| \geq 0$ for every real number x, and $|x| = 0$ only if $x = 0$. People sometimes find it confusing to say that $|x| = -x$ when x is negative, but this is correct since $-x$ is positive in that case. The symbol $\sqrt{a}$ always denotes the *nonnegative* square root of a, so an alternative definition of $|x|$ is $|x| = \sqrt{x^2}$.

It is important to remember that $\sqrt{a^2} = |a|$. Do not write $\sqrt{a^2} = a$ unless you already know that $a \geq 0$.

Geometrically, $|x|$ represents the (nonnegative) distance from x to 0 on the real line. More generally, $|x - y|$ represents the (nonnegative) distance between the points x and y on the real line, since this distance is the same as that from the point $x - y$ to 0 (see Figure P.6):

$$|x - y| = \begin{cases} x - y, & \text{if } x \geq y \\ y - x, & \text{if } x < y. \end{cases}$$

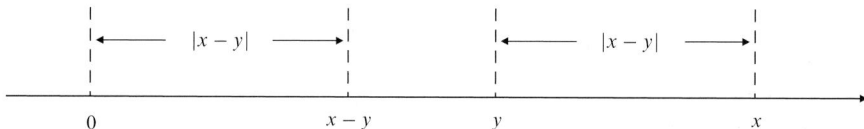

Figure P.6

$|x - y|$ = distance from x to y

The absolute value function has the following properties:

> **Properties of absolute values**
>
> 1. $|-a| = |a|$. A number and its negative have the same absolute value.
> 2. $|ab| = |a||b|$ and $\left|\dfrac{a}{b}\right| = \dfrac{|a|}{|b|}$. The absolute value of a product (or quotient) of two numbers is the product (or quotient) of their absolute values.
> 3. $|a \pm b| \leq |a| + |b|$ (the **triangle inequality**). The absolute value of a sum of or difference between numbers is less than or equal to the sum of their absolute values.

The first two of these properties can be checked by considering the cases where either of a or b is either positive or negative. The third property follows from the first two because $\pm 2ab \leq |2ab| = 2|a||b|$. Therefore, we have

$$|a \pm b|^2 = (a \pm b)^2 = a^2 \pm 2ab + b^2$$
$$\leq |a|^2 + 2|a||b| + |b|^2 = (|a| + |b|)^2,$$

and taking the (positive) square roots of both sides, we obtain $|a \pm b| \leq |a| + |b|$. This result is called the "triangle inequality" because it follows from the geometric fact that the length of any side of a triangle cannot exceed the sum of the lengths of the other two sides. For instance, if we regard the points 0, a, and b on the number line as the vertices of a degenerate "triangle," then the sides of the triangle have lengths $|a|$, $|b|$, and $|a - b|$. The triangle is degenerate since all three of its vertices lie on a straight line.

Equations and Inequalities Involving Absolute Values

The equation $|x| = D$ (where $D > 0$) has two solutions, $x = D$ and $x = -D$: the two points on the real line that lie at distance D from the origin. Equations and inequalities involving absolute values can be solved algebraically by breaking them into cases according to the definition of absolute value, but often they can also be solved geometrically by interpreting absolute values as distances. For example, the inequality $|x - a| < D$ says that the distance from x to a is less than D, so x must lie between $a - D$ and $a + D$. (Or, equivalently, a must lie between $x - D$ and $x + D$.) If D is a positive number, then

$$
\begin{aligned}
|x| = D &\iff \text{either } x = -D \text{ or } x = D \\
|x| < D &\iff -D < x < D \\
|x| \leq D &\iff -D \leq x \leq D \\
|x| > D &\iff \text{either } x < -D \text{ or } x > D
\end{aligned}
$$

More generally,

$$
\begin{aligned}
|x - a| = D &\iff \text{either } x = a - D \text{ or } x = a + D \\
|x - a| < D &\iff a - D < x < a + D \\
|x - a| \leq D &\iff a - D \leq x \leq a + D \\
|x - a| > D &\iff \text{either } x < a - D \text{ or } x > a + D
\end{aligned}
$$

EXAMPLE 7   Solve: (a) $|2x + 5| = 3$ (b) $|3x - 2| \leq 1$.

Solution

(a) $|2x + 5| = 3 \iff 2x + 5 = \pm 3$. Thus, either $2x = -3 - 5 = -8$ or $2x = 3 - 5 = -2$. The solutions are $x = -4$ and $x = -1$.

(b) $|3x - 2| \leq 1 \iff -1 \leq 3x - 2 \leq 1$. We solve this pair of inequalities:

$$
\left\{
\begin{aligned}
-1 &\leq 3x - 2 \\
-1 + 2 &\leq 3x \\
1/3 &\leq x
\end{aligned}
\right\}
\quad \text{and} \quad
\left\{
\begin{aligned}
3x - 2 &\leq 1 \\
3x &\leq 1 + 2 \\
x &\leq 1
\end{aligned}
\right\}.
$$

Thus the solutions lie in the interval $[1/3, 1]$.

Remark   Here is how part (b) of Example 7 could have been solved geometrically, by interpreting the absolute value as a distance:

$$
|3x - 2| = \left| 3\left(x - \frac{2}{3} \right) \right| = 3 \left| x - \frac{2}{3} \right|.
$$

Thus, the given inequality says that

$$
3 \left| x - \frac{2}{3} \right| \leq 1 \quad \text{or} \quad \left| x - \frac{2}{3} \right| \leq \frac{1}{3}.
$$

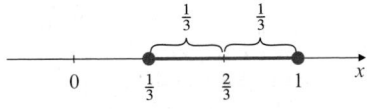

Figure P.7   The solution set for Example 7(b)

This says that the distance from x to $2/3$ does not exceed $1/3$. The solutions for x therefore lie between $1/3$ and 1, including both of these endpoints. (See Figure P.7.)

EXAMPLE 8 Solve the equation $|x + 1| = |x - 3|$.

Solution The equation says that x is equidistant from -1 and 3. Therefore, x is the point halfway between -1 and 3; $x = (-1 + 3)/2 = 1$. Alternatively, the given equation says that either $x + 1 = x - 3$ or $x + 1 = -(x - 3)$. The first of these equations has no solutions; the second has the solution $x = 1$.

EXAMPLE 9 What values of x satisfy the inequality $\left|5 - \dfrac{2}{x}\right| < 3$?

Solution We have

$$\left|5 - \frac{2}{x}\right| < 3 \quad\Longleftrightarrow\quad -3 < 5 - \frac{2}{x} < 3 \qquad \text{Subtract 5 from each member.}$$

$$-8 < -\frac{2}{x} < -2 \qquad \text{Divide each member by } -2.$$

$$4 > \frac{1}{x} > 1 \qquad \text{Take reciprocals.}$$

$$\frac{1}{4} < x < 1.$$

In this calculation we manipulated a system of two inequalities simultaneously, rather than split it up into separate inequalities as we have done in previous examples. Note how the various rules for inequalities were used here. Multiplying an inequality by a negative number reverses the inequality. So does taking reciprocals of an inequality in which both sides are positive. The given inequality holds for all x in the open interval $(1/4, 1)$.

EXERCISES P.1

In Exercises 1–2, express the given rational number as a repeating decimal. Use a bar to indicate the repeating digits.

1. $\dfrac{2}{9}$

2. $\dfrac{1}{11}$

In Exercises 3–4, express the given repeating decimal as a quotient of integers in lowest terms.

3. $0.\overline{12}$

4. $3.2\overline{7}$

5. Express the rational numbers $1/7, 2/7, 3/7$, and $4/7$ as repeating decimals. (Use a calculator to give as many decimal digits as possible.) Do you see a pattern? Guess the decimal expansions of $5/7$ and $6/7$ and check your guesses.

6. Can two different decimals represent the same number? What number is represented by $0.999\ldots = 0.\overline{9}$?

In Exercises 7–12, express the set of all real numbers x satisfying the given conditions as an interval or a union of intervals.

7. $x \geq 0$ and $x \leq 5$

8. $x < 2$ and $x \geq -3$

9. $x > -5$ or $x < -6$

10. $x \leq -1$

11. $x > -2$

12. $x < 4$ or $x \geq 2$

In Exercises 13–26, solve the given inequality, giving the solution set as an interval or union of intervals.

13. $-2x > 4$

14. $3x + 5 \leq 8$

15. $5x - 3 \leq 7 - 3x$

16. $\dfrac{6 - x}{4} \geq \dfrac{3x - 4}{2}$

17. $3(2 - x) < 2(3 + x)$

18. $x^2 < 9$

19. $\dfrac{1}{2 - x} < 3$

20. $\dfrac{x + 1}{x} \geq 2$

21. $x^2 - 2x \leq 0$

22. $6x^2 - 5x \leq -1$

23. $x^3 > 4x$

24. $x^2 - x \leq 2$

25. $\dfrac{x}{2} \geq 1 + \dfrac{4}{x}$

26. $\dfrac{3}{x - 1} < \dfrac{2}{x + 1}$

Solve the equations in Exercises 27–32.

27. $|x| = 3$

28. $|x - 3| = 7$

29. $|2t + 5| = 4$

30. $|1 - t| = 1$

31. $|8 - 3s| = 9$

32. $\left|\dfrac{s}{2} - 1\right| = 1$

In Exercises 33–40, write the interval defined by the given inequality.

33. $|x| < 2$

34. $|x| \leq 2$

35. $|s - 1| \leq 2$

36. $|t + 2| < 1$

37. $|3x - 7| < 2$ **38.** $|2x + 5| < 1$

39. $\left|\dfrac{x}{2} - 1\right| \le 1$ **40.** $\left|2 - \dfrac{x}{2}\right| < \dfrac{1}{2}$

In Exercises 41–42, solve the given inequality by interpreting it as a statement about distances on the real line.

41. $|x + 1| > |x - 3|$ **42.** $|x - 3| < 2|x|$

❷ **43.** Do not fall into the trap $|-a| = a$. For what real numbers a is

this equation true? For what numbers is it false?

44. Solve the equation $|x - 1| = 1 - x$.

❷ **45.** Show that the inequality

$$|a - b| \ge \Big||a| - |b|\Big|$$

holds for all real numbers a and b.

P.2   Cartesian Coordinates in the Plane

The positions of all points in a plane can be measured with respect to two perpendicular real lines in the plane intersecting at the 0-point of each. These lines are called **coordinate axes** in the plane. Usually (but not always) we call one of these axes the x-axis and draw it horizontally with numbers x on it increasing to the right; then we call the other the y-axis, and draw it vertically with numbers y on it increasing upward. The point of intersection of the coordinate axes (the point where x and y are both zero) is called the **origin** and is often denoted by the letter O.

If P is any point in the plane, we can draw a line through P perpendicular to the x-axis. If a is the value of x where that line intersects the x-axis, we call a the **x-coordinate** of P. Similarly, the **y-coordinate** of P is the value of y where a line through P perpendicular to the y-axis meets the y-axis. The **ordered pair** (a, b) is called the **coordinate pair**, or the **Cartesian coordinates**, of the point P. We refer to the point as $P(a, b)$ to indicate both the name P of the point and its coordinates (a, b). (See Figure P.8.) Note that the x-coordinate appears first in a coordinate pair. Coordinate pairs are in one-to-one correspondence with points in the plane; each point has a unique coordinate pair, and each coordinate pair determines a unique point. We call such a set of coordinate axes and the coordinate pairs they determine a **Cartesian coordinate system** in the plane, after the seventeenth-century philosopher René Descartes, who created analytic (coordinate) geometry. When equipped with such a coordinate system, a plane is called a **Cartesian plane**. Note that we are using the same notation (a, b) for the Cartesian coordinates of a point in the plane as we use for an open interval on the real line. However, this should not cause any confusion because the intended meaning will be clear from the context.

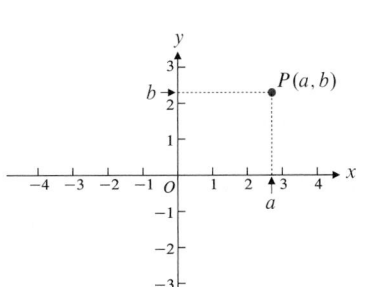

Figure P.8   The coordinate axes and the point P with coordinates (a, b)

Figure P.9 shows the coordinates of some points in the plane. Note that all points on the x-axis have y-coordinate 0. We usually just write the x-coordinates to label such points. Similarly, points on the y-axis have $x = 0$, and we can label such points using their y-coordinates only.

The coordinate axes divide the plane into four regions called **quadrants**. These quadrants are numbered I to IV, as shown in Figure P.10. The **first quadrant** is the upper right one; both coordinates of any point in that quadrant are positive numbers. Both coordinates are negative in quadrant III; only y is positive in quadrant II; only x is positive in quadrant IV.

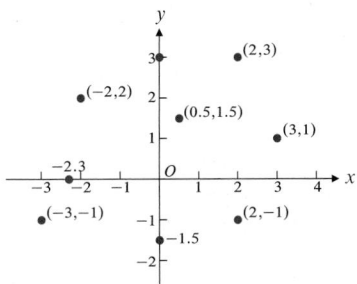

Figure P.9   Some points with their coordinates

Axis Scales

When we plot data in the coordinate plane or graph formulas whose variables have different units of measure, we do not need to use the same scale on the two axes. If, for example, we plot height versus time for a falling rock, there is no reason to place the mark that shows 1 m on the height axis the same distance from the origin as the mark that shows 1 s on the time axis.

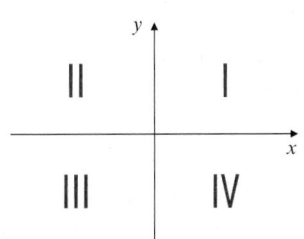

Figure P.10   The four quadrants

When we graph functions whose variables do not represent physical measurements and when we draw figures in the coordinate plane to study their geometry or trigonometry, we usually make the scales identical. A vertical unit of distance then looks the same as a horizontal unit. As on a surveyor's map or a scale drawing, line segments that are supposed to have the same length will look as if they do, and angles that are supposed to be equal will look equal. Some of the geometric results we obtain later, such as the relationship between the slopes of perpendicular lines, are valid only if equal scales are used on the two axes.

Computer and calculator displays are another matter. The vertical and horizontal scales on machine-generated graphs usually differ, with resulting distortions in distances, slopes, and angles. Circles may appear elliptical, and squares may appear rectangular or even as parallelograms. Right angles may appear as acute or obtuse. Circumstances like these require us to take extra care in interpreting what we see. High-quality computer software for drawing Cartesian graphs usually allows the user to compensate for such scale problems by adjusting the *aspect ratio* (the ratio of vertical to horizontal scale). Some computer screens also allow adjustment within a narrow range. When using graphing software, try to adjust your particular software/hardware configuration so that the horizontal and vertical diameters of a drawn circle appear to be equal.

Increments and Distances

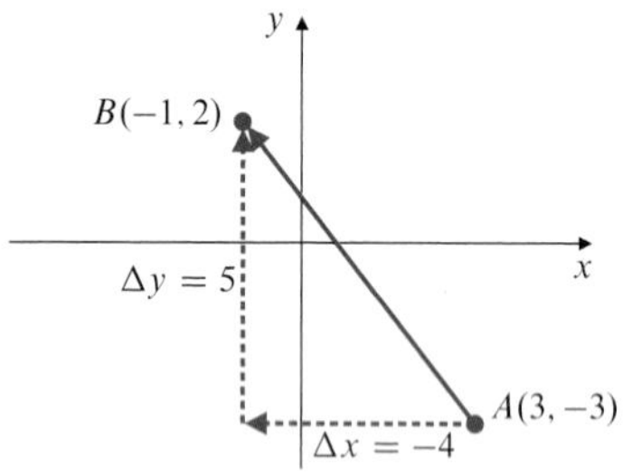

When a particle moves from one point to another, the net changes in its coordinates are called increments. They are calculated by subtracting the coordinates of the starting point from the coordinates of the ending point. An **increment** in a variable is the net change in the value of the variable. If x changes from x_1 to x_2, then the increment in x is $\Delta x = x_2 - x_1$. (Here Δ is the upper case Greek letter delta.)

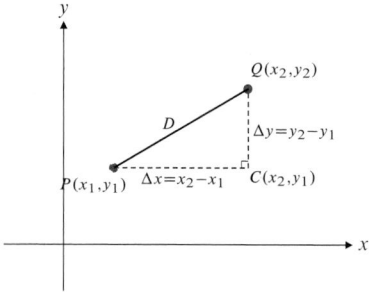

Figure P.11 Increments in x and y

EXAMPLE 1 Find the increments in the coordinates of a particle that moves from $A(3, -3)$ to $B(-1, 2)$.

Solution The increments (see Figure P.11) are:

$$\Delta x = -1 - 3 = -4 \qquad \text{and} \qquad \Delta y = 2 - (-3) = 5.$$

If $P(x_1, y_1)$ and $Q(x_2, y_2)$ are two points in the plane, the straight line segment PQ is the hypotenuse of a right triangle PCQ, as shown in Figure P.12. The sides PC and CQ of the triangle have lengths

$$|\Delta x| = |x_2 - x_1| \qquad \text{and} \qquad |\Delta y| = |y_2 - y_1|.$$

These are the *horizontal distance* and *vertical distance* between P and Q. By the Pythagorean Theorem, the length of PQ is the square root of the sum of the squares of these lengths.

Figure P.12 The distance from P to Q is $D = \sqrt{(x_2 - x_1)^2 + (y_2 - y_1)^2}$

> **Distance formula for points in the plane**
>
> The distance D between $P(x_1, y_1)$ and $Q(x_2, y_2)$ is
>
> $$D = \sqrt{(\Delta x)^2 + (\Delta y)^2} = \sqrt{(x_2 - x_1)^2 + (y_2 - y_1)^2}.$$

EXAMPLE 2 The distance between $A(3, -3)$ and $B(-1, 2)$ in Figure P.11 is

$$\sqrt{(-1 - 3)^2 + (2 - (-3))^2} = \sqrt{(-4)^2 + 5^2} = \sqrt{41} \text{ units.}$$

EXAMPLE 3   The distance from the origin $O(0,0)$ to a point $P(x, y)$ is

$$\sqrt{(x - 0)^2 + (y - 0)^2} = \sqrt{x^2 + y^2}.$$

Graphs

The **graph** of an equation (or inequality) involving the variables x and y is the set of all points $P(x, y)$ whose coordinates satisfy the equation (or inequality).

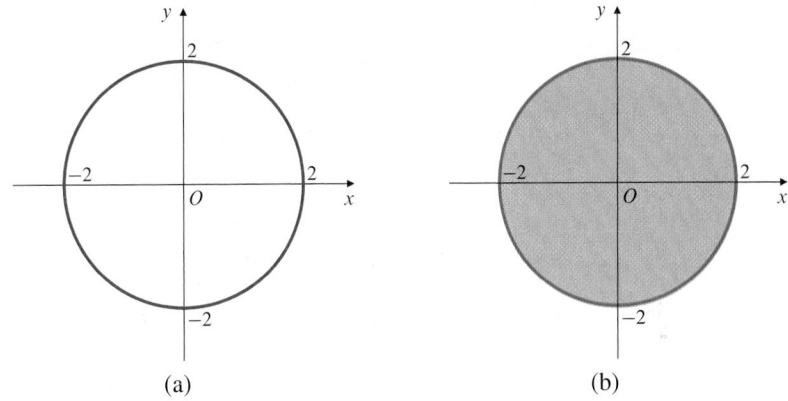

(a)   (b)

Figure P.13

(a) The circle $x^2 + y^2 = 4$

(b) The disk $x^2 + y^2 \le 4$

EXAMPLE 4   The equation $x^2 + y^2 = 4$ represents all points $P(x, y)$ whose distance from the origin is $\sqrt{x^2 + y^2} = \sqrt{4} = 2$. These points lie on the **circle** of radius 2 centred at the origin. This circle is the graph of the equation $x^2 + y^2 = 4$. (See Figure P.13(a).)

EXAMPLE 5   Points (x, y) whose coordinates satisfy the inequality $x^2 + y^2 \le 4$ all have distance ≤ 2 from the origin. The graph of the inequality is therefore the disk of radius 2 centred at the origin. (See Figure P.13(b).)

EXAMPLE 6   Consider the equation $y = x^2$. Some points whose coordinates satisfy this equation are $(0, 0)$, $(1, 1)$, $(-1, 1)$, $(2, 4)$, and $(-2, 4)$. These points (and all others satisfying the equation) lie on a smooth curve called a **parabola**. (See Figure P.14.)

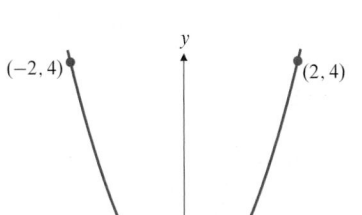

Figure P.14   The parabola $y = x^2$

Straight Lines

Given two points $P_1(x_1, y_1)$ and $P_2(x_2, y_2)$ in the plane, we call the increments $\Delta x = x_2 - x_1$ and $\Delta y = y_2 - y_1$, respectively, the **run** and the **rise** between P_1 and P_2. Two such points always determine a unique **straight line** (usually called simply a **line**) passing through them both. We call the line $P_1 P_2$.

Any nonvertical line in the plane has the property that the ratio

$$m = \frac{\text{rise}}{\text{run}} = \frac{\Delta y}{\Delta x} = \frac{y_2 - y_1}{x_2 - x_1}$$

has the *same value* for every choice of two distinct points $P_1(x_1, y_1)$ and $P_2(x_2, y_2)$ on the line. (See Figure P.15.) The constant $m = \Delta y / \Delta x$ is called the **slope** of the nonvertical line.

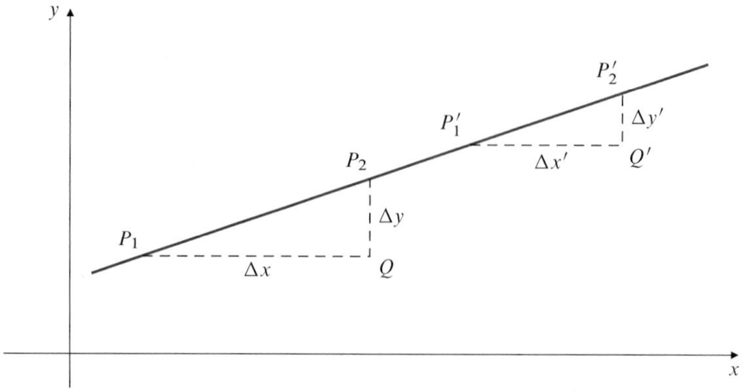

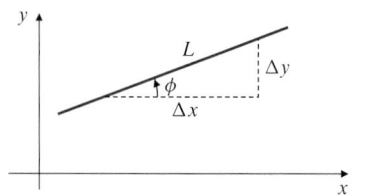

Figure P.15 $\Delta y/\Delta x = \Delta y'/\Delta x'$ because triangles $P_1 Q P_2$ and $P_1' Q' P_2'$ are similar

---

EXAMPLE 7 The slope of the line joining $A\,(3, -3)$ and $B\,(-1, 2)$ is

$$m = \frac{\Delta y}{\Delta x} = \frac{2 - (-3)}{-1 - 3} = \frac{5}{-4} = -\frac{5}{4}.$$

The slope tells us the direction and steepness of a line. A line with positive slope rises uphill to the right; one with negative slope falls downhill to the right. The greater the absolute value of the slope, the steeper the rise or fall. Since the run Δx is zero for a vertical line, we cannot form the ratio m; the slope of a vertical line is *undefined*.

The direction of a line can also be measured by an angle. The **inclination** of a line is the smallest counterclockwise angle from the positive direction of the x-axis to the line. In Figure P.16 the angle ϕ (the Greek letter "phi") is the inclination of the line L. The inclination ϕ of any line satisfies $0° \le \phi < 180°$. The inclination of a horizontal line is $0°$ and that of a vertical line is $90°$.

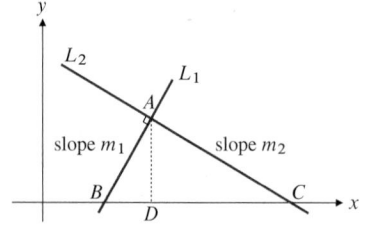

Figure P.16 Line L has inclination ϕ

Provided equal scales are used on the coordinate axes, the relationship between the slope m of a nonvertical line and its inclination ϕ is shown in Figure P.16:

$$m = \frac{\Delta y}{\Delta x} = \tan \phi.$$

(The trigonometric function tan is defined in Section P.7.)

Parallel lines have the same inclination. If they are not vertical, they must therefore have the same slope. Conversely, lines with equal slopes have the same inclination and so are parallel.

If two nonvertical lines, L_1 and L_2, are perpendicular, their slopes m_1 and m_2 satisfy $m_1 m_2 = -1$, so each slope is the *negative reciprocal* of the other:

$$m_1 = -\frac{1}{m_2} \quad \text{and} \quad m_2 = -\frac{1}{m_1}.$$

(This result also assumes equal scales on the two coordinate axes.) To see this, observe in Figure P.17 that

$$m_1 = \frac{AD}{BD} \quad \text{and} \quad m_2 = -\frac{AD}{DC}.$$

Figure P.17 $\triangle ABD$ is similar to $\triangle CAD$

Since $\triangle ABD$ is similar to $\triangle CAD$, we have $\dfrac{AD}{BD} = \dfrac{DC}{AD}$, and so

$$m_1 m_2 = \left(\frac{DC}{AD}\right)\left(-\frac{AD}{DC}\right) = -1.$$

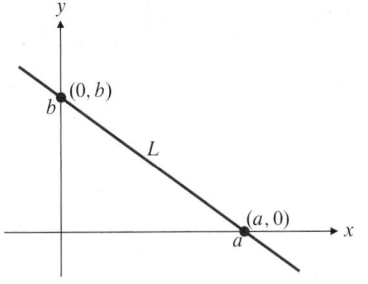

Figure P.18 The lines $y = 1$ and $x = 3$

Equations of Lines

Straight lines are particularly simple graphs, and their corresponding equations are also simple. All points on the vertical line through the point a on the x-axis have their x-coordinates equal to a. Thus $x = a$ is the equation of the line. Similarly, $y = b$ is the equation of the horizontal line meeting the y-axis at b.

EXAMPLE 8 The horizontal and vertical lines passing through the point $(3, 1)$ (Figure P.18) have equations $y = 1$ and $x = 3$, respectively.

To write an equation for a nonvertical straight line L, it is enough to know its slope m and the coordinates of one point $P_1(x_1, y_1)$ on it. If $P(x, y)$ is any other point on L, then

$$\frac{y - y_1}{x - x_1} = m,$$

so that

$$y - y_1 = m(x - x_1) \qquad \text{or} \qquad y = m(x - x_1) + y_1.$$

The equation

$$y = m(x - x_1) + y_1$$

is the **point-slope equation** of the line that passes through the point (x_1, y_1) and has slope m.

EXAMPLE 9 Find an equation of the line that has slope -2 and passes through the point $(1, 4)$.

Solution We substitute $x_1 = 1$, $y_1 = 4$, and $m = -2$ into the point-slope form of the equation and obtain

$$y = -2(x - 1) + 4 \qquad \text{or} \qquad y = -2x + 6.$$

EXAMPLE 10 Find an equation of the line through the points $(1, -1)$ and $(3, 5)$.

Solution The slope of the line is $m = \dfrac{5 - (-1)}{3 - 1} = 3$. We can use this slope with either of the two points to write an equation of the line. If we use $(1, -1)$ we get

$$y = 3(x - 1) - 1, \qquad \text{which simplifies to} \quad y = 3x - 4.$$

If we use $(3, 5)$ we get

$$y = 3(x - 3) + 5, \qquad \text{which also simplifies to} \quad y = 3x - 4.$$

Either way, $y = 3x - 4$ is an equation of the line.

The y-coordinate of the point where a nonvertical line intersects the y-axis is called the **y-intercept** of the line. (See Figure P.19.) Similarly, the **x-intercept** of a non-horizontal line is the x-coordinate of the point where it crosses the x-axis. A line with slope m and y-intercept b passes through the point $(0, b)$, so its equation is

$$y = m(x - 0) + b \qquad \text{or, more simply,} \quad y = mx + b.$$

Figure P.19 Line L has x-intercept a and y-intercept b

A line with slope m and x-intercept a passes through $(a, 0)$, and so its equation is

$$y = m(x - a).$$

> The equation $y = mx + b$ is called the **slope–y-intercept equation** of the line with slope m and y-intercept b.
>
> The equation $y = m(x - a)$ is called the **slope–x-intercept equation** of the line with slope m and x-intercept a.

EXAMPLE 11 Find the slope and the two intercepts of the line with equation $8x + 5y = 20$.

Solution Solving the equation for y we get

$$y = \frac{20 - 8x}{5} = -\frac{8}{5}x + 4.$$

Comparing this with the general form $y = mx + b$ of the slope–y-intercept equation, we see that the slope of the line is $m = -8/5$, and the y-intercept is $b = 4$. To find the x-intercept, put $y = 0$ and solve for x, obtaining $8x = 20$, or $x = 5/2$. The x-intercept is $a = 5/2$.

The equation $Ax + By = C$ (where A and B are not both zero) is called the **general linear equation** in x and y because its graph always represents a straight line, and every line has an equation in this form.

Many important quantities are related by linear equations. Once we know that a relationship between two variables is linear, we can find it from any two pairs of corresponding values, just as we find the equation of a line from the coordinates of two points.

EXAMPLE 12 The relationship between Fahrenheit temperature (F) and Celsius temperature (C) is given by a linear equation of the form $F = mC + b$. The freezing point of water is $F = 32°$ or $C = 0°$, while the boiling point is $F = 212°$ or $C = 100°$. Thus,

$$32 = 0m + b \qquad \text{and} \qquad 212 = 100m + b,$$

so $b = 32$ and $m = (212 - 32)/100 = 9/5$. The relationship is given by the linear equation

$$F = \frac{9}{5}C + 32 \qquad \text{or} \qquad C = \frac{5}{9}(F - 32).$$

EXERCISES P.2

In Exercises 1–4, a particle moves from A to B. Find the net increments Δx and Δy in the particle's coordinates. Also find the distance from A to B.

1. $A(0, 3)$, $B(4, 0)$

2. $A(-1, 2)$, $B(4, -10)$

3. $A(3, 2)$, $B(-1, -2)$

4. $A(0.5, 3)$, $B(2, 3)$

5. A particle starts at $A(-2, 3)$ and its coordinates change by $\Delta x = 4$ and $\Delta y = -7$. Find its new position.

6. A particle arrives at the point $(-2, -2)$ after its coordinates experience increments $\Delta x = -5$ and $\Delta y = 1$. From where did it start?

Describe the graphs of the equations and inequalities in Exercises 7–12.

7. $x^2 + y^2 = 1$

8. $x^2 + y^2 = 2$

9. $x^2 + y^2 \le 1$

10. $x^2 + y^2 = 0$

11. $y \ge x^2$

12. $y < x^2$

In Exercises 13–14, find an equation for (a) the vertical line and (b) the horizontal line through the given point.

13. $(-2, 5/3)$

14. $(\sqrt{2}, -1.3)$

In Exercises 15–18, write an equation for the line through P with slope m.

15. $P(-1, 1), \quad m = 1$

16. $P(-2, 2), \quad m = 1/2$

17. $P(0, b), \quad m = 2$

18. $P(a, 0), \quad m = -2$

In Exercises 19–20, does the given point P lie on, above, or below the given line?

19. $P(2, 1), \quad 2x + 3y = 6$

20. $P(3, -1), \quad x - 4y = 7$

In Exercises 21–24, write an equation for the line through the two points.

21. $(0, 0), \quad (2, 3)$

22. $(-2, 1), \quad (2, -2)$

23. $(4, 1), \quad (-2, 3)$

24. $(-2, 0), \quad (0, 2)$

In Exercises 25–26, write an equation for the line with slope m and y-intercept b.

25. $m = -2, \quad b = \sqrt{2}$

26. $m = -1/2, \quad b = -3$

In Exercises 27–30, determine the x- and y-intercepts and the slope of the given lines, and sketch their graphs.

27. $3x + 4y = 12$

28. $x + 2y = -4$

29. $\sqrt{2}x - \sqrt{3}y = 2$

30. $1.5x - 2y = -3$

In Exercises 31–32, find equations for the lines through P that are (a) parallel to and (b) perpendicular to the given line.

31. $P(2, 1), \quad y = x + 2$

32. $P(-2, 2), \quad 2x + y = 4$

33. Find the point of intersection of the lines $3x + 4y = -6$ and $2x - 3y = 13$.

34. Find the point of intersection of the lines $2x + y = 8$ and $5x - 7y = 1$.

35. **(Two-intercept equations)** If a line is neither horizontal nor vertical and does not pass through the origin, show that its equation can be written in the form $\dfrac{x}{a} + \dfrac{y}{b} = 1$, where a is its x-intercept and b is its y-intercept.

36. Determine the intercepts and sketch the graph of the line $\dfrac{x}{2} - \dfrac{y}{3} = 1$.

37. Find the y-intercept of the line through the points $(2, 1)$ and $(3, -1)$.

38. A line passes through $(-2, 5)$ and $(k, 1)$ and has x-intercept 3. Find k.

39. The cost of printing x copies of a pamphlet is $\$C$, where $C = Ax + B$ for certain constants A and B. If it costs $\$5,000$ to print 10,000 copies and $\$6,000$ to print 15,000 copies, how much will it cost to print 100,000 copies?

40. **(Fahrenheit versus Celsius)** In the FC-plane, sketch the graph of the equation $C = \dfrac{5}{9}(F - 32)$ linking Fahrenheit and Celsius temperatures found in Example 12. On the same graph sketch the line with equation $C = F$. Is there a temperature at which a Celsius thermometer gives the same numerical reading as a Fahrenheit thermometer? If so, find that temperature.

Geometry

41. By calculating the lengths of its three sides, show that the triangle with vertices at the points $A(2, 1)$, $B(6, 4)$, and $C(5, -3)$ is isosceles.

42. Show that the triangle with vertices $A(0, 0)$, $B(1, \sqrt{3})$, and $C(2, 0)$ is equilateral.

43. Show that the points $A(2, -1)$, $B(1, 3)$, and $C(-3, 2)$ are three vertices of a square and find the fourth vertex.

44. Find the coordinates of the midpoint on the line segment $P_1 P_2$ joining the points $P_1(x_1, y_1)$ and $P_2(x_2, y_2)$.

45. Find the coordinates of the point of the line segment joining the points $P_1(x_1, y_1)$ and $P_2(x_2, y_2)$ that is two-thirds of the way from P_1 to P_2.

46. The point P lies on the x-axis and the point Q lies on the line $y = -2x$. The point $(2, 1)$ is the midpoint of PQ. Find the coordinates of P.

In Exercises 47–48, interpret the equation as a statement about distances, and hence determine the graph of the equation.

47. $\sqrt{(x - 2)^2 + y^2} = 4$

48. $\sqrt{(x - 2)^2 + y^2} = \sqrt{x^2 + (y - 2)^2}$

49. For what value of k is the line $2x + ky = 3$ perpendicular to the line $4x + y = 1$? For what value of k are the lines parallel?

50. Find the line that passes through the point $(1, 2)$ and through the point of intersection of the two lines $x + 2y = 3$ and $2x - 3y = -1$.

P.3 Graphs of Quadratic Equations

This section reviews circles, parabolas, ellipses, and hyperbolas, the graphs that are represented by quadratic equations in two variables.

Circles and Disks

The **circle** having **centre** C and **radius** a is the set of all points in the plane that are at distance a from the point C.

The distance from $P(x, y)$ to the point $C(h, k)$ is $\sqrt{(x - h)^2 + (y - k)^2}$, so that the equation of the circle of radius $a > 0$ with centre at $C(h, k)$ is

$$\sqrt{(x - h)^2 + (y - k)^2} = a.$$

A simpler form of this equation is obtained by squaring both sides.

Standard equation of a circle

The circle with centre (h, k) and radius $a \geq 0$ has equation

$$(x - h)^2 + (y - k)^2 = a^2.$$

In particular, the circle with centre at the origin $(0, 0)$ and radius a has equation

$$x^2 + y^2 = a^2.$$

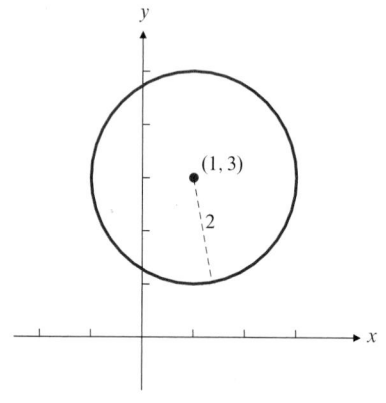

Figure P.20   Circle
$(x - 1)^2 + (y - 3)^2 = 4$

EXAMPLE 1   The circle with radius 2 and centre $(1, 3)$ (Figure P.20) has equation $(x - 1)^2 + (y - 3)^2 = 4$.

EXAMPLE 2   The circle having equation $(x + 2)^2 + (y - 1)^2 = 7$ has centre at the point $(-2, 1)$ and radius $\sqrt{7}$. (See Figure P.21.)

If the squares in the standard equation $(x - h)^2 + (y - k)^2 = a^2$ are multiplied out, and all constant terms collected on the right-hand side, the equation becomes

$$x^2 - 2hx + y^2 - 2ky = a^2 - h^2 - k^2.$$

A quadratic equation of the form

$$x^2 + y^2 + 2ax + 2by = c$$

must represent a circle, which can be a single point if the radius is 0, or no points at all. To identify the graph, we complete the squares on the left side of the equation. Since $x^2 + 2ax$ are the first two terms of the square $(x + a)^2 = x^2 + 2ax + a^2$, we add a^2 to both sides to complete the square of the x terms. (Note that a^2 is *the square of half the coefficient of* x.) Similarly, add b^2 to both sides to complete the square of the y terms. The equation then becomes

$$(x + a)^2 + (y + b)^2 = c + a^2 + b^2.$$

If $c + a^2 + b^2 > 0$, the graph is a circle with centre $(-a, -b)$ and radius $\sqrt{c + a^2 + b^2}$. If $c + a^2 + b^2 = 0$, the graph consists of the single point $(-a, -b)$. If $c + a^2 + b^2 < 0$, no points lie on the graph.

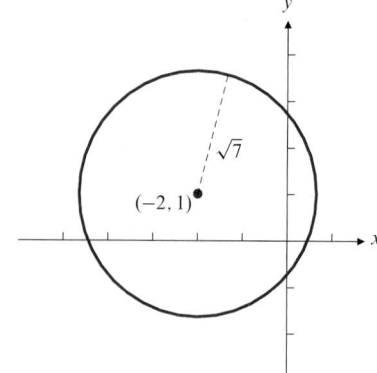

Figure P.21   Circle
$(x + 2)^2 + (y - 1)^2 = 7$

EXAMPLE 3   Find the centre and radius of the circle $x^2 + y^2 - 4x + 6y = 3$.

Solution   Observe that $x^2 - 4x$ are the first two terms of the binomial square $(x - 2)^2 = x^2 - 4x + 4$, and $y^2 + 6y$ are the first two terms of the square $(y + 3)^2 = y^2 + 6y + 9$. Hence, we add $4 + 9$ to both sides of the given equation and obtain

$$x^2 - 4x + 4 + y^2 + 6y + 9 = 3 + 4 + 9 \quad \text{or} \quad (x - 2)^2 + (y + 3)^2 = 16.$$

This is the equation of a circle with centre $(2, -3)$ and radius 4.

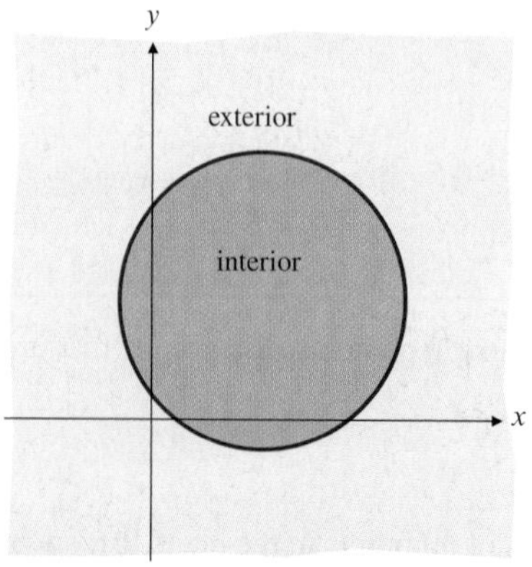

The set of all points *inside* a circle is called the **interior** of the circle; it is also called an **open disk**. The set of all points *outside* the circle is called the **exterior** of the circle. (See Figure P.22.) The interior of a circle together with the circle itself is called a **closed disk**, or simply a **disk**. The inequality

$$(x - h)^2 + (y - k)^2 \leq a^2$$

represents the disk of radius $|a|$ centred at (h, k).

Figure P.22   The interior (green) of a circle (red) and the exterior (blue)

EXAMPLE 4 Identify the graphs of

(a) $x^2 + 2x + y^2 \leq 8$ (b) $x^2 + 2x + y^2 < 8$ (c) $x^2 + 2x + y^2 > 8$.

Solution We can complete the square in the equation $x^2 + y^2 + 2x = 8$ as follows:

$$x^2 + 2x + 1 + y^2 = 8 + 1$$
$$(x + 1)^2 + y^2 = 9.$$

Thus the equation represents the circle of radius 3 with centre at $(-1, 0)$. Inequality (a) represents the (closed) disk with the same radius and centre. (See Figure P.23.) Inequality (b) represents the interior of the circle (or the open disk). Inequality (c) represents the exterior of the circle.

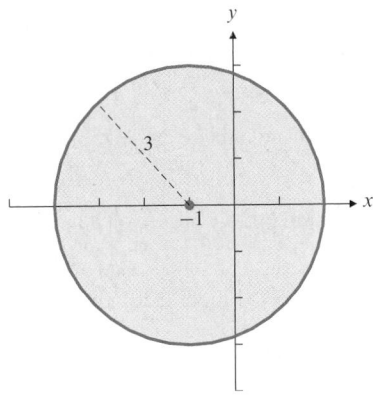

Figure P.23   The disk $x^2 + y^2 + 2x \leq 8$

Equations of Parabolas

A **parabola** is a plane curve whose points are equidistant from a fixed point F and a fixed straight line L that does not pass through F. The point F is the **focus** of the parabola; the line L is the parabola's **directrix**. The line through F perpendicular to L is the parabola's **axis**. The point V where the axis meets the parabola is the parabola's **vertex**.

Observe that the vertex V of a parabola is halfway between the focus F and the point on the directrix L that is closest to F. If the directrix is either horizontal or vertical, and the vertex is at the origin, then the parabola will have a particularly simple equation.

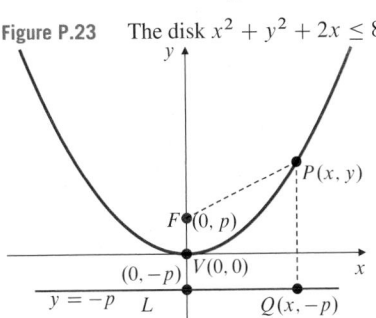

Figure P.24   The parabola $4py = x^2$ with focus $F(0, p)$ and directrix $y = -p$

EXAMPLE 5 Find an equation of the parabola having the point $F(0, p)$ as focus and the line L with equation $y = -p$ as directrix.

Solution If $P(x, y)$ is any point on the parabola, then (see Figure P.24) the distances from P to F and to (the closest point Q on) the line L are given by

$$PF = \sqrt{(x - 0)^2 + (y - p)^2} = \sqrt{x^2 + y^2 - 2py + p^2}$$
$$PQ = \sqrt{(x - x)^2 + (y - (-p))^2} = \sqrt{y^2 + 2py + p^2}.$$

Since P is on the parabola, $PF = PQ$ and so the squares of these distances are also equal:

$$x^2 + y^2 - 2py + p^2 = y^2 + 2py + p^2,$$

or, after simplifying,

$$x^2 = 4py \qquad \text{or} \qquad y = \frac{x^2}{4p} \qquad \text{(called **standard forms**)}.$$

Figure P.24 shows the situation for $p > 0$; the parabola opens upward and is symmetric about its axis, the y-axis. If $p < 0$, the focus $(0, p)$ will lie below the origin and the directrix $y = -p$ will lie above the origin. In this case the parabola will open downward instead of upward.

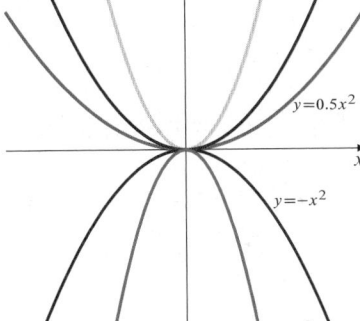

Figure P.25   Some parabolas $y = ax^2$

Figure P.25 shows several parabolas with equations of the form $y = ax^2$ for positive and negative values of a.

EXAMPLE 6 An equation for the parabola with focus $(0, 1)$ and directrix $y = -1$ is $y = x^2/4$, or $x^2 = 4y$. (We took $p = 1$ in the standard equation.)

EXAMPLE 7   Find the focus and directrix of the parabola $y = -x^2$.

Solution The given equation matches the standard form $y = x^2/(4p)$ provided $4p = -1$. Thus $p = -1/4$. The focus is $(0, -1/4)$, and the directrix is the line $y = 1/4$.

───────────────────────────────●

Interchanging the roles of x and y in the derivation of the standard equation above shows that the equation

$$y^2 = 4px \qquad \text{or} \qquad x = \frac{y^2}{4p} \qquad \text{(standard equation)}$$

represents a parabola with focus at $(p, 0)$ and vertical directrix $x = -p$. The axis is the x-axis.

Reflective Properties of Parabolas

One of the chief applications of parabolas is their use as reflectors of light and radio waves. Rays originating from the focus of a parabola will be reflected in a beam parallel to the axis, as shown in Figure P.26. Similarly, all the rays in a beam striking a parabola parallel to its axis will reflect through the focus. This property is the reason why telescopes and spotlights use parabolic mirrors and radio telescopes and microwave antennas are parabolic in shape. We will examine this property of parabolas more carefully in Section 8.1.

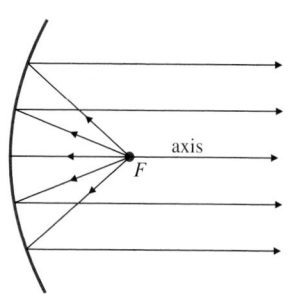

Figure P.26   Reflection by a parabola

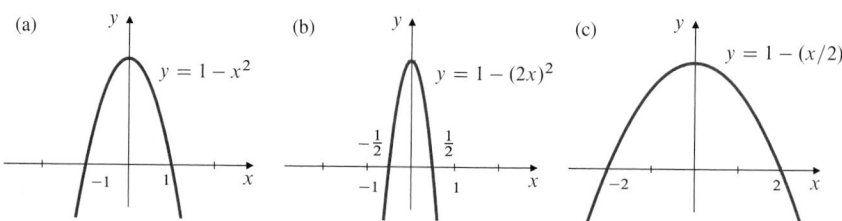

Figure P.27   Horizontal scaling:
(a) the graph $y = 1 - x^2$
(b) graph of (a) compressed horizontally
(c) graph of (a) expanded horizontally

Scaling a Graph

The graph of an equation can be compressed or expanded horizontally by replacing x with a multiple of x. If a is a positive number, replacing x with ax in an equation multiplies horizontal distances in the graph of the equation by a factor $1/a$. (See Figure P.27.) Replacing y with ay will multiply vertical distances in a similar way.

You may find it surprising that, like circles, all parabolas are *similar* geometric figures; they may have different sizes, but they all have the same shape. We can change the *size* while preserving the shape of a curve represented by an equation in x and y by scaling both the coordinates by the same amount. If we scale the equation $4py = x^2$ by replacing x and y with $4px$ and $4py$, respectively, we get $4p(4py) = (4px)^2$, or $y = x^2$. Thus, the general parabola $4py = x^2$ has the same shape as the specific parabola $y = x^2$, as shown in Figure P.28.

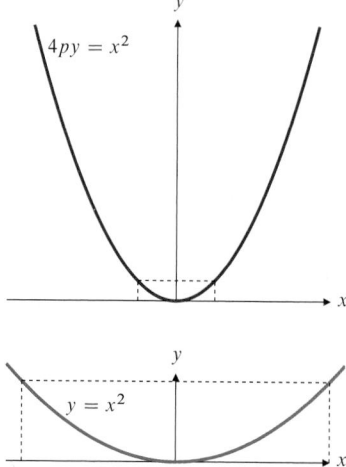

Figure P.28   The two parabolas are similar. Compare the parts inside the rectangles

Shifting a Graph

The graph of an equation (or inequality) can be shifted c units horizontally by replacing x with $x - c$ or vertically by replacing y with $y - c$.

> **Shifts**
>
> To shift a graph c units to the right, replace x in its equation or inequality with $x - c$. (If $c < 0$, the shift will be to the left.)
>
> To shift a graph c units upward, replace y in its equation or inequality with $y - c$. (If $c < 0$, the shift will be downward.)

EXAMPLE 8   The graph of $y = (x-3)^2$ (blue) is the parabola $y = x^2$ (red) shifted 3 units to the right. The graph of $y = (x+1)^2$ (green) is the parabola $y = x^2$ shifted 1 unit to the left. (See Figure P.29(a).)

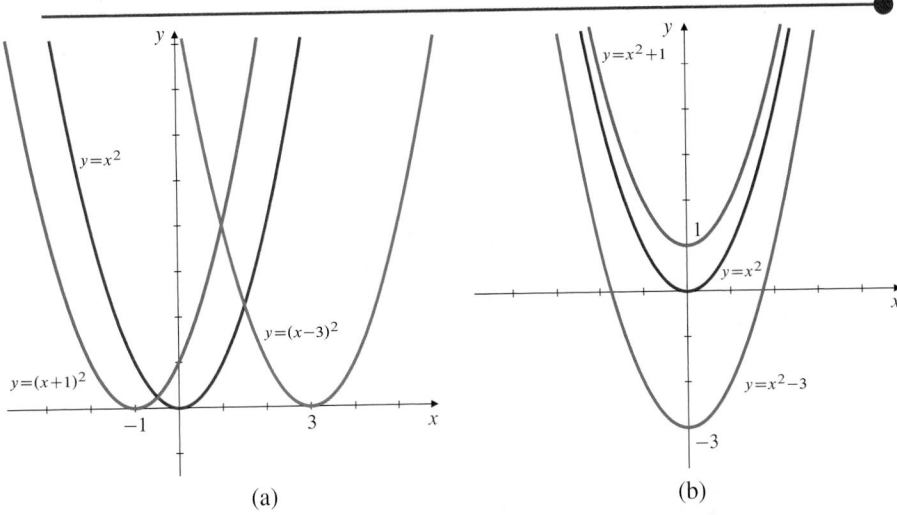

Figure P.29

(a) Horizontal shifts of $y = x^2$

(b) Vertical shifts of $y = x^2$

(a)   (b)

EXAMPLE 9   The graph of $y = x^2 + 1$ (or $y - 1 = x^2$) (green in Figure P.29(b)) is the parabola $y = x^2$ (red) shifted up 1 unit. The graph of $y = x^2 - 3$ (or $y - (-3) = x^2$) (blue) is the parabola $y = x^2$ shifted down 3 units.

EXAMPLE 10   The circle with equation $(x-h)^2 + (y-k)^2 = a^2$ having centre (h,k) and radius a can be obtained by shifting the circle $x^2 + y^2 = a^2$ of radius a centred at the origin h units to the right and k units upward. These shifts correspond to replacing x with $x - h$ and y with $y - k$.

The graph of $y = ax^2 + bx + c$ is a parabola whose axis is parallel to the y-axis. The parabola opens upward if $a > 0$ and downward if $a < 0$. We can complete the square and write the equation in the form $y = a(x-h)^2 + k$ to find the vertex (h,k).

EXAMPLE 11   Describe the graph of $y = x^2 - 4x + 3$.

Solution   The equation $y = x^2 - 4x + 3$ represents a parabola, opening upward. To find its vertex and axis we can complete the square:

$$y = x^2 - 4x + 4 - 1 = (x-2)^2 - 1, \quad \text{so} \quad y - (-1) = (x-2)^2.$$

This curve is the parabola $y = x^2$ shifted to the right 2 units and down 1 unit. Therefore, its vertex is $(2,-1)$, and its axis is the line $x = 2$. Since $y = x^2$ has focus $(0, 1/4)$, the focus of this parabola is $(0+2, (1/4)-1)$, or $(2, -3/4)$. (See Figure P.30.)

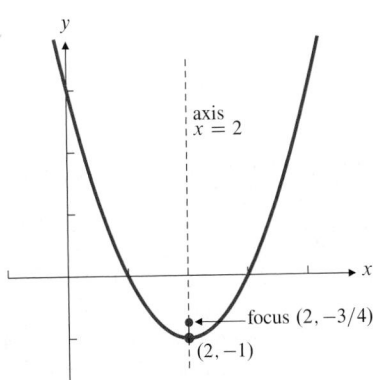

Figure P.30   The parabola $y = x^2 - 4x + 3$

Ellipses and Hyperbolas

If a and b are positive numbers, the equation

$$\frac{x^2}{a^2} + \frac{y^2}{b^2} = 1$$

represents a curve called an **ellipse** that lies wholly within the rectangle $-a \le x \le a$, $-b \le y \le b$. (Why?) If $a = b$, the ellipse is just the circle of radius a centred at the origin. If $a \ne b$, the ellipse is a circle that has been squashed by scaling it by different amounts in the two coordinate directions.

The ellipse has centre at the origin, and it passes through the four points $(a, 0)$, $(0, b)$, $(-a, 0)$, and $(0, -b)$. (See Figure P.31.) The line segments from $(-a, 0)$ to $(a, 0)$ and from $(0, -b)$ to $(0, b)$ are called the **principal axes** of the ellipse; the longer of the two is the **major axis**, and the shorter is the **minor axis**.

EXAMPLE 12 The equation $\dfrac{x^2}{9} + \dfrac{y^2}{4} = 1$ represents an ellipse with major axis from $(-3, 0)$ to $(3, 0)$ and minor axis from $(0, -2)$ to $(0, 2)$.

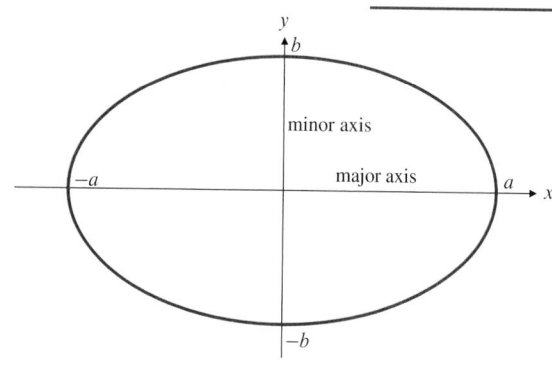

Figure P.31   The ellipse $\dfrac{x^2}{a^2} + \dfrac{y^2}{b^2} = 1$

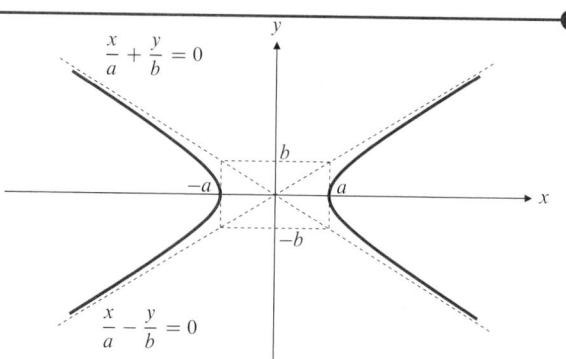

Figure P.32   The hyperbola $\dfrac{x^2}{a^2} - \dfrac{y^2}{b^2} = 1$ and its asymptotes

The equation

$$\frac{x^2}{a^2} - \frac{y^2}{b^2} = 1$$

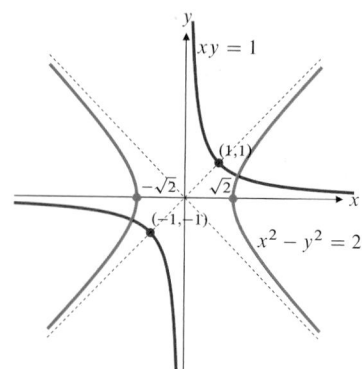

Figure P.33   Two rectangular hyperbolas

represents a curve called a **hyperbola** that has centre at the origin and passes through the points $(-a, 0)$ and $(a, 0)$. (See Figure P.32.) The curve is in two parts (called **branches**). Each branch approaches two straight lines (called **asymptotes**) as it recedes far away from the origin. The asymptotes have equations

$$\frac{x}{a} - \frac{y}{b} = 0 \qquad \text{and} \qquad \frac{x}{a} + \frac{y}{b} = 0.$$

The equation $xy = 1$ also represents a hyperbola. This one passes through the points $(-1, -1)$ and $(1, 1)$ and has the coordinate axes as its asymptotes. It is, in fact, the hyperbola $x^2 - y^2 = 2$ rotated $45°$ counterclockwise about the origin. (See Figure P.33.) These hyperbolas are called **rectangular hyperbolas**, since their asymptotes intersect at right angles.

We will study ellipses and hyperbolas in more detail in Chapter 8.

EXERCISES P.3

In Exercises 1–4, write an equation for the circle with centre C and radius r.

1. $C(0, 0)$, $r = 4$

2. $C(0, 2)$, $r = 2$

3. $C(-2, 0)$, $r = 3$

4. $C(3, -4)$, $r = 5$

In Exercises 5–8, find the centre and radius of the circle having the given equation.

5. $x^2 + y^2 - 2x = 3$

6. $x^2 + y^2 + 4y = 0$

7. $x^2 + y^2 - 2x + 4y = 4$

8. $x^2 + y^2 - 2x - y + 1 = 0$

Describe the regions defined by the inequalities and pairs of inequalities in Exercises 9–16.

9. $x^2 + y^2 > 1$

10. $x^2 + y^2 < 4$

11. $(x + 1)^2 + y^2 \le 4$

12. $x^2 + (y - 2)^2 \le 4$

13. $x^2 + y^2 > 1$, $x^2 + y^2 < 4$

14. $x^2 + y^2 \le 4$, $(x + 2)^2 + y^2 \le 4$

15. $x^2 + y^2 < 2x$, $x^2 + y^2 < 2y$

16. $x^2 + y^2 - 4x + 2y > 4$, $x + y > 1$

17. Write an inequality that describes the interior of the circle with centre $(-1, 2)$ and radius $\sqrt{6}$.

18. Write an inequality that describes the exterior of the circle with centre $(2, -3)$ and radius 4.

19. Write a pair of inequalities that describe that part of the interior of the circle with centre $(0, 0)$ and radius $\sqrt{2}$ lying on or to the right of the vertical line through $(1, 0)$.

20. Write a pair of inequalities that describe the points that lie outside the circle with centre $(0, 0)$ and radius 2, and inside the circle with centre $(1, 3)$ that passes through the origin.

In Exercises 21–24, write an equation of the parabola having the given focus and directrix.

21. Focus: $(0, 4)$ Directrix: $y = -4$

22. Focus: $(0, -1/2)$ Directrix: $y = 1/2$

23. Focus: $(2, 0)$ Directrix: $x = -2$

24. Focus: $(-1, 0)$ Directrix: $x = 1$

In Exercises 25–28, find the parabola's focus and directrix, and make a sketch showing the parabola, focus, and directrix.

25. $y = x^2/2$ **26.** $y = -x^2$

27. $x = -y^2/4$ **28.** $x = y^2/16$

29. Figure P.34 shows the graph $y = x^2$ and four shifted versions of it. Write equations for the shifted versions.

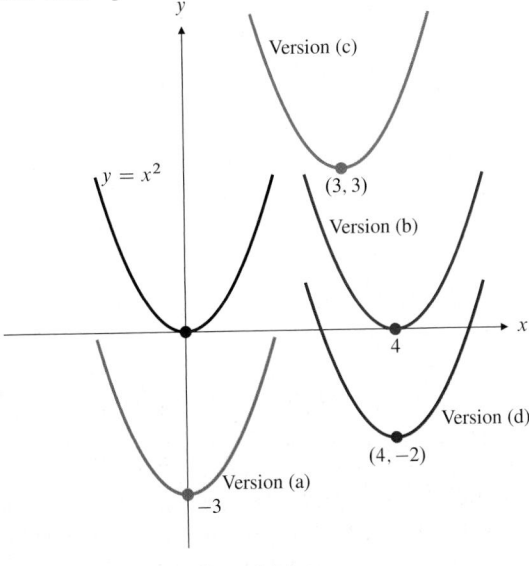

Figure P.34

30. What equations result from shifting the line $y = mx$
  (a) horizontally to make it pass through the point (a, b)
  (b) vertically to make it pass through (a, b)?

In Exercises 31–34, the graph of $y = \sqrt{x + 1}$ is to be scaled in the indicated way. Give the equation of the graph that results from the scaling where

31. horizontal distances are multiplied by 3.

32. vertical distances are divided by 4.

33. horizontal distances are multiplied by 2/3.

34. horizontal distances are divided by 4 and vertical distances are multiplied by 2.

In Exercises 35–38, write an equation for the graph obtained by shifting the graph of the given equation as indicated.

35. $y = 1 - x^2$ down 1, left 1

36. $x^2 + y^2 = 5$ up 2, left 4

37. $y = (x - 1)^2 - 1$ down 1, right 1

38. $y = \sqrt{x}$ down 2, left 4

Find the points of intersection of the pairs of curves in Exercises 39–42.

39. $y = x^2 + 3$, $y = 3x + 1$

40. $y = x^2 - 6$, $y = 4x - x^2$

41. $x^2 + y^2 = 25$, $3x + 4y = 0$

42. $2x^2 + 2y^2 = 5$, $xy = 1$

In Exercises 43–50, identify and sketch the curve represented by the given equation.

43. $\dfrac{x^2}{4} + y^2 = 1$ **44.** $9x^2 + 16y^2 = 144$

45. $\dfrac{(x - 3)^2}{9} + \dfrac{(y + 2)^2}{4} = 1$ **46.** $(x - 1)^2 + \dfrac{(y + 1)^2}{4} = 4$

47. $\dfrac{x^2}{4} - y^2 = 1$ **48.** $x^2 - y^2 = -1$

49. $xy = -4$ **50.** $(x - 1)(y + 2) = 1$

51. What is the effect on the graph of an equation in x and y of
  (a) replacing x with $-x$?
  (b) replacing y with $-y$?

52. What is the effect on the graph of an equation in x and y of replacing x with $-x$ and y with $-y$ simultaneously?

53. Sketch the graph of $|x| + |y| = 1$.

P.4 Functions and Their Graphs

The area of a circle depends on its radius. The temperature at which water boils depends on the altitude above sea level. The interest paid on a cash investment depends on the length of time for which the investment is made.

Whenever one quantity depends on another quantity, we say that the former quantity is a function of the latter. For instance, the area A of a circle depends on the radius r according to the formula

$$A = \pi r^2,$$

so we say that the area is a function of the radius. The formula is a *rule* that tells us how to calculate a *unique* (single) output value of the area A for each possible input value of the radius r.

The set of all possible input values for the radius is called the **domain** of the function. The set of all output values of the area is the **range** of the function. Since circles cannot have negative radii or areas, the domain and range of the circular area function are both the interval $[0, \infty)$ consisting of all nonnegative real numbers.

The domain and range of a mathematical function can be any sets of objects; they do not have to consist of numbers. Throughout much of this book, however, the domains and ranges of functions we consider will be sets of real numbers.

In calculus we often want to refer to a generic function without having any particular formula in mind. To denote that y is a function of x we write

$$y = f(x),$$

which we read as "y equals f of x." In this notation, due to the eighteenth-century mathematician Leonhard Euler, the function is represented by the symbol f. Also, x, called the **independent variable**, represents an input value from the domain of f, and y, the **dependent variable**, represents the corresponding output value $f(x)$ in the range of f.

DEFINITION

1

> A **function** f on a set D into a set S is a rule that assigns a *unique* element $f(x)$ in S to each element x in D.

In this definition $D = \mathcal{D}(f)$ (read "D of f") is the domain of the function f. The range $\mathcal{R}(f)$ of f is the subset of S consisting of all *values $f(x)$* of the function. Think of a function f as a kind of machine (Figure P.35) that produces an output value $f(x)$ in its range whenever we feed it an input value x from its domain.

There are several ways to represent a function symbolically. The squaring function that converts any input real number x into its square x^2 can be denoted:

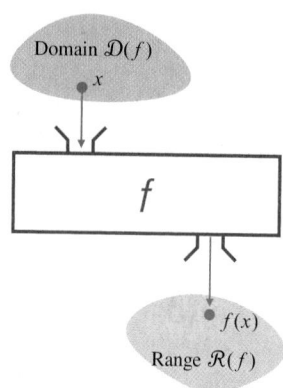

Domain $\mathcal{D}(f)$

x

f

$f(x)$

Range $\mathcal{R}(f)$

Figure P.35   A function machine

(a) by a formula such as $y = x^2$, which uses a dependent variable y to denote the value of the function;

(b) by a formula such as $f(x) = x^2$, which defines a function symbol f to name the function; or

(c) by a mapping rule such as $x \longrightarrow x^2$. (Read this as "x goes to x^2.")

In this book we will usually use either (a) or (b) to define functions. Strictly speaking, we should call a function f and not $f(x)$, since the latter denotes the value of the function at the point x. However, as is common usage, we will often refer to the function as $f(x)$ in order to name the variable on which f depends. Sometimes it is convenient to use the same letter to denote both a dependent variable and a function symbol; the circular area function can be written $A = f(r) = \pi r^2$ or as $A = A(r) = \pi r^2$. In the latter case we are using A to denote both the dependent variable and the name of the function.

EXAMPLE 1   The volume of a ball of radius r is given by the function

$$V(r) = \frac{4}{3} \pi r^3$$

for $r \geq 0$. Thus the volume of a ball of radius 3 ft is

$$V(3) = \frac{4}{3} \pi (3)^3 = 36\pi \ \text{ft}^3.$$

Note how the variable r is replaced by the special value 3 in the formula defining the function to obtain the value of the function at $r = 3$.

EXAMPLE 2 A function F is defined for all real numbers t by

$$F(t) = 2t + 3.$$

Find the output values of F that correspond to the input values 0, 2, $x + 2$, and $F(2)$.

Solution In each case we substitute the given input for t in the definition of F:

$$F(0) = 2(0) + 3 = 0 + 3 = 3$$
$$F(2) = 2(2) + 3 = 4 + 3 = 7$$
$$F(x + 2) = 2(x + 2) + 3 = 2x + 7$$
$$F(F(2)) = F(7) = 2(7) + 3 = 17.$$

The Domain Convention

A function is not properly defined until its domain is specified. For instance, the function $f(x) = x^2$ defined for all real numbers $x \geq 0$ is different from the function $g(x) = x^2$ defined for all real x because they have different domains, even though they have the same values at every point where both are defined. In Chapters 1–9 we will be dealing with real functions (functions whose input and output values are real numbers). When the domain of such a function is not specified explicitly, we will assume that the domain is the largest set of real numbers to which the function assigns real values. Thus, if we talk about the function x^2 without specifying a domain, we mean the function $g(x)$ above.

> **The domain convention**
>
> When a function f is defined without specifying its domain, we assume that the domain consists of all real numbers x for which the value $f(x)$ of the function is a real number.

In practice, it is often easy to determine the domain of a function $f(x)$ given by an explicit formula. We just have to exclude those values of x that would result in dividing by 0 or taking even roots of negative numbers.

EXAMPLE 3 **The square root function.** The domain of $f(x) = \sqrt{x}$ is the interval $[0, \infty)$, since negative numbers do not have real square roots. We have $f(0) = 0$, $f(4) = 2$, $f(10) \approx 3.16228$. Note that, although there are *two* numbers whose square is 4, namely, -2 and 2, only *one* of these numbers, 2, is the square root of 4. (Remember that a function assigns a *unique* value to each element in its domain; it cannot assign two different values to the same input.) The **square root function** $\sqrt{x}$ always denotes the *nonnegative* square root of x. The two solutions of the equation $x^2 = 4$ are $x = \sqrt{4} = 2$ and $x = -\sqrt{4} = -2$.

EXAMPLE 4 The domain of the function $h(x) = \dfrac{x}{x^2 - 4}$ consists of all real numbers except $x = -2$ and $x = 2$. Expressed in terms of intervals,

$$\mathcal{D}(h) = (-\infty, -2) \cup (-2, 2) \cup (2, \infty).$$

Most of the functions we encounter will have domains that are either intervals or unions of intervals.

EXAMPLE 5 The domain of $S(t) = \sqrt{1 - t^2}$ consists of all real numbers t for which $1 - t^2 \geq 0$. Thus we require that $t^2 \leq 1$, or $-1 \leq t \leq 1$. The domain is the closed interval $[-1, 1]$.

Graphs of Functions

An old maxim states that "a picture is worth a thousand words." This is certainly true in mathematics; the behaviour of a function is best described by drawing its graph.

The **graph of a function** f is just the graph of the *equation* $y = f(x)$. It consists of those points in the Cartesian plane whose coordinates (x, y) are pairs of input–output values for f. Thus (x, y) lies on the graph of f provided x is in the domain of f and $y = f(x)$.

Drawing the graph of a function f sometimes involves making a table of coordinate pairs $(x, f(x))$ for various values of x in the domain of f, then plotting these points and connecting them with a "smooth curve."

EXAMPLE 6 Graph the function $f(x) = x^2$.

Solution Make a table of (x, y) pairs that satisfy $y = x^2$. (See Table 1.) Now plot the points and join them with a smooth curve. (See Figure P.36(a).)

Table 1.

x	$y = f(x)$
-2	4
-1	1
0	0
1	1
2	4

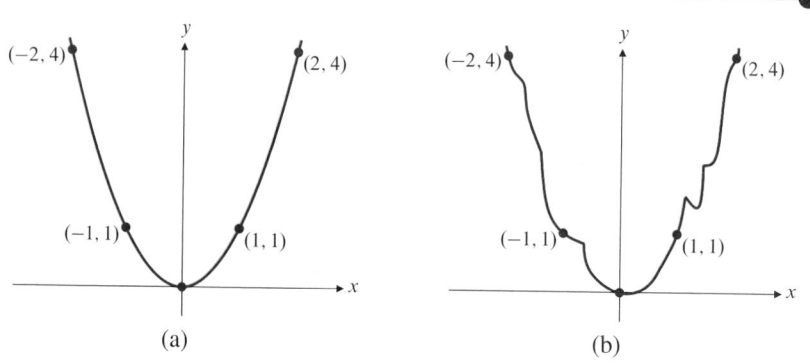

Figure P.36

(a) Correct graph of $f(x) = x^2$

(b) Incorrect graph of $f(x) = x^2$

How do we know the graph is smooth and doesn't do weird things between the points we have calculated, for example, as shown in Figure P.36(b)? We could, of course, plot more points, spaced more closely together, but how do we know how the graph behaves between the points we have plotted? In Chapter 4, calculus will provide useful tools for answering these questions.

Some functions occur often enough in applications that you should be familiar with their graphs. Some of these are shown in Figures P.37–P.46. Study them for a while; they are worth remembering. Note, in particular, the graph of the **absolute value function**, $f(x) = |x|$, shown in Figure P.46. It is made up of the two half-lines $y = -x$ for $x < 0$ and $y = x$ for $x \geq 0$.

If you know the effects of vertical and horizontal shifts on the equations representing graphs (see Section P.3), you can easily sketch some graphs that are shifted versions of the ones in Figures P.37–P.46.

EXAMPLE 7 Sketch the graph of $y = 1 + \sqrt{x - 4}$.

Solution This is just the graph of $y = \sqrt{x}$ in Figure P.40 shifted to the right 4 units (because x is replaced by $x - 4$) and up 1 unit. See Figure P.47.

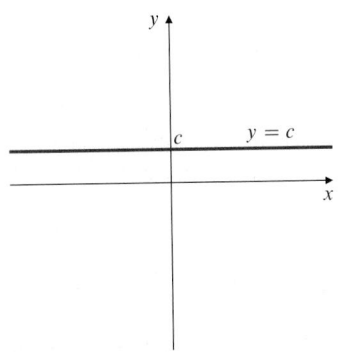

Figure P.37 The graph of a constant function $f(x) = c$

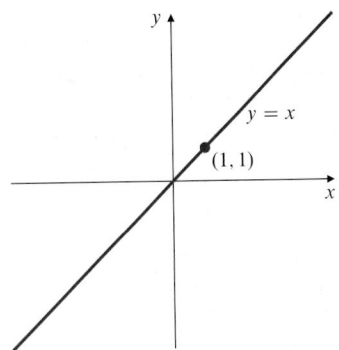

Figure P.38 The graph of $f(x) = x$

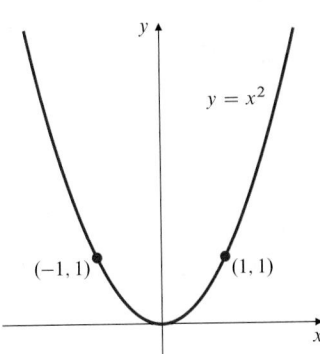

Figure P.39 The graph of $f(x) = x^2$

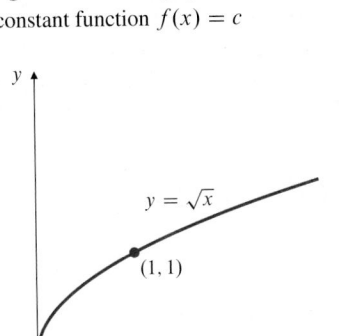

Figure P.40 The graph of $f(x) = \sqrt{x}$

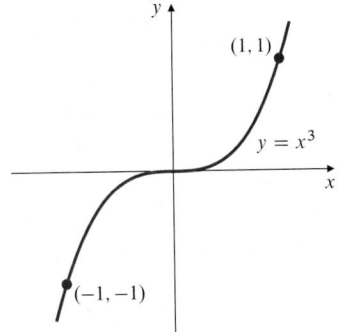

Figure P.41 The graph of $f(x) = x^3$

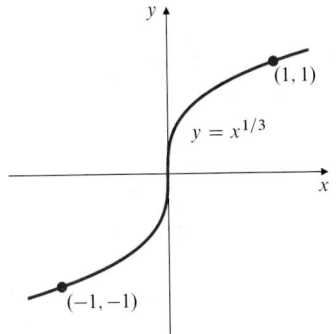

Figure P.42 The graph of $f(x) = x^{1/3}$

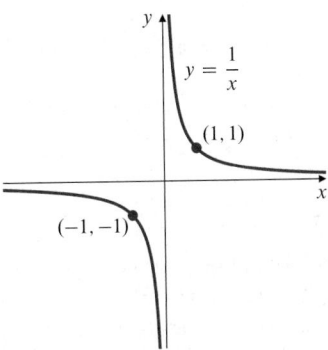

Figure P.43 The graph of $f(x) = 1/x$

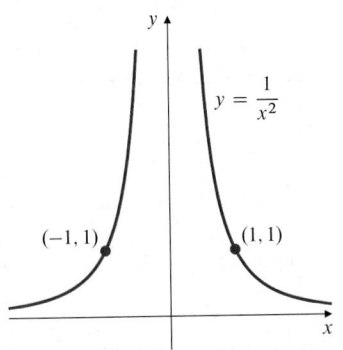

Figure P.44 The graph of $f(x) = 1/x^2$

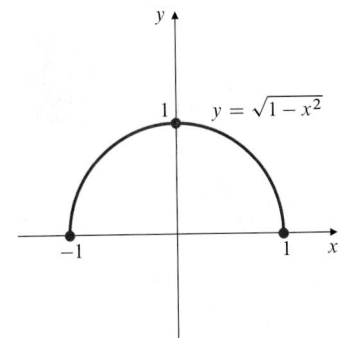

Figure P.45 The graph of $f(x) = \sqrt{1 - x^2}$

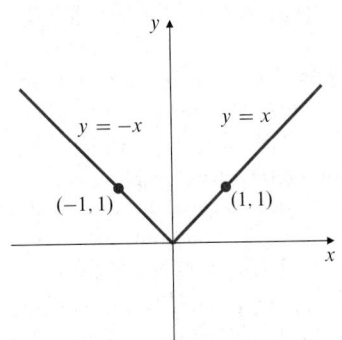

Figure P.46 The graph of $f(x) = |x|$

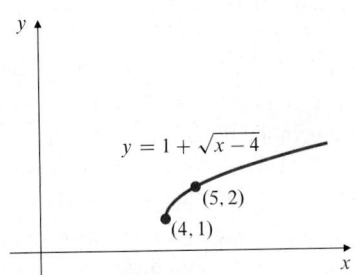

Figure P.47 The graph of $y = \sqrt{x}$ shifted right 4 units and up 1 unit

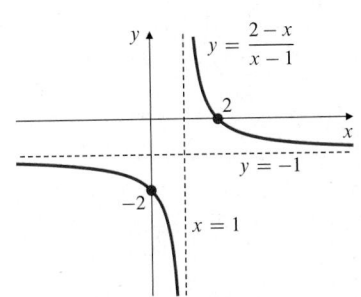

Figure P.48 The graph of $\dfrac{2 - x}{x - 1}$

EXAMPLE 8 Sketch the graph of the function $f(x) = \dfrac{2-x}{x-1}$.

Solution It is not immediately obvious that this graph is a shifted version of a known graph. To see that it is, we can divide $x-1$ into $2-x$ to get a quotient of -1 and a remainder of 1:

$$\frac{2-x}{x-1} = \frac{-x+1+1}{x-1} = \frac{-(x-1)+1}{x-1} = -1 + \frac{1}{x-1}.$$

Thus, the graph is that of $1/x$ from Figure P.43 shifted to the right 1 unit and down 1 unit. See Figure P.48.

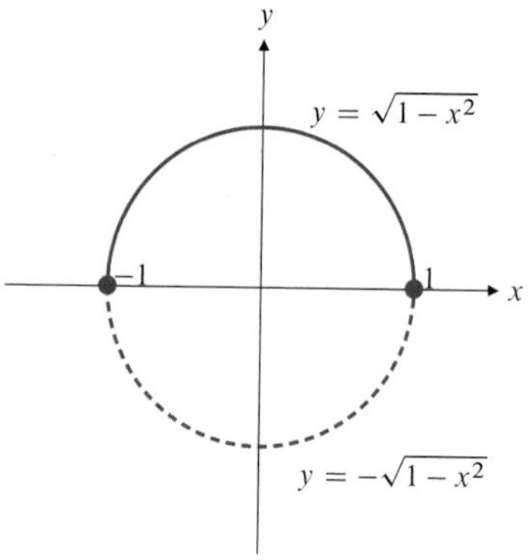

Figure P.49 The circle $x^2 + y^2 = 1$ is not the graph of a function

Not every curve you can draw is the graph of a function. A function f can have only one value $f(x)$ for each x in its domain, so no *vertical line* can intersect the graph of a function at more than one point. If a is in the domain of function f, then the vertical line $x = a$ will intersect the graph of f at the single point $(a, f(a))$. The circle $x^2 + y^2 = 1$ in Figure P.49 cannot be the graph of a function since some vertical lines intersect it twice. It is, however, the union of the graphs of two functions, namely,

$$y = \sqrt{1 - x^2} \qquad \text{and} \qquad y = -\sqrt{1 - x^2},$$

which are, respectively, the upper and lower halves (semicircles) of the given circle.

Even and Odd Functions; Symmetry and Reflections

It often happens that the graph of a function will have certain kinds of symmetry. The simplest kinds of symmetry relate the values of a function at x and $-x$.

DEFINITION

2

Even and odd functions

Suppose that $-x$ belongs to the domain of f whenever x does. We say that f is an **even function** if

$$f(-x) = f(x) \qquad \text{for every } x \text{ in the domain of } f.$$

We say that f is an **odd function** if

$$f(-x) = -f(x) \qquad \text{for every } x \text{ in the domain of } f.$$

The names *even* and *odd* come from the fact that even powers such as $x^0 = 1, x^2, x^4, \ldots, x^{-2}, x^{-4}, \ldots$ are even functions, and odd powers such as $x^1 = x, x^3, \ldots, x^{-1}, x^{-3}, \ldots$ are odd functions. For example, $(-x)^4 = x^4$ and $(-x)^{-3} = -x^{-3}$.

Since $(-x)^2 = x^2$, any function that depends only on x^2 is even. For instance, the absolute value function $y = |x| = \sqrt{x^2}$ is even.

The graph of an even function is *symmetric about the y-axis*. A horizontal straight line drawn from a point on the graph to the y-axis and continued an equal distance on the other side of the y-axis comes to another point on the graph. (See Figure P.50(a).)

The graph of an odd function is *symmetric about the origin*. A straight line drawn from a point on the graph to the origin will, if continued an equal distance on the other side of the origin, come to another point on the graph. If an odd function f is defined at $x = 0$, then its value must be zero there: $f(0) = 0$. (See Figure P.50(b).)

If $f(x)$ is even (or odd), then so is any constant multiple of $f(x)$, such as $2f(x)$ or $-5f(x)$. Sums and differences of even (or odd) functions are even (or odd). For example, $f(x) = 3x^4 - 5x^2 - 1$ is even, since it is the sum of three even functions: $3x^4$, $-5x^2$, and $-1 = -x^0$. Similarly, $4x^3 - (2/x)$ is odd. The function $g(x) = x^2 - 2x$ is the sum of an even function and an odd function and is itself neither even nor odd.

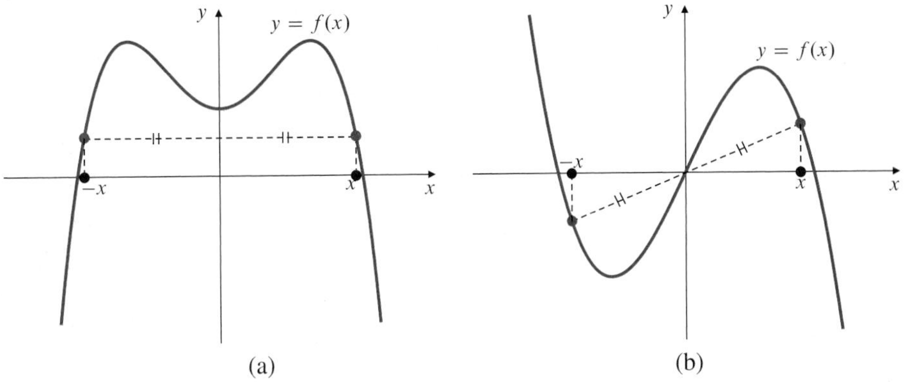

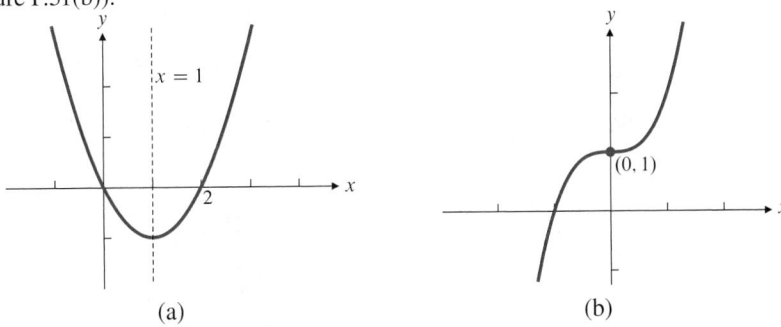

Figure P.50

(a) The graph of an even function is symmetric about the y-axis

(b) The graph of an odd function is symmetric about the origin

Other kinds of symmetry are also possible. For example, consider the function $g(x) = x^2 - 2x$, which can be written in the form $g(x) = (x-1)^2 - 1$. This shows that the values of $g(1 \pm u)$ are equal, so the graph (Figure P.51(a)) is symmetric about the vertical line $x = 1$; it is the parabola $y = x^2$ shifted 1 unit to the right and 1 unit down. Similarly, the graph of $h(x) = x^3 + 1$ is symmetric about the point $(0, 1)$ (Figure P.51(b)).

Figure P.51

(a) The graph of $g(x) = x^2 - 2x$ is symmetric about $x = 1$

(b) The graph of $y = h(x) = x^3 + 1$ is symmetric about $(0, 1)$

Reflections in Straight Lines

The image of an object reflected in a plane mirror appears to be as far behind the mirror as the object is in front of it. Thus, the mirror bisects at right angles the line from a point in the object to the corresponding point in the image. Given a line L and a point P not on L, we call a point Q the **reflection**, or the **mirror image**, of P in L if L is the right bisector of the line segment PQ. The reflection of any graph G in L is the graph consisting of the reflections of all the points of G.

Some reflections of graphs are easily described in terms of the equations of the graphs:

Reflections in special lines

1. Substituting $-x$ in place of x in an equation in x and y corresponds to reflecting the graph of the equation in the y-axis.

2. Substituting $-y$ in place of y in an equation in x and y corresponds to reflecting the graph of the equation in the x-axis.

3. Substituting $a - x$ in place of x in an equation in x and y corresponds to reflecting the graph of the equation in the line $x = a/2$.

4. Substituting $b - y$ in place of y in an equation in x and y corresponds to reflecting the graph of the equation in the line $y = b/2$.

5. Interchanging x and y in an equation in x and y corresponds to reflecting the graph of the equation in the line $y = x$.

EXAMPLE 9   Describe and sketch the graph of $y = \sqrt{2 - x} - 3$.

Solution   The graph of $y = \sqrt{2 - x}$ (green in Figure P.52(a)) is the reflection of the graph of $y = \sqrt{x}$ (blue) in the vertical line $x = 1$. The graph of $y = \sqrt{2-x} - 3$ (red) is the result of lowering this reflection by 3 units.

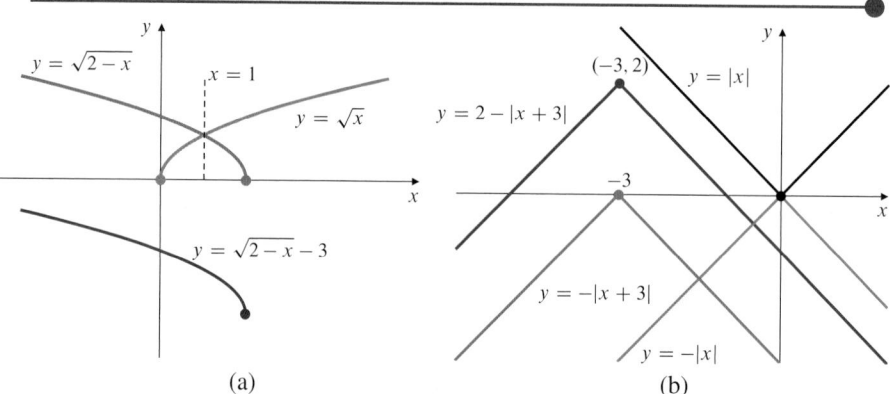

Figure P.52

(a) Constructing the graph of $y = \sqrt{2-x} - 3$

(b) Transforming $y = |x|$ to produce the coloured graph

EXAMPLE 10 Express the equation of the red graph in Figure P.52(b) in terms of the absolute value function $|x|$.

Solution We can get the red graph by first reflecting the graph of $|x|$ (black in Figure P.52(b)) in the x-axis to get the blue graph and then shifting that reflection left 3 units to get the green graph, and then shifting that graph up 2 units. The reflection of $y = |x|$ in the x-axis has equation $-y = |x|$, or $y = -|x|$. Shifting this left 3 units gives $y = -|x + 3|$. Finally, shifting up 2 units gives $y = 2 - |x + 3|$, which is the desired equation.

Defining and Graphing Functions with Maple

Many of the calculations and graphs encountered in studying calculus can be produced using a computer algebra system such as Maple or Mathematica. Here and there, throughout this book, we will include examples illustrating how to get Maple to perform such tasks. (The examples were done with Maple 10, but most of them will work with earlier or later versions of Maple as well.)

We begin with an example showing how to define a function in Maple and then plot its graph. We show in magenta the input you type into Maple and in cyan Maple's response. Let us define the function $f(x) = x^3 - 2x^2 - 12x + 1$.

```
>   f := x -> x^3-2*x^2-12*x+1; <enter>
```

$$f := x \longrightarrow x^3 - 2x^2 - 12x + 1$$

Note the use of := to indicate the symbol to the left is being defined and the use of -> to indicate the rule for the construction of $f(x)$ from x. Also note that Maple uses the asterisk * to indicate multiplication and the caret ^ to indicate an exponent. A Maple instruction should end with a semicolon ; (or a colon : if no output is desired) before the Enter key is pressed. Hereafter we will not show the `<enter>` in our input.

We can now use f as an ordinary function:

```
>   f(t)+f(1);
```

$$t^3 - 2t^2 - 12t - 11$$

The following command results in a plot of the graph of f on the interval $[-4, 5]$ shown in Figure P.53.

```
>   plot(f(x), x=-4..5);
```

We could have specified the expression x^3-2*x^2-12*x+1 directly in the plot command instead of first defining the function $f(x)$. Note the use of two dots .. to separate the left and right endpoints of the plot interval. Other options can be included in the plot command; all such options are separated with commas. You can specify the range of values of y in addition to that for x (which is required), and you can specify `scaling=CONSTRAINED` if you want equal unit distances on both axes. (This would be a bad idea for the graph of our $f(x)$. Why?)

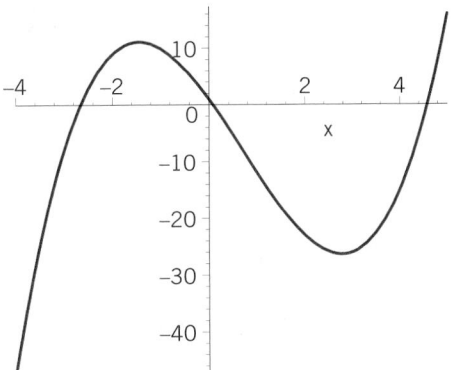

Figure P.53   A Maple plot

⚠ When using a graphing calculator or computer graphing software, things can go horribly wrong in some circumstances. The following example illustrates the catastrophic effects that **round-off error** can have.

EXAMPLE 11   Consider the function $g(x) = \dfrac{|1 + x| - 1}{x}$.

If $x > -1$, then $|1 + x| = 1 + x$, so the formula for $g(x)$ simplifies to $g(x) = \dfrac{(1 + x) - 1}{x} = \dfrac{x}{x} = 1$, at least provided $x \neq 0$. Thus the graph of g on an interval lying to the right of $x = -1$ should be the horizontal line $y = 1$, possibly with a hole in it at $x = 0$. The Maple commands

```
> g := x -> (abs(1+x)-1)/x: plot(g(x), x=-0.5..0.5);
```

lead, as expected, to the graph in Figure P.54. But plotting the same function on a very tiny interval near $x = 0$ leads to quite a different graph. The command

```
> plot([g(x),1],x=-7*10^(-16)..5*10^(-16),
    style=[point,line],numpoints=4000);
```

produces the graph in Figure P.55.

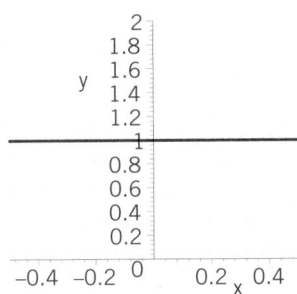

Figure P.54   The graph of $y = g(x)$ on the interval $[-0.5, 0.5]$

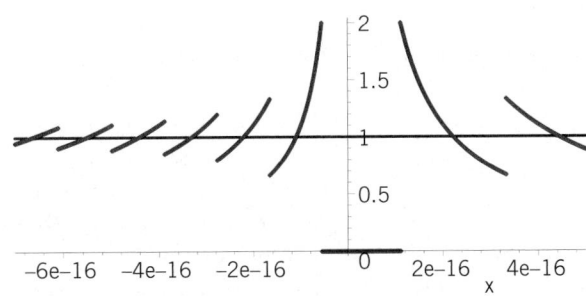

Figure P.55   The graphs of $y = g(x)$ (colour) and $y = 1$ (black) on the interval $[-7 \times 10^{-16}, 5 \times 10^{-16}]$

The coloured arcs and short line through the origin are the graph of $y = g(x)$ plotted as 4,000 individual points over the interval from -7×10^{-16} to 5×10^{-16}. For comparison sake, the black horizontal line $y = 1$ is also plotted. What makes the graph of g so strange on this interval is the fact that Maple can only represent finitely many real numbers in its finite memory. If the number x is too close to zero, Maple cannot tell the difference between $1 + x$ and 1, so it calculates $1 - 1 = 0$ for the numerator and uses $g(x) = 0$ in the plot. This seems to happen between about -0.5×10^{-16} and

0.8×10^{-16} (the coloured horizontal line). As we move further away from the origin, Maple can tell the difference between $1 + x$ and 1, but loses most of the significant figures in the representation of x when it adds 1, and these remain lost when it subtracts 1 again. Thus the numerator remains constant over short intervals while the denominator increases as x moves away from 0. In those intervals the fraction behaves like constant/x, so the arcs are hyperbolas, sloping downward away from the origin. The effect diminishes the farther x moves away from 0, as more of its significant figures are retained by Maple. It should be noted that the reason we used the absolute value of $1 + x$ instead of just $1 + x$ is that this forced Maple to add the x to the 1 before subtracting the second 1. (If we had used $(1 + x) - 1$ as the numerator for $g(x)$, Maple would have simplified it algebraically and obtained $g(x) = 1$ before using any values of x for plotting.)

In later chapters we will encounter more such strange behaviour (which we call **numerical monsters** and denote by the symbol ⚠) in the context of calculator and computer calculations with floating point (i.e., real) numbers. They are a necessary consequence of the limitations of such hardware and software and are not restricted to Maple, though they may show up somewhat differently with other software. It is necessary to be aware of how calculators and computers do arithmetic in order to be able to use them effectively without falling into errors that you do not recognize as such.

One final comment about Figure P.55: the graph of $y = g(x)$ was plotted as individual points, rather than a line, as was $y = 1$, in order to make the jumps between consecutive arcs more obvious. Had we omitted the `style=[point,line]` option in the plot command, the default line style would have been used for both graphs and the arcs in the graph of g would have been connected with vertical line segments. Note how the command called for the plotting of two different functions by listing them within square brackets, and how the corresponding styles were correspondingly listed.

EXERCISES P.4

In Exercises 1–6, find the domain and range of each function.

1. $f(x) = 1 + x^2$

2. $f(x) = 1 - \sqrt{x}$

3. $G(x) = \sqrt{8 - 2x}$

4. $F(x) = 1/(x - 1)$

5. $h(t) = \dfrac{t}{\sqrt{2 - t}}$

6. $g(x) = \dfrac{1}{1 - \sqrt{x - 2}}$

7. Which of the graphs in Figure P.56 are graphs of functions $y = f(x)$? Why?

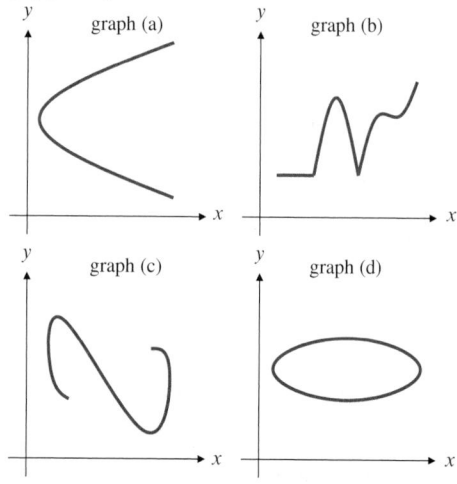

Figure P.56

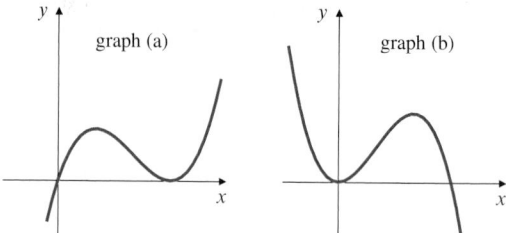

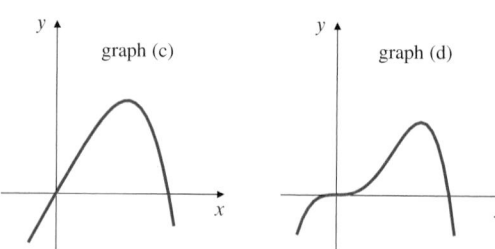

Figure P.57

8. Figure P.57 shows the graphs of the functions: (i) $x - x^4$, (ii) $x^3 - x^4$, (iii) $x(1 - x)^2$, (iv) $x^2 - x^3$. Which graph corresponds to which function?

In Exercises 9–10, sketch the graph of the function f by first making a table of values of $f(x)$ at $x = 0$, $x = \pm 1/2$, $x = \pm 1$, $x = \pm 3/2$, and $x = \pm 2$.

9. $f(x) = x^4$

10. $f(x) = x^{2/3}$

In Exercises 11–22, what (if any) symmetry does the graph of f possess? In particular, is f even or odd?

11. $f(x) = x^2 + 1$

12. $f(x) = x^3 + x$

13. $f(x) = \dfrac{x}{x^2 - 1}$

14. $f(x) = \dfrac{1}{x^2 - 1}$

15. $f(x) = \dfrac{1}{x - 2}$

16. $f(x) = \dfrac{1}{x + 4}$

17. $f(x) = x^2 - 6x$

18. $f(x) = x^3 - 2$

19. $f(x) = |x^3|$

20. $f(x) = |x + 1|$

21. $f(x) = \sqrt{2x}$

22. $f(x) = \sqrt{(x - 1)^2}$

Sketch the graphs of the functions in Exercises 23–38.

23. $f(x) = -x^2$

24. $f(x) = 1 - x^2$

25. $f(x) = (x - 1)^2$

26. $f(x) = (x - 1)^2 + 1$

27. $f(x) = 1 - x^3$

28. $f(x) = (x + 2)^3$

29. $f(x) = \sqrt{x} + 1$

30. $f(x) = \sqrt{x + 1}$

31. $f(x) = -|x|$

32. $f(x) = |x| - 1$

33. $f(x) = |x - 2|$

34. $f(x) = 1 + |x - 2|$

35. $f(x) = \dfrac{2}{x + 2}$

36. $f(x) = \dfrac{1}{2 - x}$

37. $f(x) = \dfrac{x}{x + 1}$

38. $f(x) = \dfrac{x}{1 - x}$

In Exercises 39–46, f refers to the function with domain $[0, 2]$ and range $[0, 1]$, whose graph is shown in Figure P.58. Sketch the graphs of the indicated functions and specify their domains and ranges.

39. $f(x) + 2$

40. $f(x) - 1$

41. $f(x + 2)$

42. $f(x - 1)$

43. $-f(x)$

44. $f(-x)$

45. $f(4 - x)$

46. $1 - f(1 - x)$

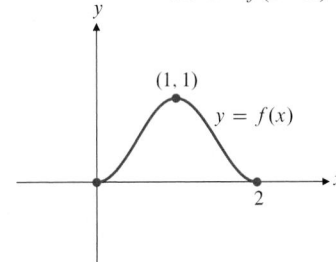

Figure P.58

It is often quite difficult to determine the range of a function exactly. In Exercises 47–48, use a graphing utility (calculator or computer) to graph the function f, and by zooming in on the graph, determine the range of f with accuracy of 2 decimal places.

47. $f(x) = \dfrac{x + 2}{x^2 + 2x + 3}$

48. $f(x) = \dfrac{x - 1}{x^2 + x}$

In Exercises 49–52, use a graphing utility to plot the graph of the given function. Examine the graph (zooming in or out as necessary) for symmetries. About what lines and/or points are the graphs symmetric? Try to verify your conclusions algebraically.

49. $f(x) = x^4 - 6x^3 + 9x^2 - 1$

50. $f(x) = \dfrac{3 - 2x + x^2}{2 - 2x + x^2}$

51. $f(x) = \dfrac{x - 1}{x - 2}$

52. $f(x) = \dfrac{2x^2 + 3x}{x^2 + 4x + 5}$

❷ 53. What function $f(x)$, defined on the real line $\mathbb{R}$, is both even and odd?

P.5

Combining Functions to Make New Functions

Functions can be combined in a variety of ways to produce new functions. We begin by examining algebraic means of combining functions, that is, addition, subtraction, multiplication, and division.

Sums, Differences, Products, Quotients, and Multiples

Like numbers, functions can be added, subtracted, multiplied, and divided (except where the denominator is zero) to produce new functions.

DEFINITION

3

If f and g are functions, then for every x that belongs to the domains of both f and g we define functions $f + g$, $f - g$, fg, and f/g by the formulas:

$$(f + g)(x) = f(x) + g(x)$$
$$(f - g)(x) = f(x) - g(x)$$
$$(fg)(x) = f(x)g(x)$$
$$\left(\frac{f}{g}\right)(x) = \frac{f(x)}{g(x)}, \qquad \text{where } g(x) \neq 0.$$

A special case of the rule for multiplying functions shows how functions can be multiplied by constants. If c is a real number, then the function cf is defined for all x

in the domain of f by

$$(cf)(x) = c \cdot f(x).$$

EXAMPLE 1 Figure P.59(a) shows the graphs of $f(x) = x^2$, $g(x) = x - 1$, and their sum $(f + g)(x) = x^2 + x - 1$. Observe that the height of the graph of $f + g$ at any point x is the sum of the heights of the graphs of f and g at that point.

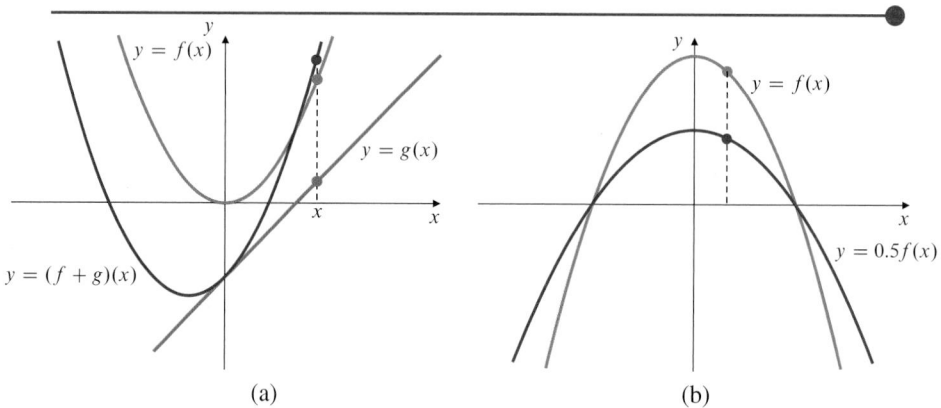

Figure P.59

(a) $(f + g)(x) = f(x) + g(x)$

(b) $g(x) = (0.5) f(x)$

(a) (b)

EXAMPLE 2 Figure P.59(b) shows the graphs of $f(x) = 2 - x^2$ and the multiple $g(x) = (0.5) f(x)$. Note how the height of the graph of g at any point x is half the height of the graph of f there.

EXAMPLE 3 The functions f and g are defined by the formulas

$$f(x) = \sqrt{x} \quad \text{and} \quad g(x) = \sqrt{1 - x}.$$

Find formulas for the values of $3f$, $f + g$, $f - g$, fg, f/g, and g/f at x, and specify the domains of each of these functions.

Solution The information is collected in Table 2:

Table 2. Combinations of f and g and their domains

Function	Formula	Domain
f	$f(x) = \sqrt{x}$	$[0, \infty)$
g	$g(x) = \sqrt{1 - x}$	$(-\infty, 1]$
$3f$	$(3f)(x) = 3\sqrt{x}$	$[0, \infty)$
$f + g$	$(f + g)(x) = f(x) + g(x) = \sqrt{x} + \sqrt{1 - x}$	$[0, 1]$
$f - g$	$(f - g)(x) = f(x) - g(x) = \sqrt{x} - \sqrt{1 - x}$	$[0, 1]$
fg	$(fg)(x) = f(x)g(x) = \sqrt{x(1 - x)}$	$[0, 1]$
f/g	$\dfrac{f}{g}(x) = \dfrac{f(x)}{g(x)} = \sqrt{\dfrac{x}{1 - x}}$	$[0, 1)$
g/f	$\dfrac{g}{f}(x) = \dfrac{g(x)}{f(x)} = \sqrt{\dfrac{1 - x}{x}}$	$(0, 1]$

Note that most of the combinations of f and g have domains

$$[0, \infty) \ \cap \ (-\infty, 1] = [0, 1],$$

the intersection of the domains of f and g. However, the domains of the two quotients f/g and g/f had to be restricted further to remove points where the denominator was zero.

Composite Functions

There is another method, called **composition**, by which two functions can be combined to form a new function.

DEFINITION

4

> **Composite functions**
>
> If f and g are two functions, the **composite** function $f \circ g$ is defined by
>
> $$f \circ g(x) = f(g(x)).$$
>
> The domain of $f \circ g$ consists of those numbers x in the domain of g for which $g(x)$ is in the domain of f. In particular, if the range of g is contained in the domain of f, then the domain of $f \circ g$ is just the domain of g.

As shown in Figure P.60, forming $f \circ g$ is equivalent to arranging "function machines" g and f in an "assembly line" so that the output of g becomes the input of f.

In calculating $f \circ g(x) = f(g(x))$, we first calculate $g(x)$ and then calculate f of the result. We call g the *inner* function and f the *outer* function of the composition. We can, of course, also calculate the composition $g \circ f(x) = g(f(x))$, where f is the inner function, the one that gets calculated first, and g is the outer function, which gets calculated last. The functions $f \circ g$ and $g \circ f$ are usually quite different, as the following example shows.

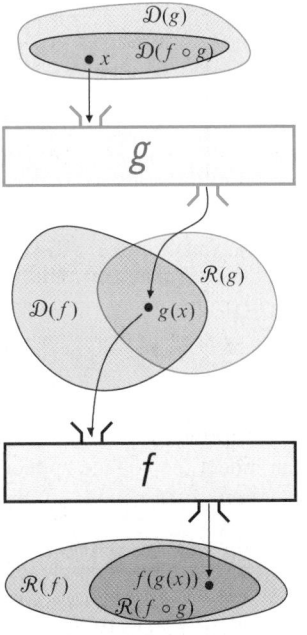

Figure P.60 $f \circ g(x) = f(g(x))$

EXAMPLE 4   Given $f(x) = \sqrt{x}$ and $g(x) = x+1$, calculate the four composite functions $f \circ g(x)$, $g \circ f(x)$, $f \circ f(x)$, and $g \circ g(x)$, and specify the domain of each.

Solution   Again, we collect the results in a table. (See Table 3.)

Table 3.   Composites of f and g and their domains

Function	Formula	Domain
f	$f(x) = \sqrt{x}$	$[0, \infty)$
g	$g(x) = x + 1$	$\mathbb{R}$
$f \circ g$	$f \circ g(x) = f(g(x)) = f(x+1) = \sqrt{x+1}$	$[-1, \infty)$
$g \circ f$	$g \circ f(x) = g(f(x)) = g(\sqrt{x}) = \sqrt{x} + 1$	$[0, \infty)$
$f \circ f$	$f \circ f(x) = f(f(x)) = f(\sqrt{x}) = \sqrt{\sqrt{x}} = x^{1/4}$	$[0, \infty)$
$g \circ g$	$g \circ g(x) = g(g(x)) = g(x+1) = (x+1)+1 = x+2$	$\mathbb{R}$

To see why, for example, the domain of $f \circ g$ is $[-1, \infty)$, observe that $g(x) = x + 1$ is defined for all real x but belongs to the domain of f only if $x + 1 \geq 0$, that is, if $x \geq -1$.

EXAMPLE 5   If $G(x) = \dfrac{1-x}{1+x}$, calculate $G \circ G(x)$ and specify its domain.

Solution   We calculate

$$G \circ G(x) = G(G(x)) = G\left(\frac{1-x}{1+x}\right) = \frac{1 - \dfrac{1-x}{1+x}}{1 + \dfrac{1-x}{1+x}} = \frac{1+x-1+x}{1+x+1-x} = x.$$

Because the resulting function, x, is defined for all real x, we might be tempted to say that the domain of $G \circ G$ is $\mathbb{R}$. This is wrong! To belong to the domain of $G \circ G$, x must satisfy two conditions:

(i) x must belong to the domain of G, and

(ii) $G(x)$ must belong to the domain of G.

The domain of G consists of all real numbers *except* $x = -1$. If we exclude $x = -1$ from the domain of $G \circ G$, condition (i) will be satisfied. Now observe that the equation $G(x) = -1$ has no solution x, since it can be rewritten in the form $1 - x = -(1 + x)$, or $1 = -1$. Therefore, all numbers $G(x)$ belong to the domain of G, and condition (ii) is satisfied with no further restrictions on x. The domain of $G \circ G$ is $(-\infty, -1) \cup (-1, \infty)$, that is, all real numbers except -1.

Piecewise Defined Functions

Sometimes it is necessary to define a function by using different formulas on different parts of its domain. One example is the absolute value function

$$|x| = \begin{cases} x & \text{if } x \geq 0 \\ -x & \text{if } x < 0. \end{cases}$$

Another would be the tax rates applied to various levels of income. Here are some other examples. (Note how we use solid and hollow dots in the graphs to indicate, respectively, which endpoints do or do not lie on various parts of the graph.)

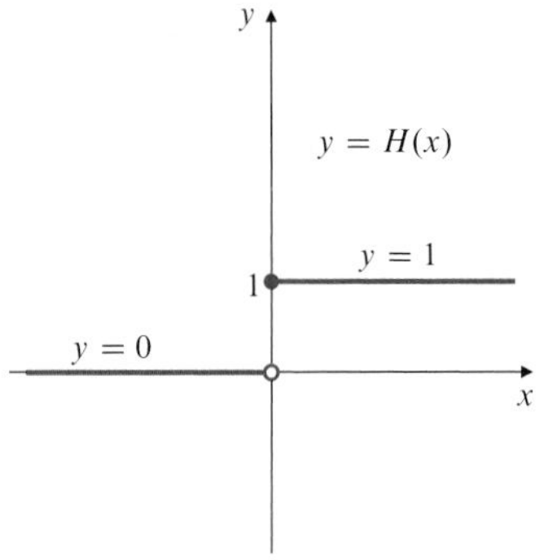

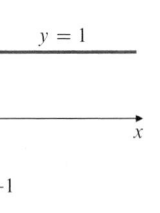

Figure P.61   The Heaviside function

EXAMPLE 6   **The Heaviside function.** The Heaviside function (or unit step function) (Figure P.61) is defined by

$$H(x) = \begin{cases} 1 & \text{if } x \geq 0 \\ 0 & \text{if } x < 0. \end{cases}$$

For instance, if t represents time, the function $6H(t)$ can model the voltage applied to an electric circuit by a 6-volt battery if a switch in the circuit is turned on at time $t = 0$.

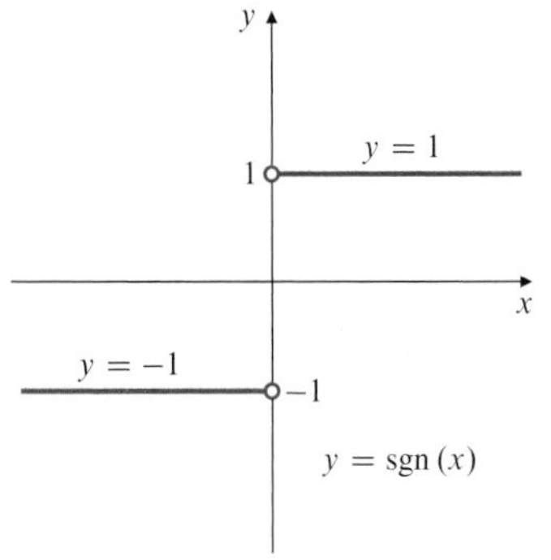

EXAMPLE 7   **The signum function.** The signum function (Figure P.62) is defined as follows:

$$\text{sgn}(x) = \frac{x}{|x|} = \begin{cases} 1 & \text{if } x > 0, \\ -1 & \text{if } x < 0, \\ \text{undefined} & \text{if } x = 0. \end{cases}$$

Figure P.62   The signum function

The name *signum* is the Latin word meaning "sign." The value of the $\text{sgn}(x)$ tells whether x is positive or negative. Since 0 is neither positive nor negative, $\text{sgn}(0)$ is not defined. The signum function is an odd function.

EXAMPLE 8   The function

$$f(x) = \begin{cases} (x+1)^2 & \text{if } x < -1, \\ -x & \text{if } -1 \leq x < 1, \\ \sqrt{x-1} & \text{if } x \geq 1, \end{cases}$$

is defined on the whole real line but has values given by three different formulas depending on the position of x. Its graph is shown in Figure P.63(a).

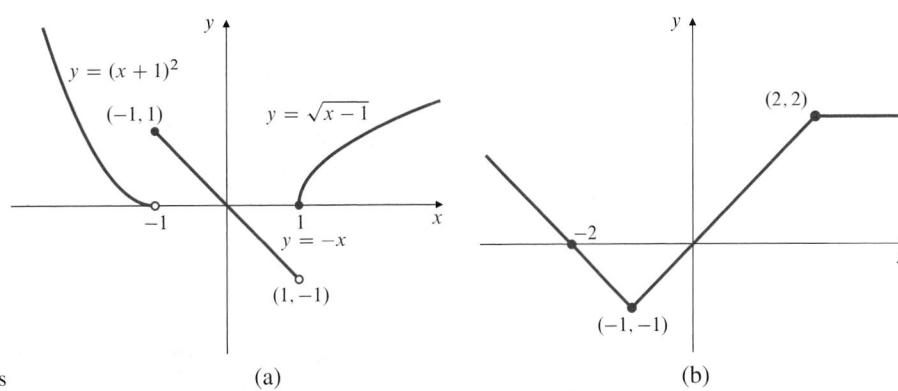

Figure P.63   Piecewise defined functions

(a)   (b)

EXAMPLE 9   Find a formula for function $g(x)$ graphed in Figure P.63(b).

Solution   The graph consists of parts of three lines. For the part $x < -1$, the line has slope -1 and x-intercept -2, so its equation is $y = -(x + 2)$. The middle section is the line $y = x$ for $-1 \le x \le 2$. The right section is $y = 2$ for $x > 2$. Combining these formulas, we write

$$g(x) = \begin{cases} -(x + 2) & \text{if } x < -1 \\ x & \text{if } -1 \le x \le 2 \\ 2 & \text{if } x > 2. \end{cases}$$

Unlike the previous example, it does not matter here which of the two possible formulas we use to define $g(-1)$, since both give the same value. The same is true for $g(2)$.

The following two functions could be defined by different formulas on every interval between consecutive integers, but we will use an easier way to define them.

EXAMPLE 10   **The greatest integer function.** The function whose value at any number x is the *greatest integer less than or equal to* x is called the **greatest integer function**, or the **integer floor function**. It is denoted $\lfloor x \rfloor$, or, in some books, $[x]$ or $[[x]]$. The graph of $y = \lfloor x \rfloor$ is given in Figure P.64(a). Observe that

$$\lfloor 2.4 \rfloor = 2, \qquad \lfloor 1.9 \rfloor = 1, \qquad \lfloor 0 \rfloor = 0, \qquad \lfloor -1.2 \rfloor = -2,$$
$$\lfloor 2 \rfloor = 2, \qquad \lfloor 0.2 \rfloor = 0, \qquad \lfloor -0.3 \rfloor = -1, \qquad \lfloor -2 \rfloor = -2.$$

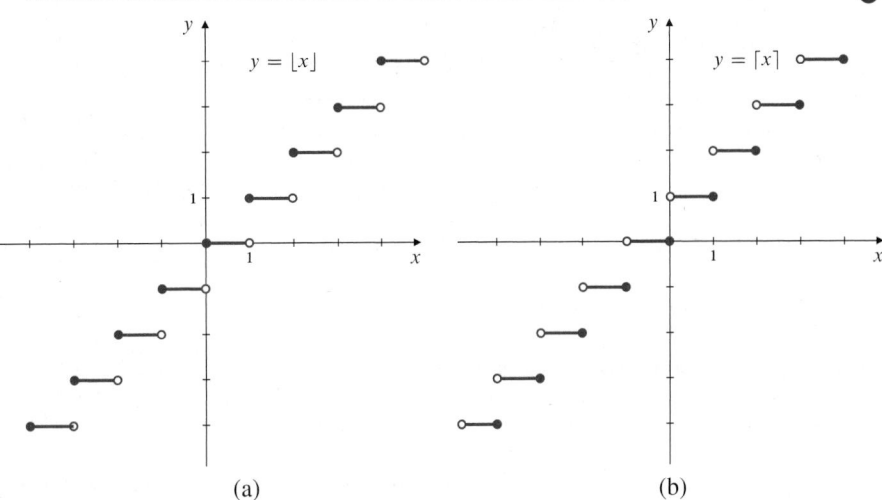

Figure P.64

(a) The greatest integer function $\lfloor x \rfloor$

(b) The least integer function $\lceil x \rceil$

(a)   (b)

EXAMPLE 11   **The least integer function.** The function whose value at any number x is the *smallest integer greater than or equal to x* is called the **least integer function**, or the **integer ceiling function**. It is denoted $\lceil x \rceil$. Its graph is given in Figure P.64(b). For positive values of x, this function might represent, for example, the cost of parking x hours in a parking lot that charges \$1 for each hour or part of an hour.

EXERCISES P.5

In Exercises 1–2, find the domains of the functions $f + g$, $f - g$, fg, f/g, and g/f, and give formulas for their values.

1. $f(x) = x$, $g(x) = \sqrt{x - 1}$

2. $f(x) = \sqrt{1 - x}$, $g(x) = \sqrt{1 + x}$

Sketch the graphs of the functions in Exercises 3–6 by combining the graphs of simpler functions from which they are built up.

3. $x - x^2$ **4.** $x^3 - x$

5. $x + |x|$ **6.** $|x| + |x - 2|$

7. If $f(x) = x + 5$ and $g(x) = x^2 - 3$, find the following:

(a) $f \circ g(0)$ (b) $g(f(0))$
(c) $f(g(x))$ (d) $g \circ f(x)$
(e) $f \circ f(-5)$ (f) $g(g(2))$
(g) $f(f(x))$ (h) $g \circ g(x)$

In Exercises 8–10, construct the following composite functions and specify the domain of each.

(a) $f \circ f(x)$ (b) $f \circ g(x)$
(c) $g \circ f(x)$ (d) $g \circ g(x)$

8. $f(x) = 2/x$, $g(x) = x/(1 - x)$

9. $f(x) = 1/(1 - x)$, $g(x) = \sqrt{x - 1}$

10. $f(x) = (x + 1)/(x - 1)$, $g(x) = \operatorname{sgn}(x)$

Find the missing entries in Table 4 (Exercises 11–16).

Table 4.

	$f(x)$	$g(x)$	$f \circ g(x)$
11.	x^2	$x + 1$	
12.		$x + 4$	x
13.	$\sqrt{x}$		$|x|$
14.		$x^{1/3}$	$2x + 3$
15.	$(x + 1)/x$		x
16.		$x - 1$	$1/x^2$

⊞ 17. Use a graphing utility to examine in order the graphs of the functions

$$y = \sqrt{x}, \qquad\qquad y = 2 + \sqrt{x},$$
$$y = 2 + \sqrt{3 + x}, \qquad y = 1/(2 + \sqrt{3 + x}).$$

Describe the effect on the graph of the change made in the function at each stage.

⊞ 18. Repeat the previous exercise for the functions

$$y = 2x, \qquad y = 2x - 1, \qquad y = 1 - 2x,$$
$$y = \sqrt{1 - 2x}, \qquad y = \frac{1}{\sqrt{1 - 2x}}, \qquad y = \frac{1}{\sqrt{1 - 2x}} - 1.$$

In Exercises 19–24, f refers to the function with domain $[0, 2]$ and range $[0, 1]$, whose graph is shown in Figure P.65. Sketch the graphs of the indicated functions, and specify their domains and ranges.

19. $2f(x)$ **20.** $-(1/2)f(x)$

21. $f(2x)$ **22.** $f(x/3)$

23. $1 + f(-x/2)$ **24.** $2f((x - 1)/2)$

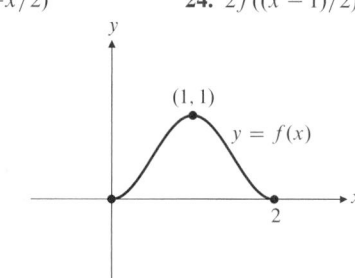

Figure P.65

In Exercises 25–26, sketch the graphs of the given functions.

25. $f(x) = \begin{cases} x & \text{if } 0 \le x \le 1 \\ 2 - x & \text{if } 1 < x \le 2 \end{cases}$

26. $g(x) = \begin{cases} \sqrt{x} & \text{if } 0 \le x \le 1 \\ 2 - x & \text{if } 1 < x \le 2 \end{cases}$

27. Find all real values of the constants A and B for which the function $F(x) = Ax + B$ satisfies:

(a) $F \circ F(x) = F(x)$ for all x.
(b) $F \circ F(x) = x$ for all x.

Greatest and least integer functions

28. For what values of x is (a) $\lfloor x \rfloor = 0$? (b) $\lceil x \rceil = 0$?

29. What real numbers x satisfy the equation $\lfloor x \rfloor = \lceil x \rceil$?

30. True or false: $\lceil -x \rceil = -\lfloor x \rfloor$ for all real x?

31. Sketch the graph of $y = x - \lfloor x \rfloor$.

32. Sketch the graph of the function

$$f(x) = \begin{cases} \lfloor x \rfloor & \text{if } x \ge 0 \\ \lceil x \rceil & \text{if } x < 0. \end{cases}$$

Why is $f(x)$ called *the integer part of x*?

Even and odd functions

❷ **33.** Assume that f is an even function, g is an odd function, and both f and g are defined on the whole real line $\mathbb{R}$. Is each of the following functions even, odd, or neither?

$$f + g, \quad fg, \quad f/g, \quad g/f, \quad f^2 = ff, \quad g^2 = gg$$

$$f \circ g, \quad g \circ f, \quad f \circ f, \quad g \circ g$$

❷ **34.** If f is both an even and an odd function, show that $f(x) = 0$ at every point of its domain.

❷ **35.** Let f be a function whose domain is symmetric about the origin, that is, $-x$ belongs to the domain whenever x does.

(a) Show that f is the sum of an even function and an odd function:

$$f(x) = E(x) + O(x),$$

where E is an even function and O is an odd function. *Hint:* Let $E(x) = (f(x) + f(-x))/2$. Show that $E(-x) = E(x)$, so that E is even. Then show that $O(x) = f(x) - E(x)$ is odd.

(b) Show that there is only one way to write f as the sum of an even and an odd function. *Hint:* One way is given in part (a). If also $f(x) = E_1(x) + O_1(x)$, where E_1 is even and O_1 is odd, show that $E - E_1 = O_1 - O$ and then use Exercise 34 to show that $E = E_1$ and $O = O_1$.

P.6 Polynomials and Rational Functions

Among the easiest functions to deal with in calculus are polynomials. These are sums of terms each of which is a constant multiple of a nonnegative integer power of the variable of the function.

DEFINITION

5

A **polynomial** is a function P whose value at x is

$$P(x) = a_n x^n + a_{n-1} x^{n-1} + \cdots + a_2 x^2 + a_1 x + a_0,$$

where $a_n, a_{n-1}, \ldots, a_2, a_1,$ and a_0, called the **coefficients** of the polynomial, are constants and, if $n > 0$, then $a_n \neq 0$. The number n, the degree of the highest power of x in the polynomial, is called the **degree** of the polynomial. (The degree of the zero polynomial is not defined.)

For example,

$$3 \qquad \text{is a polynomial of degree 0.}$$
$$2 - x \qquad \text{is a polynomial of degree 1.}$$
$$2x^3 - 17x + 1 \qquad \text{is a polynomial of degree 3.}$$

Generally, we assume that the polynomials we deal with are **real polynomials**; that is, their coefficients are real numbers rather than more general complex numbers. Often the coefficients will be integers or rational numbers. Polynomials play a role in the study of functions somewhat analogous to the role played by integers in the study of numbers. For instance, just as we always get an integer result if we add, subtract, or multiply two integers, we always get a polynomial result if we add, subtract, or multiply two polynomials. Adding or subtracting polynomials produces a polynomial whose degree does not exceed the larger of the two degrees of the polynomials being combined. Multiplying two polynomials of degrees m and n produces a product polynomial of degree $m + n$. For instance, for the product

$$(x^2 + 1)(x^3 - x - 2) = x^5 - 2x^2 - x - 2,$$

the two factors have degrees 2 and 3, so the result has degree 5.

The following definition is analogous to the definition of a rational number as the quotient of two integers.

DEFINITION

6

If $P(x)$ and $Q(x)$ are two polynomials and $Q(x)$ is not the zero polynomial, then the function

$$R(x) = \frac{P(x)}{Q(x)}$$

is called a **rational function**. By the domain convention, the domain of $R(x)$ consists of all real numbers x except those for which $Q(x) = 0$.

Two examples of rational functions and their domains are

$$R(x) = \frac{2x^3 - 3x^2 + 3x + 4}{x^2 + 1} \quad \text{with domain } \mathbb{R}, \text{ all real numbers.}$$

$$S(x) = \frac{1}{x^2 - 4} \quad \text{with domain all real numbers except } \pm 2.$$

Remark If the numerator and denominator of a rational function have a common factor, that factor can be cancelled out just as with integers. However, the resulting simpler rational function may not have the same domain as the original one, so it should be regarded as a different rational function even though it is equal to the original one at all points of the original domain. For instance,

$$\frac{x^2 - x}{x^2 - 1} = \frac{x(x - 1)}{(x + 1)(x - 1)} = \frac{x}{x + 1} \quad \text{only if } x \neq \pm 1,$$

even though $x = 1$ is in the domain of $x/(x + 1)$.

When we divide a positive integer a by a smaller positive integer b, we can obtain an integer quotient q and an integer remainder r satisfying $0 \leq r < b$ and hence write the fraction a/b (in a unique way) as the sum of the integer q and another fraction whose numerator (the remainder r) is smaller than its denominator b. For instance,

$$\frac{7}{3} = 2 + \frac{1}{3}; \quad \text{the quotient is 2, the remainder is 1.}$$

Similarly, if A_m and B_n are polynomials having degrees m and n, respectively, and if $m > n$, then we can express the rational function A_m/B_n (in a unique way) as the sum of a quotient polynomial Q_{m-n} of degree $m - n$ and another rational function R_k/B_n where the numerator polynomial R_k (the remainder in the division) is either zero or has degree $k < n$:

$$\frac{A_m(x)}{B_n(x)} = Q_{m-n}(x) + \frac{R_k(x)}{B_n(x)}. \quad \textbf{(The Division Algorithm)}$$

We calculate the quotient and remainder polynomials by using long division or an equivalent method.

EXAMPLE 1 Write the division algorithm for $\dfrac{2x^3 - 3x^2 + 3x + 4}{x^2 + 1}$.

Solution **METHOD I.** Use long division:

$$
\begin{array}{r}
2x \;\;\; - \;\; 3 \\
x^2 + 1 \,\overline{\big)\, 2x^3 - 3x^2 + 3x + 4} \\
\underline{2x^3 + 2x } \\
-3x^2 + x + 4 \\
\underline{-3x^2 - 3} \\
x + 7
\end{array}
$$

Thus,

$$\frac{2x^3 - 3x^2 + 3x + 4}{x^2 + 1} = 2x - 3 + \frac{x + 7}{x^2 + 1}.$$

The quotient is $2x - 3$, and the remainder is $x + 7$.

METHOD II. Use short division; add appropriate lower-degree terms to the terms of the numerator that have degrees not less than the degree of the denominator to enable factoring out the denominator, and then subtract those terms off again.

$$
\begin{aligned}
& 2x^3 - 3x^2 + 3x + 4 \\
= \; & 2x^3 + 2x - 3x^2 - 3 + 3x + 4 - 2x + 3 \\
= \; & 2x(x^2 + 1) - 3(x^2 + 1) + x + 7,
\end{aligned}
$$

from which it follows at once that

$$\frac{2x^3 - 3x^2 + 3x + 4}{x^2 + 1} = 2x - 3 + \frac{x + 7}{x^2 + 1}.$$

Roots, Zeros, and Factors

A number r is called a **root** or **zero** of the polynomial P if $P(r) = 0$. For example, $P(x) = x^3 - 4x$ has three roots: 0, 2, and -2; substituting any of these numbers for x makes $P(x) = 0$. In this context the terms "root" and "zero" are often used interchangeably. It is technically more correct to call a number r satisfying $P(r) = 0$ a *zero* of the polynomial *function* P and a *root* of the *equation* $P(x) = 0$, and later in this book we will follow this convention more closely. But for now, to avoid confusion with the *number* zero, we will prefer to use "root" rather than "zero" even when referring to the polynomial P rather than the equation $P(x) = 0$.

The **Fundamental Theorem of Algebra** (see Appendix II) states that every polynomial of degree at least 1 has a root (although the root might be a complex number). For example, the linear (degree 1) polynomial $ax + b$ has the root $-b/a$ since $a(-b/a) + b = 0$. A constant polynomial (one of degree zero) cannot have any roots unless it is the zero polynomial, in which case every number is a root.

Real polynomials do not always have real roots; the polynomial $x^2 + 4$ is never zero for any real number x, but it is zero if x is either of the two complex numbers $2i$ and $-2i$, where i is the so-called imaginary unit satisfying $i^2 = -1$. (See Appendix I for a discussion of complex numbers.) The numbers $2i$ and $-2i$ are *complex conjugates of each other*. Any complex roots of a real polynomial must occur in conjugate pairs. (See Appendix II for a proof of this fact.)

In our study of calculus we will often find it useful to factor polynomials into products of polynomials of lower degree, especially degree 1 or 2 (linear or quadratic polynomials). The following theorem shows the connection between linear factors and roots.

THEOREM

1

The Factor Theorem

The number r is a root of the polynomial P of degree not less than 1 if and only if $x - r$ is a factor of $P(x)$.

PROOF By the division algorithm there exists a quotient polynomial Q having degree one less than that of P and a remainder polynomial of degree 0 (i.e., a constant c) such that

$$\frac{P(x)}{x - r} = Q(x) + \frac{c}{x - r}.$$

Thus $P(x) = (x - r)Q(x) + c$, and $P(r) = 0$ if and only if $c = 0$, in which case $P(x) = (x - r)Q(x)$ and $x - r$ is a factor of $P(x)$.

It follows from Theorem 1 and the Fundamental Theorem of Algebra that every polynomial of degree $n \geq 1$ has n roots. (If P has degree $n \geq 2$, then P has a zero r and $P(x) = (x - r)Q(x)$, where Q is a polynomial of degree $n - 1 \geq 1$, which in turn has a root, etc.) Of course, the roots of a polynomial need not all be different. The 4th-degree polynomial $P(x) = x^4 - 3x^3 + 3x^2 - x = x(x - 1)^3$ has four roots; one is 0 and the other three are each equal to 1. We say that the root 1 has **multiplicity** 3 because we can divide $P(x)$ by $(x - 1)^3$ and still get zero remainder.

If P is a real polynomial having a complex root $r_1 = u + iv$, where u and v are real and $v \neq 0$, then, as asserted above, the complex conjugate of r_1, namely, $r_2 = u - iv$, will also be a root of P. (Moreover, r_1 and r_2 will have the same multiplicity.) Thus, both $x - u - iv$ and $x - u + iv$ are factors of $P(x)$, and so, therefore, is their product

$$(x - u - iv)(x - u + iv) = (x - u)^2 + v^2 = x^2 - 2ux + u^2 + v^2,$$

which is a quadratic polynomial having no real roots. It follows that every real polynomial can be factored into a product of real (possibly repeated) linear factors and real (also possibly repeated) quadratic factors having no real zeros.

EXAMPLE 2
What is the degree of $P(x) = x^3(x^2 + 2x + 5)^2$? What are the roots of P, and what is the multiplicity of each root?

Solution If P is expanded, the highest power of x present in the expansion is $x^3(x^2)^2 = x^7$, so P has degree 7. The factor $x^3 = (x - 0)^3$ indicates that 0 is a root of P having multiplicity 3. The remaining four roots will be the two roots of $x^2 + 2x + 5$, each having multiplicity 2. Now

$$\left[x^2 + 2x + 5\right]^2 = \left[(x + 1)^2 + 4\right]^2$$
$$= \left[(x + 1 + 2i)(x + 1 - 2i)\right]^2.$$

Hence the seven roots of P are:

$$\begin{cases} 0,\ 0,\ 0 & \text{0 has multiplicity 3,} \\ -1 - 2i,\ -1 - 2i & \text{$-1 - 2i$ has multiplicity 2,} \\ -1 + 2i,\ -1 + 2i & \text{$-1 + 2i$ has multiplicity 2.} \end{cases}$$

Roots and Factors of Quadratic Polynomials

There is a well-known formula for finding the roots of a quadratic polynomial.

The Quadratic Formula

The two solutions of the quadratic equation

$$Ax^2 + Bx + C = 0,$$

where A, B, and C are constants and $A \neq 0$, are given by

$$x = \frac{-B \pm \sqrt{B^2 - 4AC}}{2A}.$$

To see this, just divide the equation by A and complete the square for the terms in x:

$$x^2 + \frac{B}{A}x + \frac{C}{A} = 0$$

$$x^2 + \frac{2B}{2A}x + \frac{B^2}{4A^2} = \frac{B^2}{4A^2} - \frac{C}{A}$$

$$\left(x + \frac{B}{2A}\right)^2 = \frac{B^2 - 4AC}{4A^2}$$

$$x + \frac{B}{2A} = \pm\frac{\sqrt{B^2 - 4AC}}{2A}.$$

The quantity $D = B^2 - 4AC$ that appears under the square root in the quadratic formula is called the **discriminant** of the quadratic equation or polynomial. The nature of the roots of the quadratic depends on the sign of this discriminant.

(a) If $D > 0$, then $D = k^2$ for some real constant k, and the quadratic has two distinct roots, $(-B + k)/(2A)$ and $(-B - k)/(2A)$.

(b) If $D = 0$, then the quadratic has only the root $-B/(2A)$, and this root has multiplicity 2. (It is called a *double root.*)

(c) If $D < 0$, then $D = -k^2$ for some real constant k, and the quadratic has two complex conjugate roots, $(-B + ki)/(2A)$ and $(-B - ki)/(2A)$.

EXAMPLE 3   Find the roots of these quadratic polynomials and thereby factor the polynomials into linear factors:

(a) $x^2 + x - 1$ (b) $9x^2 - 6x + 1$ (c) $2x^2 + x + 1$.

Solution   We use the quadratic formula to solve the corresponding quadratic equations to find the roots of the three polynomials.

(a) $A = 1$, $B = 1$, $C = -1$

$$x = \frac{-1 \pm \sqrt{1 + 4}}{2} = -\frac{1}{2} \pm \frac{\sqrt{5}}{2}$$

$$x^2 + x - 1 = \left(x + \frac{1}{2} - \frac{\sqrt{5}}{2}\right)\left(x + \frac{1}{2} + \frac{\sqrt{5}}{2}\right).$$

(b) $A = 9$, $B = -6$, $C = 1$

$$x = \frac{6 \pm \sqrt{36 - 36}}{18} = \frac{1}{3} \quad \text{(double root)}$$

$$9x^2 - 6x + 1 = 9\left(x - \frac{1}{3}\right)^2 = (3x - 1)^2.$$

(c) $A = 2$, $B = 1$, $C = 1$

$$x = \frac{-1 \pm \sqrt{1 - 8}}{4} = -\frac{1}{4} \pm \frac{\sqrt{7}}{4}i$$

$$2x^2 + x + 1 = 2\left(x + \frac{1}{4} - \frac{\sqrt{7}}{4}i\right)\left(x + \frac{1}{4} + \frac{\sqrt{7}}{4}i\right).$$

Remark   There exist formulas for calculating exact roots of cubic (degree 3) and quartic (degree 4) polynomials, but, unlike the quadratic formula above, they are very complicated and almost never used. Instead, calculus will provide us with very powerful and easily used tools for approximating roots of polynomials (and solutions of much more general equations) to any desired degree of accuracy.

Miscellaneous Factorings

Some quadratic and higher-degree polynomials can be (at least partially) factored by inspection. Some simple examples include:

(a) Common Factor: $ax^2 + bx = x(ax + b)$.

(b) Difference of Squares: $x^2 - a^2 = (x - a)(x + a)$.

(c) Difference of Cubes: $x^3 - a^3 = (x - a)(x^2 + ax + a^2)$.

(d) More generally, a difference of nth powers for any positive integer n:

$$x^n - a^n = (x - a)(x^{n-1} + ax^{n-2} + a^2 x^{n-3} + \cdots + a^{n-2}x + a^{n-1}).$$

   Note that $x - a$ is a factor of $x^n - a^n$ for any positive integer n.

(e) It is also true that if n is an *odd positive integer*, then $x + a$ is a factor of $x^n + a^n$. For example,

$$x^3 + a^3 = (x + a)(x^2 - ax + a^2)$$
$$x^5 + a^5 = (x + a)(x^4 - ax^3 + a^2 x^2 - a^3 x + a^4).$$

Finally, we mention a trial-and-error method of factoring quadratic polynomials sometimes called *trinomial factoring*. Since

$$(x + p)(x + q) = x^2 + (p + q)x + pq,$$
$$(x - p)(x - q) = x^2 - (p + q)x + pq, \qquad \text{and}$$
$$(x + p)(x - q) = x^2 + (p - q)x - pq,$$

we can sometimes spot the factors of $x^2 + Bx + C$ by looking for factors of $|C|$ for which the sum or difference is B. More generally, we can sometimes factor

$$Ax^2 + Bx + C = (ax + b)(cx + d)$$

by looking for factors a and c of A and factors b and d of C for which $ad + bc = B$. Of course, if this fails you can always resort to the quadratic formula to find the roots and, therefore, the factors, of the quadratic polynomial.

EXAMPLE 4

$$
\begin{array}{ll}
x^2 - 5x + 6 = (x - 3)(x - 2) & p = 3, q = 2, pq = 6, p + q = 5 \\
x^2 + 7x + 6 = (x + 6)(x + 1) & p = 6, q = 1, pq = 6, p + q = 7 \\
x^2 + \ x - 6 = (x + 3)(x - 2) & p = 3, q = -2, pq = -6, p + q = 1 \\
2x^2 + x - 10 = (2x + 5)(x - 2) & a = 2, b = 5, c = 1, d = -2 \\
& ac = 2, bd = -10, ad + bc = 1.
\end{array}
$$

EXAMPLE 5   Find the roots of the following polynomials:

(a) $x^3 - x^2 - 4x + 4$,   (b) $x^4 + 3x^2 - 4$,   (c) $x^5 - x^4 - x^2 + x$.

Solution   (a) There is an obvious common factor:

$$x^3 - x^2 - 4x + 4 = (x - 1)(x^2 - 4) = (x - 1)(x - 2)(x + 2).$$

   The roots are 1, 2, and -2.

(b) This is a trinomial in x^2 for which there is an easy factoring:

$$x^4 + 3x^2 - 4 = (x^2 + 4)(x^2 - 1) = (x + 2i)(x - 2i)(x + 1)(x - 1).$$

The roots are $1, -1, 2i$, and $-2i$.

(c) We start with some obvious factorings:

$$x^5 - x^4 - x^2 + x = x(x^4 - x^3 - x + 1) = x(x-1)(x^3-1)$$
$$= x(x-1)^2(x^2+x+1).$$

Thus 0 is a root, and 1 is a double root. The remaining two roots must come from the quadratic factor $x^2 + x + 1$, which cannot be factored easily by inspection, so we use the formula:

$$x = \frac{-1 \pm \sqrt{1-4}}{2} = -\frac{1}{2} \pm \frac{\sqrt{3}}{2}i.$$

EXAMPLE 6 For what values of the real constant b will the product of the real polynomials $x^2 - bx + a^2$ and $x^2 + bx + a^2$ be equal to $x^4 + a^4$? Use your answer to express $x^4 + 1$ as a product of two real quadratic polynomials each having no real roots.

Solution We have

$$(x^2 - bx + a^2)(x^2 + bx + a^2) = (x^2 + a^2)^2 - b^2x^2$$
$$= x^4 + 2a^2x^2 + a^4 - b^2x^2 = x^4 + a^4,$$

provided that $b^2 = 2a^2$, that is, $b = \pm\sqrt{2}a$.
If $a = 1$, then $b = \pm\sqrt{2}$, and we have

$$x^4 + 1 = (x^2 - \sqrt{2}x + 1)(x^2 + \sqrt{2}x + 1).$$

EXERCISES P.6

Find the roots of the polynomials in Exercises 1–12. If a root is repeated, give its multiplicity. Also, write each polynomial as a product of linear factors.

1. $x^2 + 7x + 10$ **2.** $x^2 - 3x - 10$

3. $x^2 + 2x + 2$ **4.** $x^2 - 6x + 13$

5. $16x^4 - 8x^2 + 1$ **6.** $x^4 + 6x^3 + 9x^2$

7. $x^3 + 1$ **8.** $x^4 - 1$

9. $x^6 - 3x^4 + 3x^2 - 1$ **10.** $x^5 - x^4 - 16x + 16$

11. $x^5 + x^3 + 8x^2 + 8$ **12.** $x^9 - 4x^7 - x^6 + 4x^4$

In Exercises 13–16, determine the domains of the given rational functions.

13. $\dfrac{3x+2}{x^2+2x+2}$ **14.** $\dfrac{x^2-9}{x^3-x}$

15. $\dfrac{4}{x^3+x^2}$ **16.** $\dfrac{x^3+3x^2+6}{x^2+x-1}$

In Exercises 17–20, express the given rational function as the sum of a polynomial and another rational function whose numerator is either zero or has smaller degree than the denominator.

17. $\dfrac{x^3-1}{x^2-2}$ **18.** $\dfrac{x^2}{x^2+5x+3}$

19. $\dfrac{x^3}{x^2+2x+3}$ **20.** $\dfrac{x^4+x^2}{x^3+x^2+1}$

In Exercises 21–22, express the given polynomial as a product of real quadratic polynomials with no real roots.

21. $P(x) = x^4 + 4$ **22.** $P(x) = x^4 + x^2 + 1$

23. Show that $x - 1$ is a factor of a polynomial P of positive degree if and only if the sum of the coefficients of P is zero.

24. What condition should the coefficients of a polynomial satsify to ensure that $x + 1$ is a factor of that polynomial?

25. The complex conjugate of a complex number $z = u + iv$ (where u and v are real numbers) is the complex number $\bar{z} = u - iv$. It is shown in Appendix I that the complex conjugate of a sum (or product) of complex numbers is the sum (or product) of the complex conjugates of those numbers. Use this fact to verify that if $z = u + iv$ is a complex root of a polynomial P having real coefficients, then its conjugate $\bar{z}$ is also a root of P.

❷ 26. Continuing the previous exercise, show that if $z = u + iv$ (where u and v are real numbers) is a complex root of a polynomial P with real coefficients, then P must have the real quadratic factor $x^2 - 2ux + u^2 + v^2$.

❷ 27. Use the result of Exercise 26 to show that if $z = u + iv$ (where u and v are real numbers) is a complex root of a polynomial P with real coefficients, then z and $\bar{z}$ are roots of P having the same multiplicity.

P.7 The Trigonometric Functions

Most people first encounter the quantities $\cos t$ and $\sin t$ as ratios of sides in a right-angled triangle having t as one of the acute angles. If the sides of the triangle are labelled "hyp" for hypotenuse, "adj" for the side adjacent to angle t, and "opp" for the side opposite angle t (see Figure P.66), then

$$\cos t = \frac{\text{adj}}{\text{hyp}} \quad \text{and} \quad \sin t = \frac{\text{opp}}{\text{hyp}}. \qquad (*)$$

These ratios depend only on the angle t, not on the particular triangle, since all right-angled triangles having an acute angle t are similar.

In calculus we need more general definitions of $\cos t$ and $\sin t$ as functions defined for *all real numbers t*, not just acute angles. Such definitions are phrased in terms of a circle rather than a triangle.

Let C be the circle with centre at the origin O and radius 1; its equation is $x^2 + y^2 = 1$. Let A be the point $(1, 0)$ on C. For any real number t, let P_t be the point on C at distance $|t|$ from A, measured along C in the counterclockwise direction if $t > 0$, and the clockwise direction if $t < 0$. For example, since C has circumference 2π, the point $P_{\pi/2}$ is one-quarter of the way counterclockwise around C from A; it is the point $(0, 1)$.

We will use the arc length t as a measure of the size of the angle AOP_t. See Figure P.67.

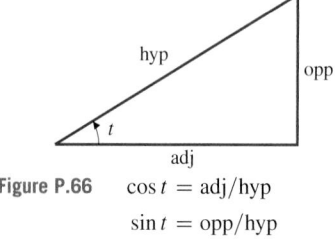

Figure P.66 $\cos t = \text{adj}/\text{hyp}$
$\sin t = \text{opp}/\text{hyp}$

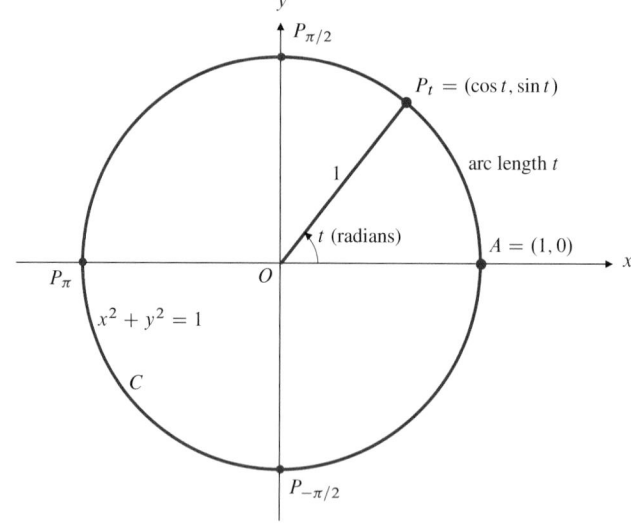

Figure P.67   If the length of arc AP_t is t units, then angle $AOP_t = t$ radians

DEFINITION

7

The **radian measure** of angle AOP_t is t radians:

$$\angle AOP_t = t \text{ radians.}$$

We are more used to measuring angles in **degrees**. Since P_π is the point $(-1, 0)$, halfway (π units of distance) around C from A, we have

$$\pi \text{ radians} = 180°.$$

To convert degrees to radians, multiply by $\pi/180$; to convert radians to degrees, multiply by $180/\pi$.

> **Angle convention**
>
> In calculus it is assumed that all angles are measured in radians unless degrees or other units are stated explicitly. When we talk about the angle $\pi/3$, we mean $\pi/3$ radians (which is $60°$), not $\pi/3$ degrees.

EXAMPLE 1 **Arc length and sector area.** An arc of a circle of radius r subtends an angle t at the centre of the circle. Find the length s of the arc and the area A of the sector lying between the arc and the centre of the circle.

Solution The length s of the arc is the same fraction of the circumference $2\pi r$ of the circle that the angle t is of a complete revolution 2π radians (or $360°$). Thus,

$$s = \frac{t}{2\pi}(2\pi r) = rt \text{ units.}$$

Similarly, the area A of the circular sector (Figure P.68) is the same fraction of the area πr^2 of the whole circle:

$$A = \frac{t}{2\pi}(\pi r^2) = \frac{r^2 t}{2} \text{ units}^2.$$

(We will show that the area of a circle of radius r is πr^2 in Section 1.1.)

Figure P.68 Arc length $s = rt$
Sector area $A = r^2 t/2$

Using the procedure described above, we can find the point P_t corresponding to any real number t, positive or negative. We define $\cos t$ and $\sin t$ to be the coordinates of P_t. (See Figure P.69.)

DEFINITION

8

> **Cosine and sine**
>
> For any real t, the **cosine** of t (abbreviated $\cos t$) and the **sine** of t (abbreviated $\sin t$) are the x- and y-coordinates of the point P_t.
>
> $$\cos t = \text{the } x\text{-coordinate of } P_t$$
> $$\sin t = \text{the } y\text{-coordinate of } P_t$$

Because they are defined this way, cosine and sine are often called the **circular functions**. Note that these definitions agree with the ones given earlier for an acute angle. (See formulas $(*)$ at the beginning of this section.) The triangle involved is $P_t O Q_t$ in Figure P.69.

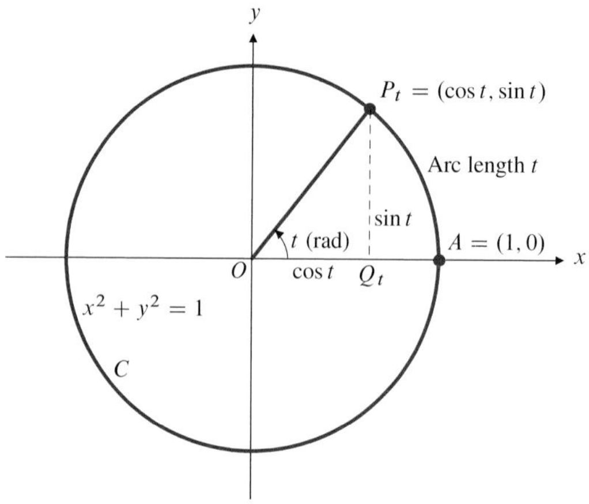

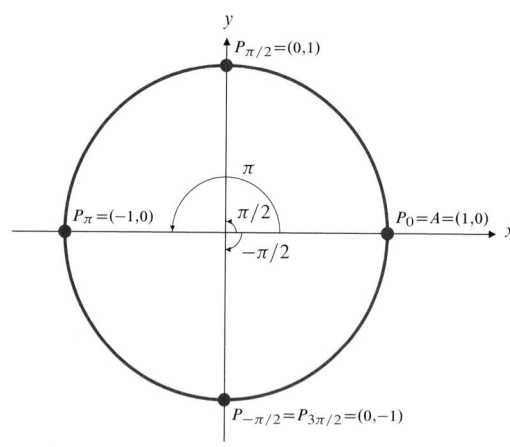

Figure P.69 The coordinates of P_t are $(\cos t, \sin t)$

Figure P.70 Some special angles

EXAMPLE 2 Examining the coordinates of $P_0 = A$, $P_{\pi/2}$, P_π, and $P_{-\pi/2} = P_{3\pi/2}$ in Figure P.70, we obtain the following values:

$$\cos 0 = 1 \quad \cos\frac{\pi}{2} = 0 \quad \cos\pi = -1 \quad \cos\left(-\frac{\pi}{2}\right) = \cos\frac{3\pi}{2} = 0$$

$$\sin 0 = 0 \quad \sin\frac{\pi}{2} = 1 \quad \sin\pi = 0 \quad \sin\left(-\frac{\pi}{2}\right) = \sin\frac{3\pi}{2} = -1$$

Some Useful Identities

Many important properties of $\cos t$ and $\sin t$ follow from the fact that they are coordinates of the point P_t on the circle C with equation $x^2 + y^2 = 1$.

The range of cosine and sine. For every real number t,

$$-1 \leq \cos t \leq 1 \quad \text{and} \quad -1 \leq \sin t \leq 1.$$

The Pythagorean identity. The coordinates $x = \cos t$ and $y = \sin t$ of P_t must satisfy the equation of the circle. Therefore, for every real number t,

$$\cos^2 t + \sin^2 t = 1.$$

(Note that $\cos^2 t$ means $(\cos t)^2$, not $\cos(\cos t)$. This is an unfortunate notation, but it is used everywhere in technical literature, so you have to get used to it!)

Periodicity. Since C has circumference 2π, adding 2π to t causes the point P_t to go one extra complete revolution around C and end up in the same place: $P_{t+2\pi} = P_t$. Thus, for every t,

$$\cos(t + 2\pi) = \cos t \qquad \text{and} \qquad \sin(t + 2\pi) = \sin t.$$

This says that cosine and sine are **periodic** with period 2π.

Cosine is an even function. Sine is an odd function. Since the circle $x^2 + y^2 = 1$ is symmetric about the x-axis, the points P_{-t} and P_t have the same x-coordinates and opposite y-coordinates (Figure P.71),

$$\cos(-t) = \cos t \qquad \text{and} \qquad \sin(-t) = -\sin t.$$

Complementary angle identities. Two angles are complementary if their sum is $\pi/2$ (or $90°$). The points $P_{(\pi/2)-t}$ and P_t are reflections of each other in the line $y = x$ (Figure P.72), so the x-coordinate of one is the y-coordinate of the other and vice versa. Thus,

$$\cos\left(\frac{\pi}{2} - t\right) = \sin t \qquad \text{and} \qquad \sin\left(\frac{\pi}{2} - t\right) = \cos t.$$

Supplementary angle identities. Two angles are supplementary if their sum is π (or $180°$). Since the circle is symmetric about the y-axis, $P_{\pi-t}$ and P_t have the same y-coordinates and opposite x-coordinates (Figure P.73). Thus,

$$\cos(\pi - t) = -\cos t \qquad \text{and} \qquad \sin(\pi - t) = \sin t.$$

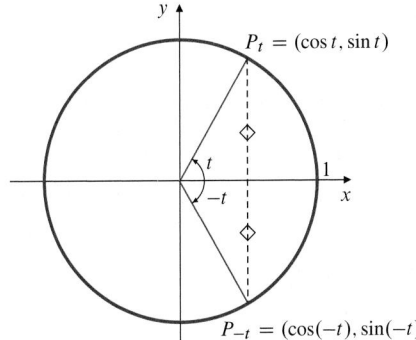

Figure P.71 $\cos(-t) = \cos t$
$\sin(-t) = -\sin t$

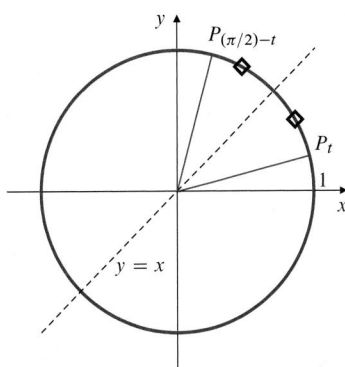

Figure P.72 $\cos((\pi/2) - t) = \sin t$
$\sin((\pi/2) - t) = \cos t$

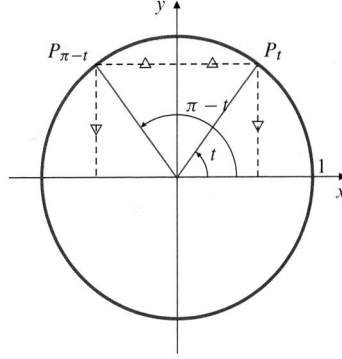

Figure P.73 $\cos(\pi - t) = -\cos t$
$\sin(\pi - t) = \sin t$

Some Special Angles

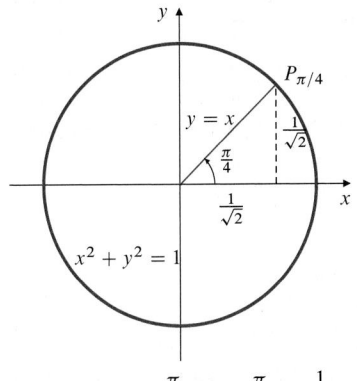

Figure P.74 $\sin\dfrac{\pi}{4} = \cos\dfrac{\pi}{4} = \dfrac{1}{\sqrt{2}}$

EXAMPLE 3   Find the sine and cosine of $\pi/4$ (i.e., $45°$).

Solution   The point $P_{\pi/4}$ lies in the first quadrant on the line $x = y$. To find its coordinates, substitute $y = x$ into the equation $x^2 + y^2 = 1$ of the circle, obtaining $2x^2 = 1$. Thus $x = y = 1/\sqrt{2}$ (see Figure P.74), and

$$\cos(45°) = \cos\frac{\pi}{4} = \frac{1}{\sqrt{2}}, \qquad \sin(45°) = \sin\frac{\pi}{4} = \frac{1}{\sqrt{2}}.$$

EXAMPLE 4   Find the values of sine and cosine of the angles $\pi/3$ (or $60°$) and $\pi/6$ (or $30°$).

Solution The point $P_{\pi/3}$ and the points $O(0,0)$ and $A(1,0)$ are the vertices of an equilateral triangle with edge length 1 (see Figure P.75). Thus, $P_{\pi/3}$ has x-coordinate $1/2$ and y-coordinate $\sqrt{1-(1/2)^2} = \sqrt{3}/2$, and

$$\cos(60°) = \cos\frac{\pi}{3} = \frac{1}{2}, \qquad \sin(60°) = \sin\frac{\pi}{3} = \frac{\sqrt{3}}{2}.$$

Since $\dfrac{\pi}{6} = \dfrac{\pi}{2} - \dfrac{\pi}{3}$, the complementary angle identities now tell us that

$$\cos(30°) = \cos\frac{\pi}{6} = \sin\frac{\pi}{3} = \frac{\sqrt{3}}{2}, \qquad \sin(30°) = \sin\frac{\pi}{6} = \cos\frac{\pi}{3} = \frac{1}{2}.$$

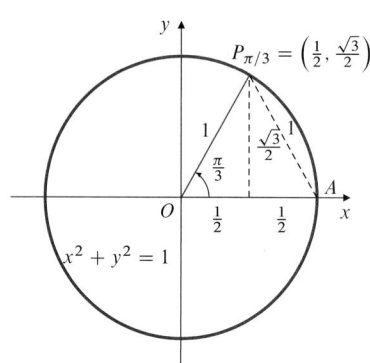

Figure P.75 $\cos\pi/3 = 1/2$
$$\sin\pi/3 = \sqrt{3}/2$$

Table 5 summarizes the values of cosine and sine at multiples of 30° and 45° between 0° and 180°. The values for 120°, 135°, and 150° were determined by using the supplementary angle identities; for example,

$$\cos(120°) = \cos\left(\frac{2\pi}{3}\right) = \cos\left(\pi - \frac{\pi}{3}\right) = -\cos\left(\frac{\pi}{3}\right) = -\cos(60°) = -\frac{1}{2}.$$

Table 5.   Cosines and sines of special angles

Degrees	0°	30°	45°	60°	90°	120°	135°	150°	180°
Radians	0	$\dfrac{\pi}{6}$	$\dfrac{\pi}{4}$	$\dfrac{\pi}{3}$	$\dfrac{\pi}{2}$	$\dfrac{2\pi}{3}$	$\dfrac{3\pi}{4}$	$\dfrac{5\pi}{6}$	π
Cosine	1	$\dfrac{\sqrt{3}}{2}$	$\dfrac{1}{\sqrt{2}}$	$\dfrac{1}{2}$	0	$-\dfrac{1}{2}$	$-\dfrac{1}{\sqrt{2}}$	$-\dfrac{\sqrt{3}}{2}$	-1
Sine	0	$\dfrac{1}{2}$	$\dfrac{1}{\sqrt{2}}$	$\dfrac{\sqrt{3}}{2}$	1	$\dfrac{\sqrt{3}}{2}$	$\dfrac{1}{\sqrt{2}}$	$\dfrac{1}{2}$	0

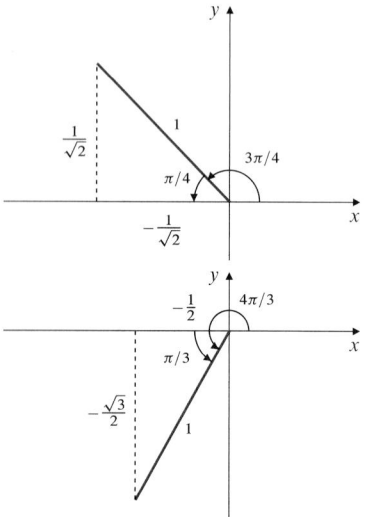

Figure P.76   Using suitably placed triangles to find trigonometric functions of special angles

EXAMPLE 5   Find: (a) $\sin(3\pi/4)$ and   (b) $\cos(4\pi/3)$.

Solution We can draw appropriate triangles in the quadrants where the angles lie to determine the required values. See Figure P.76.

(a) $\sin(3\pi/4) = \sin(\pi - (\pi/4)) = 1/\sqrt{2}$.

(b) $\cos(4\pi/3) = \cos(\pi + (\pi/3)) = -1/2$.

While decimal approximations to the values of sine and cosine can be found using a scientific calculator or mathematical tables, it is useful to remember the exact values in Table 5 for angles 0, $\pi/6$, $\pi/4$, $\pi/3$, and $\pi/2$. They occur frequently in applications.

When we treat sine and cosine as functions, we can call the variable they depend on anything we want (e.g., x, as we do with other functions), rather than t. The graphs of $\cos x$ and $\sin x$ are shown in Figures P.77 and P.78. In both graphs the pattern between $x = 0$ and $x = 2\pi$ repeats over and over to the left and right. Observe that the graph of $\sin x$ is the graph of $\cos x$ shifted to the right a distance $\pi/2$.

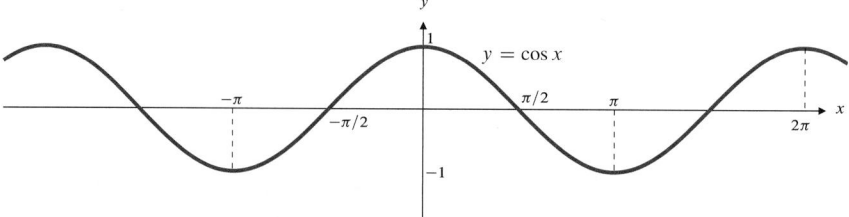

Figure P.77   The graph of $\cos x$

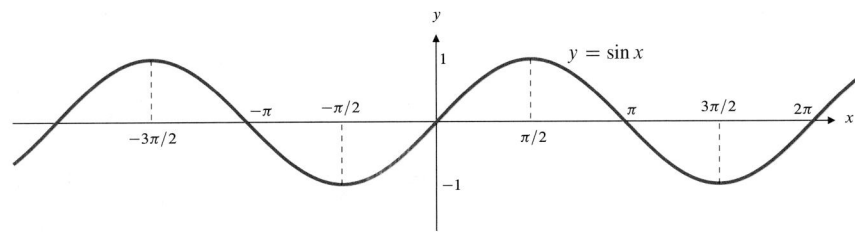

Figure P.78   The graph of $\sin x$

> **Remember this!**
>
> When using a scientific calculator to calculate any trigonometric functions, be sure you have selected the proper angular mode: degrees or radians.

The Addition Formulas

The following formulas enable us to determine the cosine and sine of a sum or difference of two angles in terms of the cosines and sines of those angles.

THEOREM

2

Addition Formulas for Cosine and Sine

$$\cos(s + t) = \cos s \cos t - \sin s \sin t$$
$$\sin(s + t) = \sin s \cos t + \cos s \sin t$$
$$\cos(s - t) = \cos s \cos t + \sin s \sin t$$
$$\sin(s - t) = \sin s \cos t - \cos s \sin t$$

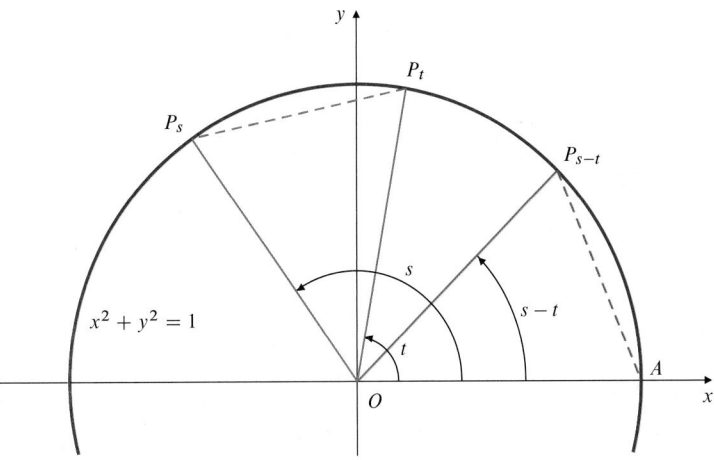

Figure P.79 $P_s P_t = P_{s-t} A$

PROOF   We prove the third of these formulas as follows: Let s and t be real numbers and consider the points

$$P_t = (\cos t, \sin t) \qquad P_{s-t} = (\cos(s - t), \sin(s - t))$$
$$P_s = (\cos s, \sin s) \qquad A = (1, 0),$$

as shown in Figure P.79.

The angle $P_t O P_s = s - t$ radians = angle AOP_{s-t}, so the distance $P_s P_t$ is equal to the distance $P_{s-t} A$. Therefore, $(P_s P_t)^2 = (P_{s-t} A)^2$. We express these squared distances in terms of coordinates and expand the resulting squares of binomials:

$$(\cos s - \cos t)^2 + (\sin s - \sin t)^2 = (\cos(s - t) - 1)^2 + \sin^2(s - t),$$

$$\cos^2 s - 2\cos s \cos t + \cos^2 t + \sin^2 s - 2\sin s \sin t + \sin^2 t$$
$$= \cos^2(s-t) - 2\cos(s-t) + 1 + \sin^2(s-t).$$

Since $\cos^2 x + \sin^2 x = 1$ for every x, this reduces to

$$\cos(s-t) = \cos s \cos t + \sin s \sin t.$$

Replacing t with $-t$ in the formula above, and recalling that $\cos(-t) = \cos t$ and $\sin(-t) = -\sin t$, we have

$$\cos(s+t) = \cos s \cos t - \sin s \sin t.$$

The complementary angle formulas can be used to obtain either of the addition formulas for sine:

$$\sin(s+t) = \cos\left(\frac{\pi}{2} - (s+t)\right)$$
$$= \cos\left(\left(\frac{\pi}{2} - s\right) - t\right)$$
$$= \cos\left(\frac{\pi}{2} - s\right)\cos t + \sin\left(\frac{\pi}{2} - s\right)\sin t$$
$$= \sin s \cos t + \cos s \sin t,$$

and the other formula again follows if we replace t with $-t$.

EXAMPLE 6 Find the value of $\cos(\pi/12) = \cos 15°$.

Solution

$$\cos\frac{\pi}{12} = \cos\left(\frac{\pi}{3} - \frac{\pi}{4}\right) = \cos\frac{\pi}{3}\cos\frac{\pi}{4} + \sin\frac{\pi}{3}\sin\frac{\pi}{4}$$
$$= \left(\frac{1}{2}\right)\left(\frac{1}{\sqrt{2}}\right) + \left(\frac{\sqrt{3}}{2}\right)\left(\frac{1}{\sqrt{2}}\right) = \frac{1+\sqrt{3}}{2\sqrt{2}}$$

From the addition formulas, we obtain as special cases certain useful formulas called **double-angle formulas**. Put $s = t$ in the addition formulas for $\sin(s+t)$ and $\cos(s+t)$ to get

$$\sin 2t = 2\sin t \cos t \qquad \text{and}$$
$$\cos 2t = \cos^2 t - \sin^2 t$$
$$= 2\cos^2 t - 1 \qquad (\text{using } \sin^2 t + \cos^2 t = 1)$$
$$= 1 - 2\sin^2 t$$

Solving the last two formulas for $\cos^2 t$ and $\sin^2 t$, we obtain

$$\cos^2 t = \frac{1 + \cos 2t}{2} \qquad \text{and} \qquad \sin^2 t = \frac{1 - \cos 2t}{2},$$

which are sometimes called **half-angle formulas** because they are used to express trigonometric functions of half of the angle $2t$. Later we will find these formulas useful when we have to integrate powers of $\cos x$ and $\sin x$.

Other Trigonometric Functions

There are four other trigonometric functions—tangent (tan), cotangent (cot), secant (sec), and cosecant (csc)—each defined in terms of cosine and sine. Their graphs are shown in Figures P.80–P.83.

DEFINITION

9

Tangent, cotangent, secant, and cosecant

$$\tan t = \frac{\sin t}{\cos t} \qquad \sec t = \frac{1}{\cos t}$$

$$\cot t = \frac{\cos t}{\sin t} = \frac{1}{\tan t} \qquad \csc t = \frac{1}{\sin t}$$

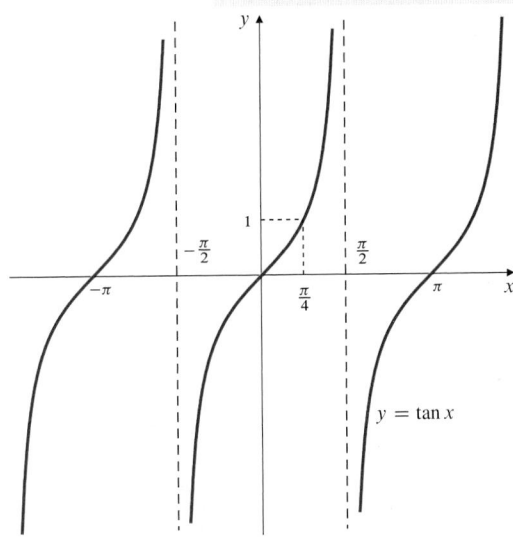

Figure P.80   The graph of $\tan x$

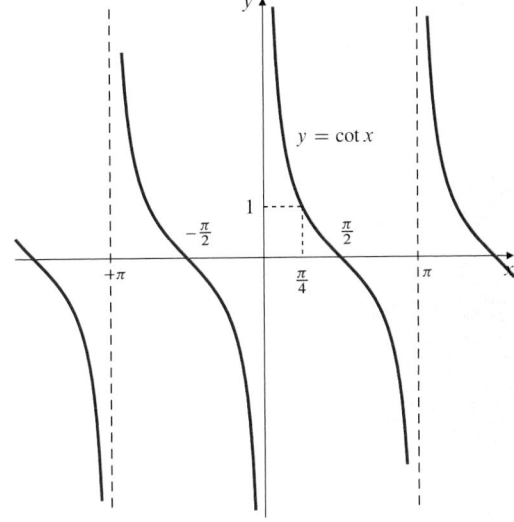

Figure P.81   The graph of $\cot x$

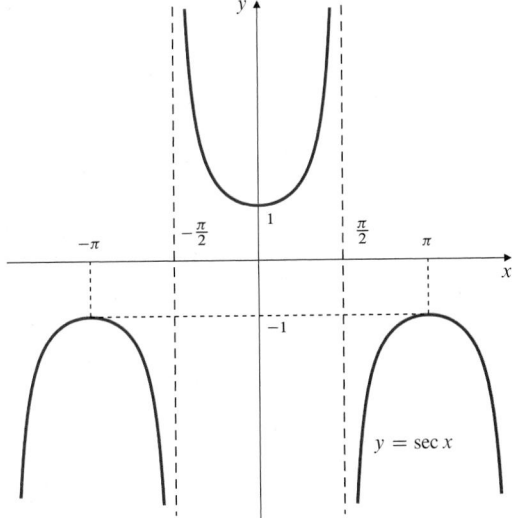

Figure P.82   The graph of $\sec x$

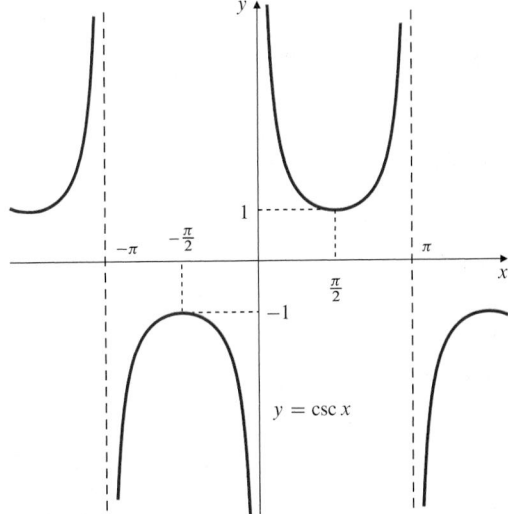

Figure P.83   The graph of $\csc x$

Observe that each of these functions is undefined (and its graph approaches vertical asymptotes) at points where the function in the denominator of its defining fraction has value 0. Observe also that tangent, cotangent, and cosecant are odd functions and secant is an even function. Since $|\sin x| \le 1$ and $|\cos x| \le 1$ for all x, $|\csc x| \ge 1$ and $|\sec x| \ge 1$ for all x where they are defined.

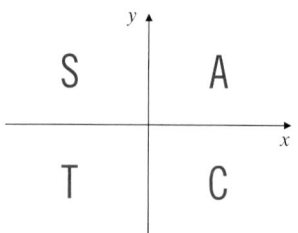

Figure P.84 The CAST rule

The three functions sine, cosine, and tangent are called **primary trigonometric functions**, while their reciprocals cosecant, secant, and cotangent are called **secondary trigonometric functions**. Scientific calculators usually just implement the primary functions; you can use the reciprocal key to find values of the corresponding secondary functions. Figure P.84 shows a useful pattern called the "CAST rule" to help you remember where the primary functions are positive. All three are positive in the first quadrant, marked A. Of the three, only sine is positive in the second quadrant S, only tangent in the third quadrant T, and only cosine in the fourth quadrant C.

EXAMPLE 7 Find the sine and tangent of the angle θ in $\left[\pi, \dfrac{3\pi}{2}\right]$ for which we have $\cos\theta = -\dfrac{1}{3}$.

Solution From the Pythagorean identity $\sin^2\theta + \cos^2\theta = 1$, we get

$$\sin^2\theta = 1 - \frac{1}{9} = \frac{8}{9}, \qquad \text{so} \quad \sin\theta = \pm\sqrt{\frac{8}{9}} = \pm\frac{2\sqrt{2}}{3}.$$

The requirement that θ should lie in $[\pi, 3\pi/2]$ makes θ a third quadrant angle. Its sine is therefore negative. We have

$$\sin\theta = -\frac{2\sqrt{2}}{3} \qquad \text{and} \qquad \tan\theta = \frac{\sin\theta}{\cos\theta} = \frac{-2\sqrt{2}/3}{-1/3} = 2\sqrt{2}.$$

Like their reciprocals cosine and sine, the functions secant and cosecant are periodic with period 2π. Tangent and cotangent, however, have period π because

$$\tan(x + \pi) = \frac{\sin(x + \pi)}{\cos(x + \pi)} = \frac{\sin x \cos\pi + \cos x \sin\pi}{\cos x \cos\pi - \sin x \sin\pi} = \frac{-\sin x}{-\cos x} = \tan x.$$

Dividing the Pythagorean identity $\sin^2 x + \cos^2 x = 1$ by $\cos^2 x$ and $\sin^2 x$, respectively, leads to two useful alternative versions of that identity:

$$1 + \tan^2 x = \sec^2 x \qquad \text{and} \qquad 1 + \cot^2 x = \csc^2 x.$$

Addition formulas for tangent and cotangent can be obtained from those for sine and cosine. For example,

$$\tan(s + t) = \frac{\sin(s + t)}{\cos(s + t)} = \frac{\sin s \cos t + \cos s \sin t}{\cos s \cos t - \sin s \sin t}.$$

Now divide the numerator and denominator of the fraction on the right by $\cos s \cos t$ to get

$$\tan(s + t) = \frac{\tan s + \tan t}{1 - \tan s \tan t}.$$

Replacing t by $-t$ leads to

$$\tan(s - t) = \frac{\tan s - \tan t}{1 + \tan s \tan t}.$$

Maple Calculations

Maple knows all six trigonometric functions and can calculate their values and manipulate them in other ways. It assumes the arguments of the trigonometric functions are in radians.

```
>  evalf(sin(30)); evalf(sin(Pi/6));
```
$$-.9880316241$$

$$.5000000000$$

Note that the constant Pi (with an uppercase P) is known to Maple. The `evalf()` function converts its argument to a number expressed as a floating point decimal with 10 significant digits. (This precision can be changed by defining a new value for the variable `Digits`.) Without it, the sine of 30 radians would have been left unexpanded because it is not an integer.

```
>  Digits := 20; evalf(100*Pi); sin(30);
```
$$Digits := 20$$

$$314.15926535897932385$$

$$sin(30)$$

It is often useful to expand trigonometric functions of multiple angles to powers of sine and cosine, and vice versa.

```
>  expand(sin(5*x));
```
$$16 \sin(x) \cos(x)^4 - 12 \sin(x) \cos(x)^2 + \sin(x)$$

```
>  combine((cos(x))^5, trig);
```
$$\frac{1}{16} \cos(5x) + \frac{5}{16} \cos(3x) + \frac{5}{8} \cos(x)$$

Other trigonometric functions can be converted to expressions involving sine and cosine.

```
>  convert(tan(4*x)*(sec(4*x))^2, sincos); combine(%,trig);
```
$$\frac{\sin(4x)}{\cos(4x)^3}$$

$$4 \frac{\sin(4x)}{\cos(12x) + 3\cos(4x)}$$

The % in the last command refers to the result of the previous calculation.

Trigonometry Review

The trigonometric functions are so called because they are often used to express the relationships between the sides and angles of a triangle. As we observed at the beginning of this section, if θ is one of the acute angles in a right-angled triangle, we can refer to the three sides of the triangle as adj (side adjacent θ), opp (side opposite θ), and hyp (hypotenuse). (See Figure P.85.) The trigonometric functions of θ can then be expressed as ratios of these sides, in particular:

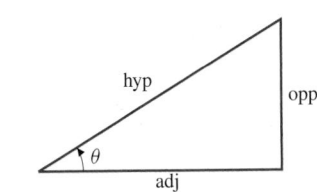

Figure P.85

$$\sin \theta = \frac{\text{opp}}{\text{hyp}}, \qquad \cos \theta = \frac{\text{adj}}{\text{hyp}}, \qquad \tan \theta = \frac{\text{opp}}{\text{adj}}.$$

EXAMPLE 8 Find the unknown sides x and y of the triangle in Figure P.86.

Solution Here, x is the side opposite and y is the side adjacent the 30° angle. The hypotenuse of the triangle is 5 units. Thus,

$$\frac{x}{5} = \sin 30° = \frac{1}{2} \qquad \text{and} \qquad \frac{y}{5} = \cos 30° = \frac{\sqrt{3}}{2},$$

$$\text{so } x = \frac{5}{2} \text{ units and } y = \frac{5\sqrt{3}}{2} \text{ units.}$$

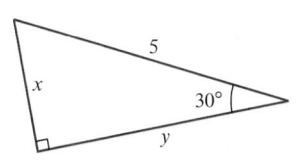

Figure P.86

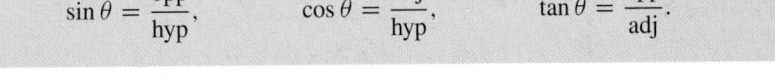

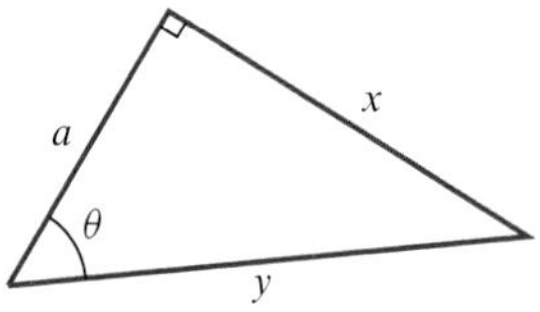

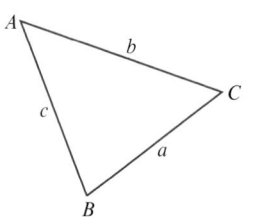

Figure P.87

EXAMPLE 9   For the triangle in Figure P.87, express sides x and y in terms of side a and angle θ.

Solution   The side x is opposite the angle θ, and the side y is the hypotenuse. The side adjacent to θ is a. Thus,

$$\frac{x}{a} = \tan\theta \qquad \text{and} \qquad \frac{a}{y} = \cos\theta.$$

Hence, $x = a\tan\theta$ and $y = \dfrac{a}{\cos\theta} = a\sec\theta$.

When dealing with general (not necessarily right-angled) triangles, it is often convenient to label the vertices with capital letters, which also denote the angles at those vertices, and refer to the sides opposite those vertices by the corresponding lowercase letters. See Figure P.88. Relationships between the sides a, b, and c and opposite angles A, B, and C of an arbitrary triangle ABC are given by the following formulas, called the **Sine Law** and the **Cosine Law**.

THEOREM

3

Sine Law: $\dfrac{\sin A}{a} = \dfrac{\sin B}{b} = \dfrac{\sin C}{c}$

Cosine Law: $a^2 = b^2 + c^2 - 2bc\cos A$

$b^2 = a^2 + c^2 - 2ac\cos B$

$c^2 = a^2 + b^2 - 2ab\cos C$

Figure P.88   In this triangle the sides are named to correspond to the opposite angles

PROOF   See Figure P.89. Let h be the length of the perpendicular from A to the side BC. From right-angled triangles (and using $\sin(\pi - t) = \sin t$ if required), we get $c\sin B = h = b\sin C$. Thus $(\sin B)/b = (\sin C)/c$. By the symmetry of the formulas (or by dropping a perpendicular to another side), both fractions must be equal to $(\sin A)/a$, so the Sine Law is proved. For the Cosine Law, observe that

$$c^2 = \begin{cases} h^2 + (a - b\cos C)^2 & \text{if } C \le \dfrac{\pi}{2} \\[2mm] h^2 + (a + b\cos(\pi - C))^2 & \text{if } C > \dfrac{\pi}{2} \end{cases}$$

$$= h^2 + (a - b\cos C)^2 \qquad (\text{since } \cos(\pi - C) = -\cos C)$$

$$= b^2\sin^2 C + a^2 - 2ab\cos C + b^2\cos^2 C$$

$$= a^2 + b^2 - 2ab\cos C.$$

The other versions of the Cosine Law can be proved in a similar way.

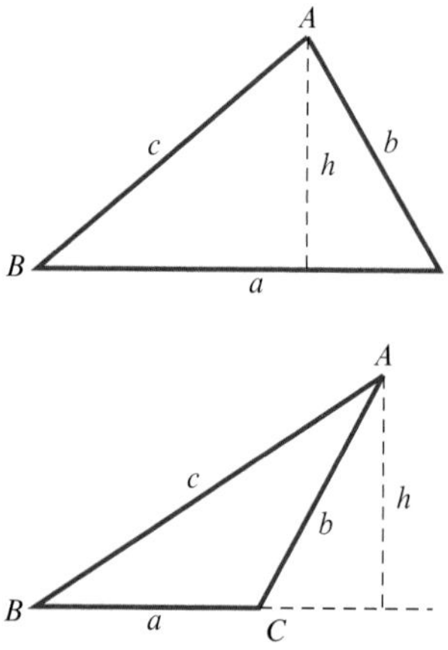

Figure P.89

EXAMPLE 10   A triangle has sides $a = 2$ and $b = 3$ and angle $C = 40°$. Find side c and the sine of angle B.

Solution   From the third version of the Cosine Law:

$$c^2 = a^2 + b^2 - 2ab\cos C = 4 + 9 - 12\cos 40° \approx 13 - 12 \times 0.766 = 3.808.$$

Side c is about $\sqrt{3.808} = 1.951$ units in length. Now using Sine Law we get

$$\sin B = b\,\frac{\sin C}{c} \approx 3 \times \frac{\sin 40°}{1.951} \approx \frac{3 \times 0.6428}{1.951} \approx 0.988.$$

A triangle is uniquely determined by any one of the following sets of data (which correspond to the known cases of congruency of triangles in classical geometry):

1. two sides and the angle contained between them (e.g., Example 10);
2. three sides, no one of which exceeds the sum of the other two in length;
3. two angles and one side; or
4. the hypotenuse and one other side of a right-angled triangle.

In such cases you can always find the unknown sides and angles by using the Pythagorean Theorem or the Sine and Cosine Laws, and the fact that the sum of the three angles of a triangle is $180°$ (or π radians).

A triangle is not determined uniquely by two sides and a noncontained angle; there may exist no triangle, one right-angled triangle, or two triangles having such data.

EXAMPLE 11   In triangle ABC, angle $B = 30°$, $b = 2$, and $c = 3$. Find a.

Solution   This is one of the ambiguous cases. By the Cosine Law,

$$b^2 = a^2 + c^2 - 2ac \cos B$$
$$4 = a^2 + 9 - 6a(\sqrt{3}/2).$$

Therefore, a must satisfy the equation $a^2 - 3\sqrt{3}a + 5 = 0$. Solving this equation using the quadratic formula, we obtain

$$a = \frac{3\sqrt{3} \pm \sqrt{27 - 20}}{2}$$
$$\approx 1.275 \quad \text{or} \quad 3.921$$

There are two triangles with the given data, as shown in Figure P.90.

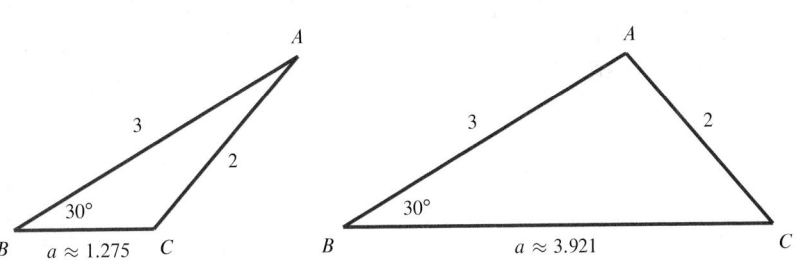

Figure P.90   Two triangles with $b = 2$, $c = 3$, $B = 30°$

EXERCISES P.7

Find the values of the quantities in Exercises 1–6 using various formulas presented in this section. Do not use tables or a calculator.

1. $\cos \dfrac{3\pi}{4}$

2. $\tan -\dfrac{3\pi}{4}$

3. $\sin \dfrac{2\pi}{3}$

4. $\sin \dfrac{7\pi}{12}$

5. $\cos \dfrac{5\pi}{12}$

6. $\sin \dfrac{11\pi}{12}$

In Exercises 7–12, express the given quantity in terms of $\sin x$ and $\cos x$.

7. $\cos(\pi + x)$

8. $\sin(2\pi - x)$

9. $\sin\left(\dfrac{3\pi}{2} - x\right)$

10. $\cos\left(\dfrac{3\pi}{2} + x\right)$

11. $\tan x + \cot x$

12. $\dfrac{\tan x - \cot x}{\tan x + \cot x}$

In Exercises 13–16, prove the given identities.

13. $\cos^4 x - \sin^4 x = \cos(2x)$

14. $\dfrac{1 - \cos x}{\sin x} = \dfrac{\sin x}{1 + \cos x} = \tan \dfrac{x}{2}$

15. $\dfrac{1 - \cos x}{1 + \cos x} = \tan^2 \dfrac{x}{2}$

16. $\dfrac{\cos x - \sin x}{\cos x + \sin x} = \sec 2x - \tan 2x$

17. Express $\sin 3x$ in terms of $\sin x$ and $\cos x$.

18. Express $\cos 3x$ in terms of $\sin x$ and $\cos x$.

In Exercises 19–22, sketch the graph of the given function. What is the period of the function?

19. $f(x) = \cos 2x$

20. $f(x) = \sin \dfrac{x}{2}$

21. $f(x) = \sin \pi x$

22. $f(x) = \cos \dfrac{\pi x}{2}$

23. Sketch the graph of $y = 2\cos\left(x - \dfrac{\pi}{3}\right)$.

24. Sketch the graph of $y = 1 + \sin\left(x + \dfrac{\pi}{4}\right)$.

In Exercises 25–30, one of $\sin\theta$, $\cos\theta$, and $\tan\theta$ is given. Find the other two if θ lies in the specified interval.

25. $\sin\theta = \dfrac{3}{5}$, θ in $\left[\dfrac{\pi}{2}, \pi\right]$

26. $\tan\theta = 2$, θ in $\left[0, \dfrac{\pi}{2}\right]$

27. $\cos\theta = \dfrac{1}{3}$, θ in $\left[-\dfrac{\pi}{2}, 0\right]$

28. $\cos\theta = -\dfrac{5}{13}$, θ in $\left[\dfrac{\pi}{2}, \pi\right]$

29. $\sin\theta = \dfrac{-1}{2}$, θ in $\left[\pi, \dfrac{3\pi}{2}\right]$

30. $\tan\theta = \dfrac{1}{2}$, θ in $\left[\pi, \dfrac{3\pi}{2}\right]$

Trigonometry Review

In Exercises 31–42, ABC is a triangle with a right angle at C. The sides opposite angles A, B, and C are a, b, and c, respectively. (See Figure P.91.)

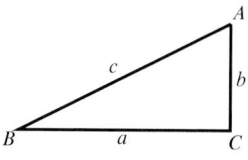

Figure P.91

31. Find a and b if $c = 2$, $B = \dfrac{\pi}{3}$.

32. Find a and c if $b = 2$, $B = \dfrac{\pi}{3}$.

33. Find b and c if $a = 5$, $B = \dfrac{\pi}{6}$.

34. Express a in terms of A and c.

35. Express a in terms of A and b.

36. Express a in terms of B and c.

37. Express a in terms of B and b.

38. Express c in terms of A and a.

39. Express c in terms of A and b.

40. Express $\sin A$ in terms of a and c.

41. Express $\sin A$ in terms of b and c.

42. Express $\sin A$ in terms of a and b.

In Exercises 43–50, ABC is an arbitrary triangle with sides a, b, and c, opposite to angles A, B, and C, respectively. (See Figure P.92.) Find the indicated quantities. Use tables or a scientific calculator if necessary.

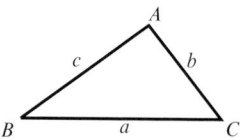

Figure P.92

43. Find $\sin B$ if $a = 4$, $b = 3$, $A = \dfrac{\pi}{4}$.

44. Find $\cos A$ if $a = 2$, $b = 2$, $c = 3$.

45. Find $\sin B$ if $a = 2$, $b = 3$, $c = 4$.

46. Find c if $a = 2$, $b = 3$, $C = \dfrac{\pi}{4}$.

47. Find a if $c = 3$, $A = \dfrac{\pi}{4}$, $B = \dfrac{\pi}{3}$.

48. Find c if $a = 2$, $b = 3$, $C = 35°$.

49. Find b if $a = 4$, $B = 40°$, $C = 70°$.

50. Find c if $a = 1$, $b = \sqrt{2}$, $A = 30°$. (There are two possible answers.)

51. Two guy wires stretch from the top T of a vertical pole to points B and C on the ground, where C is 10 m closer to the base of the pole than is B. If wire BT makes an angle of $35°$ with the horizontal, and wire CT makes an angle of $50°$ with the horizontal, how high is the pole?

52. Observers at positions A and B 2 km apart simultaneously measure the angle of elevation of a weather balloon to be $40°$ and $70°$, respectively. If the balloon is directly above a point on the line segment between A and B, find the height of the balloon.

53. Show that the area of triangle ABC is given by $(1/2)ab \sin C = (1/2)bc \sin A = (1/2)ca \sin B$.

❗ 54. Show that the area of triangle ABC is given by $\sqrt{s(s-a)(s-b)(s-c)}$, where $s = (a + b + c)/2$ is the semi-perimeter of the triangle.

❗ This symbol is used throughout the book to indicate an exercise that is somewhat more difficult than most exercises.

❷ This symbol is used throughout the book to indicate an exercise that is somewhat theoretical in nature. It does not imply difficulty.

CHAPTER 1

Limits and Continuity

> ❝ Every body continues in its state of rest, or of uniform motion in a right line, unless it is compelled to change that state by forces impressed upon it. ❞
>
> **Isaac Newton   1642–1727**
> **from *Principia Mathematica, 1687***

> ❝ It was not until Leibniz and Newton, by the discovery of the differential calculus, had dispelled the ancient darkness which enveloped the conception of the infinite, and had clearly established the conception of the continuous and continuous change, that a full productive application of the newly found mechanical conceptions made any progress. ❞
>
> **Hermann von Helmholtz   1821–1894**

Introduction

Calculus was created to describe how quantities change. It has two basic procedures that are opposites of one another, namely:

- *differentiation,* for finding the rate of change of a given function, and
- *integration,* for finding a function having a given rate of change.

Both of these procedures are based on the fundamental concept of the *limit* of a function. It is this idea of limit that distinguishes calculus from algebra, geometry, and trigonometry, which are useful for describing static situations.

In this chapter we will introduce the limit concept and develop some of its properties. We begin by considering how limits arise in some basic problems.

1.1   Examples of Velocity, Growth Rate, and Area

In this section we consider some examples of phenomena where limits arise in a natural way.

Average Velocity and Instantaneous Velocity

The position of a moving object is a function of time. The average velocity of the object over a time interval is found by dividing the change in the object's position by the length of the time interval.

___
EXAMPLE 1
___
(**The average velocity of a falling rock**) Physical experiments show that if a rock is dropped from rest near the surface of the earth, in the first t s it will fall a distance

$$y = 4.9t^2 \text{ m.}$$

(a) What is the average velocity of the falling rock during the first 2 s?

(b) What is its average velocity from $t = 1$ to $t = 2$?

Solution The *average velocity* of the falling rock over any time interval $[t_1, t_2]$ is the change Δy in the distance fallen divided by the length Δt of the time interval:

$$\text{average velocity over } [t_1, t_2] = \frac{\Delta y}{\Delta t} = \frac{4.9t_2^2 - 4.9t_1^2}{t_2 - t_1}.$$

(a) In the first 2 s (time interval $[0, 2]$), the average velocity is

$$\frac{\Delta y}{\Delta t} = \frac{4.9(2^2) - 4.9(0^2)}{2 - 0} = 9.8 \text{ m/s.}$$

(b) In the time interval $[1, 2]$, the average velocity is

$$\frac{\Delta y}{\Delta t} = \frac{4.9(2^2) - 4.9(1^2)}{2 - 1} = 14.7 \text{ m/s.}$$

●

___
EXAMPLE 2
___
How fast is the rock in Example 1 falling (a) at time $t = 1$? (b) at time $t = 2$?

Table 1. Average velocity over $[1, 1 + h]$

h	$\Delta y / \Delta t$
1	14.7000
0.1	10.2900
0.01	9.8490
0.001	9.8049
0.0001	9.8005

Table 2. Average velocity over $[2, 2 + h]$

h	$\Delta y / \Delta t$
1	24.5000
0.1	20.0900
0.01	19.6490
0.001	19.6049
0.0001	19.6005

Solution We can calculate the average velocity over any time interval, but this question asks for the *instantaneous velocity* at a given time. If the falling rock had a speedometer, what would it show at time $t = 1$? To answer this, we first write the average velocity over the time interval $[1, 1 + h]$ starting at $t = 1$ and having length h:

$$\text{Average velocity over } [1, 1 + h] = \frac{\Delta y}{\Delta t} = \frac{4.9(1 + h)^2 - 4.9(1^2)}{h}.$$

We can't calculate the instantaneous velocity at $t = 1$ by substituting $h = 0$ in this expression, because we can't divide by zero. But we can calculate the average velocities over shorter and shorter time intervals and see whether they seem to get close to a particular number. Table 1 shows the values of $\Delta y / \Delta t$ for some values of h approaching zero. Indeed, it appears that these average velocities get closer and closer to 9.8 m/s as the length of the time interval gets closer and closer to zero. This suggests that the rock is falling at a rate of 9.8 m/s one second after it is dropped.

Similarly, Table 2 shows values of the average velocities over shorter and shorter time intervals $[2, 2 + h]$ starting at $t = 2$. The values suggest that the rock is falling at 19.6 m/s two seconds after it is dropped.

●

In Example 2 the average velocity of the falling rock over the time interval $[t, t + h]$ is

$$\frac{\Delta y}{\Delta t} = \frac{4.9(t + h)^2 - 4.9t^2}{h}.$$

To find the instantaneous velocity (usually just called *the velocity*) at the instants $t = 1$ and $t = 2$, we examined the values of this average velocity for time intervals whose lengths h became smaller and smaller. We were, in fact, finding the *limit of the average velocity as h approaches zero*. This is expressed symbolically in the form

$$\text{velocity at time } t = \lim_{h \to 0} \frac{\Delta y}{\Delta t} = \lim_{h \to 0} \frac{4.9(t + h)^2 - 4.9t^2}{h}.$$

Read "$\lim_{h \to 0} \ldots$" as "the limit as h approaches zero of $\ldots$" We can't find the limit of the fraction by just substituting $h = 0$ because that would involve dividing by zero. However, we can calculate the limit by first performing some algebraic simplifications on the expression for the average velocity.

EXAMPLE 3   Simplify the expression for the average velocity of the rock over $[t, t + h]$ by first expanding $(t + h)^2$. Hence, find the velocity $v(t)$ of the falling rock at time t directly, without making a table of values.

Solution   The average velocity of the rock over time interval $[t, t + h]$ is

$$
\frac{4.9(t + h)^2 - 4.9t^2}{h} = \frac{4.9(t^2 + 2th + h^2 - t^2)}{h}
$$
$$
= \frac{4.9(2th + h^2)}{h}
$$
$$
= 9.8t + 4.9h.
$$

The final form of the expression no longer involves division by h. It approaches $9.8t + 4.9(0) = 9.8t$ as h approaches 0. Thus, t s after the rock is dropped, its velocity is $v(t) = 9.8t$ m/s. In particular, at $t = 1$ and $t = 2$ the velocities are $v(1) = 9.8$ m/s and $v(2) = 19.6$ m/s, respectively.

The Growth of an Algal Culture

In a laboratory experiment, the biomass of an algal culture was measured over a 74-day period by measuring the area in square millimetres occupied by the culture on a microscope slide. These measurements m were plotted against the time t in days and the points joined by a smooth curve $m = f(t)$, as shown in red in Figure 1.1.

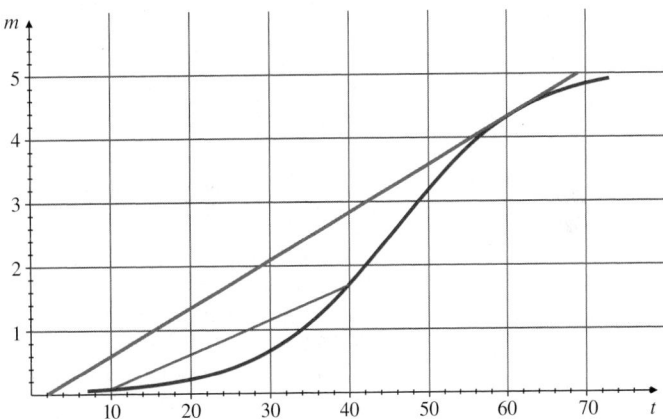

Figure 1.1   The biomass m of an algal culture after t days

Observe that the biomass was about 0.1 mm^2 on day 10 and had grown to about 1.7 mm^2 on day 40, an increase of $1.7 - 0.1 = 1.6$ mm^2 in a time interval of $40 - 10 = 30$ days. The average rate of growth over the time interval from day 10 to day 40 was therefore

$$
\frac{1.7 - 0.1}{40 - 10} = \frac{1.6}{30} \approx 0.053 \text{ mm}^2/\text{d}.
$$

This average rate is just the slope of the green line joining the points on the graph of $m = f(t)$ corresponding to $t = 10$ and $t = 40$. Similarly, the average rate of growth of the algal biomass over any time interval can be determined by measuring the slope of the line joining the points on the curve corresponding to that time interval. Such lines are called **secant lines** to the curve.

EXAMPLE 4   How fast is the biomass growing on day 60?

Solution To answer this question, we could measure the average rates of change over shorter and shorter times around day 60. The corresponding secant lines become shorter and shorter, but their slopes approach a *limit*, namely, the slope of the **tangent line** to the graph of $m = f(t)$ at the point where $t = 60$. This tangent line is sketched in blue in Figure 1.1; it seems to go through the points $(2, 0)$ and $(69, 5)$, so that its slope is

$$\frac{5 - 0}{69 - 2} \approx 0.0746 \text{ mm}^2/\text{d}.$$

This is the rate at which the biomass was growing on day 60.

The Area of a Circle

All circles are similar geometric figures; they all have the same shape and differ only in size. The ratio of the circumference C to the diameter $2r$ (twice the radius) has the same value for all circles. The number π is defined to be this common ratio:

$$\frac{C}{2r} = \pi \quad \text{or} \quad C = 2\pi r.$$

In school we are taught that the area A of a circle is this same number π times the square of the radius:

$$A = \pi r^2.$$

How can we deduce this area formula from the formula for the circumference that is the definition of π?

The answer to this question lies in regarding the circle as a "limit" of regular polygons, which are in turn made up of triangles, figures about whose geometry we know a great deal.

Suppose a regular polygon having n sides is inscribed in a circle of radius r. (See Figure 1.2.) The perimeter P_n and the area A_n of the polygon are, respectively, less than the circumference C and the area A of the circle, but if n is large, P_n is *close to* C and A_n is *close to* A. (In fact, the "circle" in Figure 1.2 was drawn by a computer as a regular polygon having 180 sides, each subtending a 2° angle at the centre of the circle. It is very difficult to distinguish this 180-sided polygon from a real circle.) We would expect P_n to approach the limit C and A_n to approach the limit A as n grows larger and larger and approaches infinity.

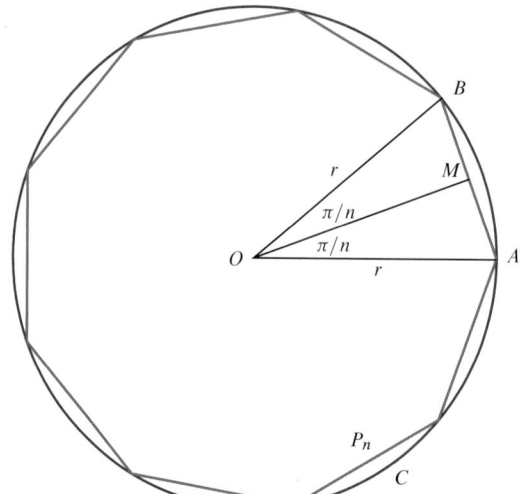

Figure 1.2 A regular polygon (green) of n sides inscribed in a red circle. Here $n = 9$

A regular polygon of n sides is the union of n nonoverlapping, congruent, isosceles triangles having a common vertex at O, the centre of the polygon. One of these triangles, $\triangle OAB$, is shown in Figure 1.2. Since the total angle around the point O is 2π radians (we are assuming that a circle of radius 1 has circumference 2π), the angle AOB is $2\pi/n$ radians. If M is the midpoint of AB, then OM bisects angle AOB. Using elementary trigonometry, we can write the length of AB and the area of triangle OAB in terms of the radius r of the circle:

$$|AB| = 2|AM| = 2r \sin \frac{\pi}{n}$$

$$\text{area } OAB = \frac{1}{2}|AB||OM| = \frac{1}{2}\left(2r \sin \frac{\pi}{n}\right)\left(r \cos \frac{\pi}{n}\right)$$

$$= r^2 \sin \frac{\pi}{n} \cos \frac{\pi}{n}.$$

The perimeter P_n and area A_n of the polygon are n times these expressions:

$$P_n = 2rn \sin \frac{\pi}{n}$$

$$A_n = r^2 n \sin \frac{\pi}{n} \cos \frac{\pi}{n}.$$

Solving the first equation for $rn \sin(\pi/n) = P_n/2$ and substituting into the second equation, we get

$$A_n = \left(\frac{P_n}{2}\right) r \cos \frac{\pi}{n}.$$

Now the angle $AOM = \pi/n$ approaches 0 as n grows large, so its cosine, $\cos(\pi/n) = |OM|/|OA|$, approaches 1. Since P_n approaches $C = 2\pi r$ as n grows large, the expression for A_n approaches $(2\pi r/2)r(1) = \pi r^2$, which must therefore be the area of the circle.

Remark There is a fundamental relationship between the problem of finding the area under the graph of a function f and the problem of finding another function g whose rate of change is f. It will be explored fully beginning in Chapter 5. As an example, for the falling rock of Example 1–Example 3, the green area A under the graph of the velocity function $v = 9.8t$ m/s and above the interval $[0, t]$ on the t-axis is the area of a triangle of base length t s and height $9.8t$ m/s, and so (see Figure 1.3) is

$$A = \frac{1}{2}(t)(9.8t) = 4.9t^2 \text{ m},$$

which is exactly the distance y that the rock falls during the first t seconds. The rate of change of the area function $A(t)$ (that is, of the distance function y) is the velocity function $v(t)$.

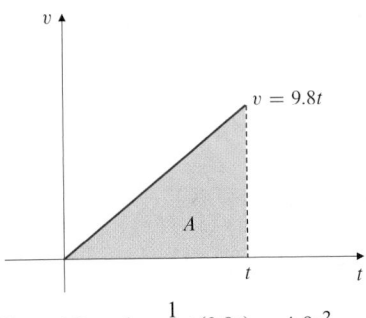

Figure 1.3 $A = \dfrac{1}{2}t\,(9.8t) = 4.9t^2$

EXERCISES 1.1

Exercises 1–4 refer to an object moving along the x-axis in such a way that at time t s its position is $x = t^2$ m to the right of the origin.

1. Find the average velocity of the object over the time interval $[t, t + h]$.

2. Make a table giving the average velocities of the object over time intervals $[2, 2 + h]$, for $h = 1, 0.1, 0.01, 0.001$, and 0.0001 s.

3. Use the results from Exercise 2 to guess the instantaneous velocity of the object at $t = 2$ s.

4. Confirm your guess in Exercise 3 by calculating the limit of the average velocity over $[2, 2 + h]$ as h approaches zero,

using the method of Example 3.

Exercises 5–8 refer to the motion of a particle moving along the x-axis so that at time t s it is at position $x = 3t^2 - 12t + 1$ m.

5. Find the average velocity of the particle over the time intervals $[1, 2]$, $[2, 3]$, and $[1, 3]$.

6. Use the method of Example 3 to find the velocity of the particle at $t = 1, t = 2$, and $t = 3$.

7. In what direction is the particle moving at $t = 1$? $t = 2$? $t = 3$?

8. Show that for any positive number k, the average velocity of the particle over the time interval $[t - k, t + k]$ is equal to its velocity at time t.

In Exercises 9–11, a weight that is suspended by a spring bobs up and down so that its height above the floor at time t s is y ft, where

$$y = 2 + \frac{1}{\pi} \sin(\pi t).$$

9. Sketch the graph of y as a function of t. How high is the weight at $t = 1$ s? In what direction is it moving at that time?

10. What is the average velocity of the weight over the time intervals $[1, 2]$, $[1, 1.1]$, $[1, 1.01]$, and $[1, 1.001]$?

11. Using the results of Exercise 10, estimate the velocity of the weight at time $t = 1$. What is the significance of the sign of your answer?

Exercises 12–13 refer to the algal biomass graphed in Figure 1.1.

12. Approximately how fast is the biomass growing on day 20?

13. On about what day is the biomass growing fastest?

14. The annual profits of a small company for each of the first five years of its operation are given in Table 3.

Table 3.

Year	Profit ($1,000s)
2011	6
2012	27
2013	62
2014	111
2015	174

(a) Plot points representing the profits as a function of year on graph paper, and join them by a smooth curve.

(b) What is the average rate of increase of the annual profits between 2013 and 2015?

(c) Use your graph to estimate the rate of increase of the profits in 2013.

1.2   Limits of Functions

In order to speak meaningfully about rates of change, tangent lines, and areas bounded by curves, we have to investigate the process of finding limits. Indeed, the concept of *limit* is the cornerstone on which the development of calculus rests. Before we try to give a definition of a limit, let us look at more examples.

EXAMPLE 1   Describe the behaviour of the function $f(x) = \dfrac{x^2 - 1}{x - 1}$ near $x = 1$.

Solution   Note that $f(x)$ is defined for all real numbers x except $x = 1$. (We can't divide by zero.) For any $x \neq 1$ we can simplify the expression for $f(x)$ by factoring the numerator and cancelling common factors:

$$f(x) = \frac{(x - 1)(x + 1)}{x - 1} = x + 1 \qquad \text{for} \quad x \neq 1.$$

The graph of f is the line $y = x + 1$ with one point removed, namely, the point $(1, 2)$. This removed point is shown as a "hole" in the graph in Figure 1.4. Even though $f(1)$ is not defined, it is clear that we can make the value of $f(x)$ *as close as we want* to 2 by choosing x *close enough* to 1. Therefore, we say that $f(x)$ approaches arbitrarily close to 2 as x approaches 1, or, more simply, $f(x)$ approaches *the limit* 2 as x approaches 1. We write this as

$$\lim_{x \to 1} f(x) = 2 \qquad \text{or} \qquad \lim_{x \to 1} \frac{x^2 - 1}{x - 1} = 2.$$

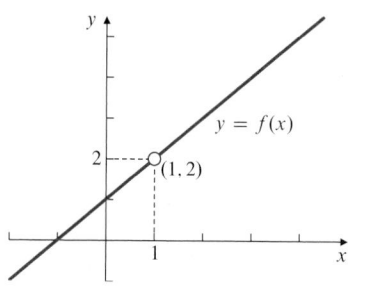

Figure 1.4   The graph of $f(x) = \dfrac{x^2 - 1}{x - 1}$

EXAMPLE 2   What happens to the function $g(x) = (1 + x^2)^{1/x^2}$ as x approaches zero?

Solution   Note that $g(x)$ is not defined at $x = 0$. In fact, for the moment it does not appear to be defined for any x whose square x^2 is not a rational number. (Recall that if $r = m/n$, where m and n are integers and $n > 0$, then x^r means the nth root of x^m.) Let us ignore for now the problem of deciding what $g(x)$ means if x^2 is irrational and consider only rational values of x. There is no obvious way to simplify the expression for $g(x)$ as we did in Example 1. However, we can use a scientific calculator to obtain approximate values of $g(x)$ for some rational values of x approaching 0. (The values in Table 4 were obtained with such a calculator.)

Except for the last value in the table, the values of $g(x)$ seem to be approaching a certain number, $2.71828\ldots$, as x gets closer and closer to 0. We will show in Section 3.4 that

$$\lim_{x\to 0} g(x) = \lim_{x\to 0} (1 + x^2)^{1/x^2} = e = 2.7\ 1828\ 1828\ 45\ 90\ 45\ldots.$$

Table 4.

x	$g(x)$
± 1.0	2.0000 00000
± 0.1	2.7048 13829
± 0.01	2.7181 45927
± 0.001	2.7182 80469
± 0.0001	2.7182 81815
± 0.00001	1.0000 00000

The number e turns out to be very important in mathematics.

Observe that the last entry in the table appears to be wrong. This is important. It is because the calculator can only represent a finite number of numbers. The calculator was unable to distinguish $1 + (0.00001)^2 = 1.0000000001$ from 1, and it therefore calculated $1^{10,000,000,000} = 1$. While for many calculations on computers this reality can be minimized, it cannot be eliminated. The wrong value warns us of something called round-off error. We can explore with computer graphics what this means for g near 0. As was the case for the *numerical monster* encountered in Section P.4, the computer can produce rich and beautiful behaviour in its failed attempt to represent g, which is very different from what g actually does. While it is possible to get computer algebra software like Maple to evaluate limits correctly (as we will see in the next section), we cannot use computer graphics or floating-point arithmetic to study many mathematical notions such as limits. In fact, we will need mathematics to understand what the computer actually does so that we can be the master of our tools.

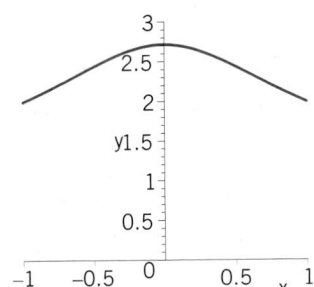

Figure 1.5   The graph of $y = g(x)$ on the interval $[-1, 1]$

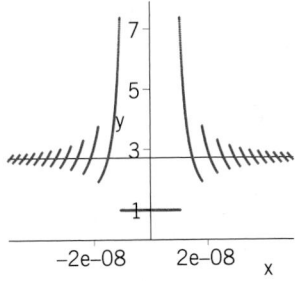

Figure 1.6   The graphs of $y = g(x)$ (colour) and $y = e \approx 2.718$ (black) on the interval $[-5 \times 10^{-8}, 5 \times 10^{-8}]$

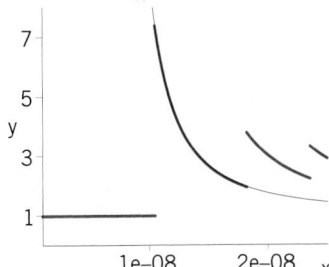

Figure 1.7   The graphs of $y = g(x)$ (colour) and $y = (1 + 2 \times 10^{-16})^{1/x^2}$ (black) on the interval $[10^{-9}, 2.5 \times 10^{-8}]$

Figures 1.5–1.7 illustrate this fascinating behaviour of g with three plots made with Maple using its default 10-significant-figure precision in representing floating-point (i.e., real) numbers. Figure 1.5 is a plot of the graph of g on the interval $[-1, 1]$. The graph starts out at height 2 at either endpoint $x = \pm 1$ and rises to height approximately $2.718 \cdots$ as x decreases in absolute value, as we would expect from Table 4. Figure 1.6 shows the graph of g restricted to the tiny interval $[-5 \times 10^{-8}, 5 \times 10^{-8}]$. It consists of many short arcs decreasing in height as $|x|$ increases, and clustering around the line $y = 2.718 \cdots$, and a horizontal part at height 1 between approximately -10^{-8} and 10^{-8}. Figure 1.7 zooms in on the part of the graph to the right of the origin up to $x = 2.5 \times 10^{-8}$. Note how the arc closest to 0 coincides with the graph of

$y = \left(1 + 2 \times 10^{-16}\right)^{1/x^2}$ (shown in black), indicating that $1 + 2 \times 10^{-16}$ may be the smallest number greater than 1 that Maple can distinguish from 1. Both figures show that the breakdown in the graph of g is not sudden, but becomes more and more pronounced as $|x|$ decreases until the breakdown is complete near $\pm 10^{-8}$.

The examples above and those in Section 1.1 suggest the following *informal* definition of limit.

DEFINITION

1

An informal definition of limit

If $f(x)$ is defined for all x near a, except possibly at a itself, and if we can ensure that $f(x)$ is as close as we want to L by taking x close enough to a, but not equal to a, we say that the function f approaches the **limit** L as x approaches a, and we write

$$\lim_{x \to a} f(x) = L \qquad \text{or} \qquad \lim_{x \to a} f(x) = L.$$

This definition is *informal* because phrases such as *close as we want* and *close enough* are imprecise; their meaning depends on the context. To a machinist manufacturing a piston, *close enough* may mean *within a few thousandths of an inch*. To an astronomer studying distant galaxies, *close enough* may mean *within a few thousand light-years*. The definition should be clear enough, however, to enable us to recognize and evaluate limits of specific functions. A more precise "formal" definition, given in Section 1.5, is needed if we want to *prove* theorems about limits like Theorems 2–4, stated later in this section.

EXAMPLE 3   Find   (a) $\lim_{x \to a} x$ and   (b) $\lim_{x \to a} c$ (where c is a constant).

Solution   In words, part (a) asks: "What does x approach as x approaches a?" The answer is surely a.

$$\lim_{x \to a} x = a.$$

Similarly, part (b) asks: "What does c approach as x approaches a?" The answer here is that c approaches c; you can't get any closer to c than by *being* c.

$$\lim_{x \to a} c = c.$$

Example 3 shows that $\lim_{x \to a} f(x)$ can *sometimes* be evaluated by just calculating $f(a)$. This will be the case if $f(x)$ is defined in an open interval containing $x = a$ and the graph of f passes unbroken through the point $(a, f(a))$. The next example shows various ways algebraic manipulations can be used to evaluate $\lim_{x \to a} f(x)$ in situations where $f(a)$ is undefined. This usually happens when $f(x)$ is a fraction with denominator equal to 0 at $x = a$.

EXAMPLE 4   Evaluate:

(a) $\displaystyle\lim_{x \to -2} \frac{x^2 + x - 2}{x^2 + 5x + 6}$,   (b) $\displaystyle\lim_{x \to a} \frac{\dfrac{1}{x} - \dfrac{1}{a}}{x - a}$,   and   (c) $\displaystyle\lim_{x \to 4} \frac{\sqrt{x} - 2}{x^2 - 16}$.

Solution   Each of these limits involves a fraction whose numerator and denominator are both 0 at the point where the limit is taken.

(a) $\displaystyle\lim_{x\to-2}\frac{x^2+x-2}{x^2+5x+6}$

fraction undefined at $x=-2$

Factor numerator and denominator. (See Section P.6.)

$\displaystyle=\lim_{x\to-2}\frac{(x+2)(x-1)}{(x+2)(x+3)}$

Cancel common factors.

$\displaystyle=\lim_{x\to-2}\frac{x-1}{x+3}$

Evaluate this limit by substituting $x=-2$.

$\displaystyle=\frac{-2-1}{-2+3}=-3.$

(b) $\displaystyle\lim_{x\to a}\frac{\dfrac{1}{x}-\dfrac{1}{a}}{x-a}$

fraction undefined at $x=a$

Simplify the numerator.

$\displaystyle=\lim_{x\to a}\frac{\dfrac{a-x}{ax}}{x-a}$

$\displaystyle=\lim_{x\to a}\frac{-(x-a)}{ax(x-a)}$

Cancel the common factor.

$\displaystyle=\lim_{x\to a}\frac{-1}{ax}=-\frac{1}{a^2}.$

(c) $\displaystyle\lim_{x\to4}\frac{\sqrt{x}-2}{x^2-16}$

fraction undefined at $x=4$

Multiply numerator and denominator by the conjugate of the expression in the numerator.

$\displaystyle=\lim_{x\to4}\frac{(\sqrt{x}-2)(\sqrt{x}+2)}{(x^2-16)(\sqrt{x}+2)}$

$\displaystyle=\lim_{x\to4}\frac{x-4}{(x-4)(x+4)(\sqrt{x}+2)}$

$\displaystyle=\lim_{x\to4}\frac{1}{(x+4)(\sqrt{x}+2)}=\frac{1}{(4+4)(2+2)}=\frac{1}{32}$

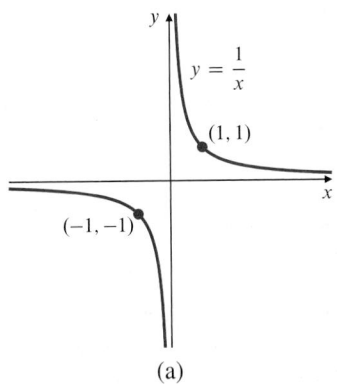

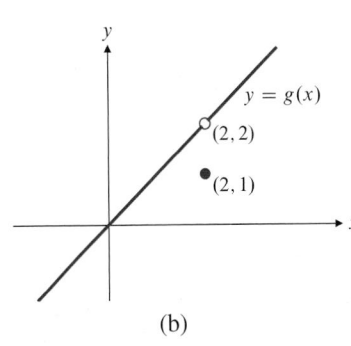

(a)

(b)

Figure 1.8

(a) $\displaystyle\lim_{x\to0}\frac{1}{x}$ does not exist

(b) $\displaystyle\lim_{x\to2}g(x)=2$, but $g(2)=1$

BEWARE! Always be aware that the existence of $\lim_{x\to a}f(x)$ does not require that $f(a)$ exist and does not depend on $f(a)$ even if $f(a)$ does exist. It depends only on the values of $f(x)$ for x near but *not equal* to a.

A function f may be defined on both sides of $x=a$ but still not have a limit at $x=a$. For example, the function $f(x)=1/x$ has no limit as x approaches 0. As can be seen in Figure 1.8(a), the values $1/x$ grow ever larger in absolute value as x approaches 0; there is no single number L that they approach.

The following example shows that even if $f(x)$ is defined at $x=a$, the limit of $f(x)$ as x approaches a may not be equal to $f(a)$.

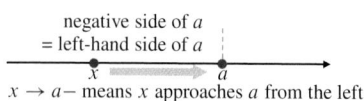

$x \to a-$ means x approaches a from the left

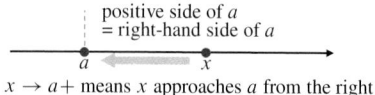

$x \to a+$ means x approaches a from the right

Figure 1.9 One-sided approach

DEFINITION

2

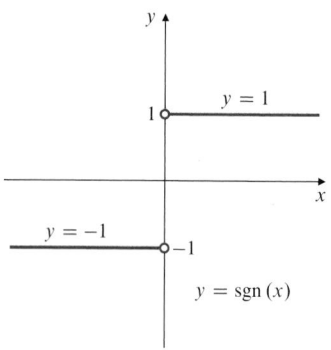

$y = \text{sgn}(x)$

Figure 1.10

$\lim\limits_{x\to 0} \text{sgn}(x)$ does not exist, because
$\lim\limits_{x\to 0-} \text{sgn}(x) = -1$, $\lim\limits_{x\to 0+} \text{sgn}(x) = 1$

THEOREM

1

EXAMPLE 5 Let $g(x) = \begin{cases} x & \text{if } x \neq 2 \\ 1 & \text{if } x = 2. \end{cases}$ (See Figure 1.8(b).) Then

$$\lim_{x\to 2} g(x) = \lim_{x\to 2} x = 2, \qquad \text{although} \quad g(2) = 1.$$

One-Sided Limits

Limits are *unique*; if $\lim_{x\to a} f(x) = L$ and $\lim_{x\to a} f(x) = M$, then $L = M$. (See Exercise 31 in Section 1.5.) Although a function f can only have one limit at any particular point, it is, nevertheless, useful to be able to describe the behaviour of functions that approach different numbers as x approaches a from one side or the other. (See Figure 1.9.)

Informal definition of left and right limits

If $f(x)$ is defined on some interval (b, a) extending to the left of $x = a$, and if we can ensure that $f(x)$ is as close as we want to L by taking x to the left of a and close enough to a, then we say $f(x)$ has **left limit** L at $x = a$, and we write

$$\lim_{x\to a-} f(x) = L.$$

If $f(x)$ is defined on some interval (a, b) extending to the right of $x = a$, and if we can ensure that $f(x)$ is as close as we want to L by taking x to the right of a and close enough to a, then we say $f(x)$ has **right limit** L at $x = a$, and we write

$$\lim_{x\to a+} f(x) = L.$$

Note the use of the suffix $+$ to denote approach from the right (the *positive* side) and the suffix $-$ to denote approach from the left (the *negative* side).

EXAMPLE 6 The signum function $\text{sgn}(x) = x/|x|$ (see Figure 1.10) has left limit -1 and right limit 1 at $x = 0$:

$$\lim_{x\to 0-} \text{sgn}(x) = -1 \qquad \text{and} \qquad \lim_{x\to 0+} \text{sgn}(x) = 1$$

because the values of $\text{sgn}(x)$ approach -1 (they *are* -1) if x is negative and approaches 0, and they approach 1 if x is positive and approaches 0. Since these left and right limits are not equal, $\lim_{x\to 0} \text{sgn}(x)$ *does not exist.*

As suggested in Example 6, the relationship between ordinary (two-sided) limits and one-sided limits can be stated as follows:

Relationship between one-sided and two-sided limits

A function $f(x)$ has limit L at $x = a$ if and only if it has both left and right limits there and these one-sided limits are both equal to L:

$$\lim_{x\to a} f(x) = L \quad \Longleftrightarrow \quad \lim_{x\to a-} f(x) = \lim_{x\to a+} f(x) = L.$$

EXAMPLE 7 If $f(x) = \dfrac{|x-2|}{x^2 + x - 6}$, find: $\lim\limits_{x\to 2+} f(x)$, $\lim\limits_{x\to 2-} f(x)$, and $\lim\limits_{x\to 2} f(x)$.

Solution Observe that $|x - 2| = x - 2$ if $x > 2$, and $|x - 2| = -(x - 2)$ if $x < 2$. Therefore,

$$
\begin{aligned}
\lim_{x \to 2+} f(x) &= \lim_{x \to 2+} \frac{x - 2}{x^2 + x - 6} & \lim_{x \to 2-} f(x) &= \lim_{x \to 2-} \frac{-(x - 2)}{x^2 + x - 6} \\
&= \lim_{x \to 2+} \frac{x - 2}{(x - 2)(x + 3)} & &= \lim_{x \to 2-} \frac{-(x - 2)}{(x - 2)(x + 3)} \\
&= \lim_{x \to 2+} \frac{1}{x + 3} = \frac{1}{5}, & &= \lim_{x \to 2-} \frac{-1}{x + 3} = -\frac{1}{5}.
\end{aligned}
$$

Since $\lim_{x \to 2-} f(x) \neq \lim_{x \to 2+} f(x)$, the limit $\lim_{x \to 2} f(x)$ does not exist.

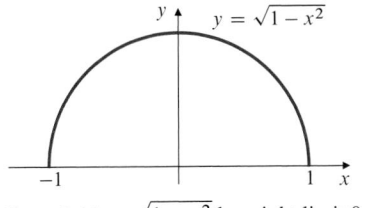

Figure 1.11 $\sqrt{1 - x^2}$ has right limit 0 at -1 and left limit 0 at 1

EXAMPLE 8 What one-sided limits does $g(x) = \sqrt{1 - x^2}$ have at $x = -1$ and $x = 1$?

Solution The domain of g is $[-1, 1]$, so $g(x)$ is defined only to the right of $x = -1$ and only to the left of $x = 1$. As can be seen in Figure 1.11,

$$
\lim_{x \to -1+} g(x) = 0 \qquad \text{and} \qquad \lim_{x \to 1-} g(x) = 0.
$$

$g(x)$ has no left limit or limit at $x = -1$ and no right limit or limit at $x = 1$.

Rules for Calculating Limits

The following theorems make it easy to calculate limits and one-sided limits of many kinds of functions when we know some elementary limits. We will not prove the theorems here. (See Section 1.5.)

THEOREM

2

Limit Rules

If $\lim_{x \to a} f(x) = L$, $\lim_{x \to a} g(x) = M$, and k is a constant, then

1. **Limit of a sum:** $\quad \lim_{x \to a} [f(x) + g(x)] = L + M$
2. **Limit of a difference:** $\quad \lim_{x \to a} [f(x) - g(x)] = L - M$
3. **Limit of a product:** $\quad \lim_{x \to a} f(x)g(x) = LM$
4. **Limit of a multiple:** $\quad \lim_{x \to a} k f(x) = kL$
5. **Limit of a quotient:** $\quad \lim_{x \to a} \dfrac{f(x)}{g(x)} = \dfrac{L}{M}, \quad$ if $M \neq 0$.

If m is an integer and n is a positive integer, then

6. **Limit of a power:** $\quad \lim_{x \to a} [f(x)]^{m/n} = L^{m/n}$, provided $L > 0$ if n is even, and $L \neq 0$ if $m < 0$.

If $f(x) \leq g(x)$ on an interval containing a in its interior, then

7. **Order is preserved:** $\quad L \leq M$

Rules 1–6 are also valid for right limits and left limits. So is Rule 7, under the assumption that $f(x) \leq g(x)$ on an open interval extending from a in the appropriate direction.

In words, rule 1 of Theorem 2 says that the limit of a sum of functions is the sum of their limits. Similarly, rule 5 says that the limit of a quotient of two functions is the quotient of their limits, provided that the limit of the denominator is not zero. Try to state the other rules in words.

We can make use of the limits (a) $\lim_{x \to a} c = c$ (where c is a constant) and (b) $\lim_{x \to a} x = a$, from Example 3, together with parts of Theorem 2 to calculate limits of many combinations of functions.

EXAMPLE 9 Find: (a) $\lim\limits_{x \to a} \dfrac{x^2 + x + 4}{x^3 - 2x^2 + 7}$ and (b) $\lim\limits_{x \to 2} \sqrt{2x + 1}$.

Solution

(a) The expression $\dfrac{x^2 + x + 4}{x^3 - 2x^2 + 7}$ is formed by combining the basic functions x and c (constant) using addition, subtraction, multiplication, and division. Theorem 2 assures us that the limit of such a combination is the same combination of the limits a and c of the basic functions, provided the denominator does not have limit zero. Thus,

$$\lim_{x \to a} \frac{x^2 + x + 4}{x^3 - 2x^2 + 7} = \frac{a^2 + a + 4}{a^3 - 2a^2 + 7} \qquad \text{provided } a^3 - 2a^2 + 7 \neq 0.$$

(b) The same argument as in (a) shows that $\lim_{x \to 2} (2x + 1) = 2(2) + 1 = 5$. Then the Power Rule (rule 6 of Theorem 2) assures us that

$$\lim_{x \to 2} \sqrt{2x + 1} = \sqrt{5}.$$

The following result is an immediate corollary of Theorem 2. (See Section P.6 for a discussion of polynomials and rational functions.)

THEOREM

3

Limits of Polynomials and Rational Functions

1. If $P(x)$ is a polynomial and a is any real number, then

$$\lim_{x \to a} P(x) = P(a).$$

2. If $P(x)$ and $Q(x)$ are polynomials and $Q(a) \neq 0$, then

$$\lim_{x \to a} \frac{P(x)}{Q(x)} = \frac{P(a)}{Q(a)}.$$

The Squeeze Theorem

The following theorem will enable us to calculate some very important limits in subsequent chapters. It is called the *Squeeze Theorem* because it refers to a function g whose values are squeezed between the values of two other functions f and h that have the same limit L at a point a. Being trapped between the values of two functions that approach L, the values of g must also approach L. (See Figure 1.12.)

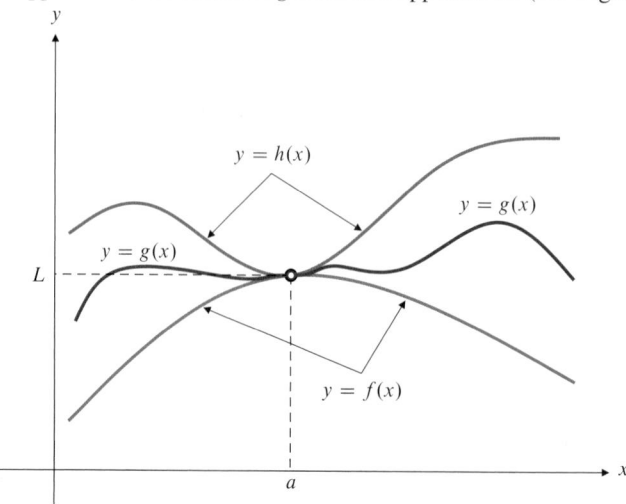

Figure 1.12 The graph of g is squeezed between those of f (blue) and h (green)

THEOREM

4

The Squeeze Theorem

Suppose that $f(x) \le g(x) \le h(x)$ holds for all x in some open interval containing a, except possibly at $x = a$ itself. Suppose also that

$$\lim_{x \to a} f(x) = \lim_{x \to a} h(x) = L.$$

Then $\lim_{x \to a} g(x) = L$ also. Similar statements hold for left and right limits.

EXAMPLE 10 Given that $3 - x^2 \le u(x) \le 3 + x^2$ for all $x \ne 0$, find $\lim_{x \to 0} u(x)$.

Solution Since $\lim_{x \to 0}(3 - x^2) = 3$ and $\lim_{x \to 0}(3 + x^2) = 3$, the Squeeze Theorem implies that $\lim_{x \to 0} u(x) = 3$.

EXAMPLE 11 Show that if $\lim_{x \to a} |f(x)| = 0$, then $\lim_{x \to a} f(x) = 0$.

Solution Since $-|f(x)| \le f(x) \le |f(x)|$, and $-|f(x)|$ and $|f(x)|$ both have limit 0 as x approaches a, so does $f(x)$ by the Squeeze Theorem.

EXERCISES 1.2

1. Find: (a) $\lim_{x \to -1} f(x)$, (b) $\lim_{x \to 0} f(x)$, and (c) $\lim_{x \to 1} f(x)$, for the function f whose graph is shown in Figure 1.13.

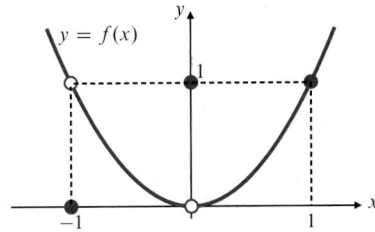

Figure 1.13

2. For the function $y = g(x)$ graphed in Figure 1.14, find each of the following limits or explain why it does not exist.
(a) $\lim_{x \to 1} g(x)$, (b) $\lim_{x \to 2} g(x)$, (c) $\lim_{x \to 3} g(x)$

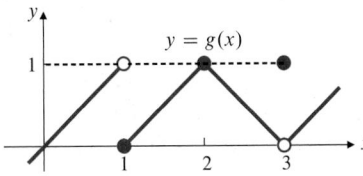

Figure 1.14

In Exercises 3–6, find the indicated one-sided limit of the function g whose graph is given in Figure 1.14.

3. $\lim_{x \to 1-} g(x)$

4. $\lim_{x \to 1+} g(x)$

5. $\lim_{x \to 3+} g(x)$

6. $\lim_{x \to 3-} g(x)$

In Exercises 7–36, evaluate the limit or explain why it does not exist.

7. $\lim_{x \to 4}(x^2 - 4x + 1)$

8. $\lim_{x \to 2} 3(1 - x)(2 - x)$

9. $\lim_{x \to 3} \dfrac{x + 3}{x + 6}$

10. $\lim_{t \to -4} \dfrac{t^2}{4 - t}$

11. $\lim_{x \to 1} \dfrac{x^2 - 1}{x + 1}$

12. $\lim_{x \to -1} \dfrac{x^2 - 1}{x + 1}$

13. $\lim_{x \to 3} \dfrac{x^2 - 6x + 9}{x^2 - 9}$

14. $\lim_{x \to -2} \dfrac{x^2 + 2x}{x^2 - 4}$

15. $\lim_{h \to 2} \dfrac{1}{4 - h^2}$

16. $\lim_{h \to 0} \dfrac{3h + 4h^2}{h^2 - h^3}$

17. $\lim_{x \to 9} \dfrac{\sqrt{x} - 3}{x - 9}$

18. $\lim_{h \to 0} \dfrac{\sqrt{4 + h} - 2}{h}$

19. $\lim_{x \to \pi} \dfrac{(x - \pi)^2}{\pi x}$

20. $\lim_{x \to -2} |x - 2|$

21. $\lim_{x \to 0} \dfrac{|x - 2|}{x - 2}$

22. $\lim_{x \to 2} \dfrac{|x - 2|}{x - 2}$

23. $\lim_{t \to 1} \dfrac{t^2 - 1}{t^2 - 2t + 1}$

24. $\lim_{x \to 2} \dfrac{\sqrt{4 - 4x + x^2}}{x - 2}$

25. $\lim_{t \to 0} \dfrac{t}{\sqrt{4 + t} - \sqrt{4 - t}}$

26. $\lim_{x \to 1} \dfrac{x^2 - 1}{\sqrt{x + 3} - 2}$

27. $\lim_{t \to 0} \dfrac{t^2 + 3t}{(t + 2)^2 - (t - 2)^2}$

28. $\lim_{s \to 0} \dfrac{(s + 1)^2 - (s - 1)^2}{s}$

29. $\lim_{y \to 1} \dfrac{y - 4\sqrt{y} + 3}{y^2 - 1}$

30. $\lim_{x \to -1} \dfrac{x^3 + 1}{x + 1}$

31. $\lim\limits_{x \to 2} \dfrac{x^4 - 16}{x^3 - 8}$

32. $\lim\limits_{x \to 8} \dfrac{x^{2/3} - 4}{x^{1/3} - 2}$

33. $\lim\limits_{x \to 2} \left(\dfrac{1}{x - 2} - \dfrac{4}{x^2 - 4} \right)$

34. $\lim\limits_{x \to 2} \left(\dfrac{1}{x - 2} - \dfrac{1}{x^2 - 4} \right)$

35. $\lim\limits_{x \to 0} \dfrac{\sqrt{2 + x^2} - \sqrt{2 - x^2}}{x^2}$

36. $\lim\limits_{x \to 0} \dfrac{|3x - 1| - |3x + 1|}{x}$

The limit $\lim\limits_{h \to 0} \dfrac{f(x + h) - f(x)}{h}$ occurs frequently in the study of calculus. (Can you guess why?) Evaluate this limit for the functions f in Exercises 37–42.

37. $f(x) = x^2$

38. $f(x) = x^3$

39. $f(x) = \dfrac{1}{x}$

40. $f(x) = \dfrac{1}{x^2}$

41. $f(x) = \sqrt{x}$

42. $f(x) = 1/\sqrt{x}$

Examine the graphs of $\sin x$ and $\cos x$ in Section P.7 to determine the limits in Exercises 43–46.

43. $\lim\limits_{x \to \pi/2} \sin x$

44. $\lim\limits_{x \to \pi/4} \cos x$

45. $\lim\limits_{x \to \pi/3} \cos x$

46. $\lim\limits_{x \to 2\pi/3} \sin x$

47. Make a table of values of $f(x) = (\sin x)/x$ for a sequence of values of x approaching 0, say $\pm 1.0, \pm 0.1, \pm 0.01, \pm 0.001,$ $\pm 0.0001,$ and $\pm 0.00001.$ Make sure your calculator is set in *radian mode* rather than degree mode. Guess the value of $\lim\limits_{x \to 0} f(x).$

48. Repeat Exercise 47 for $f(x) = \dfrac{1 - \cos x}{x^2}.$

In Exercises 49–60, find the indicated one-sided limit or explain why it does not exist.

49. $\lim\limits_{x \to 2-} \sqrt{2 - x}$

50. $\lim\limits_{x \to 2+} \sqrt{2 - x}$

51. $\lim\limits_{x \to -2-} \sqrt{2 - x}$

52. $\lim\limits_{x \to -2+} \sqrt{2 - x}$

53. $\lim\limits_{x \to 0} \sqrt{x^3 - x}$

54. $\lim\limits_{x \to 0-} \sqrt{x^3 - x}$

55. $\lim\limits_{x \to 0+} \sqrt{x^3 - x}$

56. $\lim\limits_{x \to 0+} \sqrt{x^2 - x^4}$

57. $\lim\limits_{x \to a-} \dfrac{|x - a|}{x^2 - a^2}$

58. $\lim\limits_{x \to a+} \dfrac{|x - a|}{x^2 - a^2}$

59. $\lim\limits_{x \to 2-} \dfrac{x^2 - 4}{|x + 2|}$

60. $\lim\limits_{x \to 2+} \dfrac{x^2 - 4}{|x + 2|}$

Exercises 61–64 refer to the function

$$f(x) = \begin{cases} x - 1 & \text{if } x \le -1 \\ x^2 + 1 & \text{if } -1 < x \le 0 \\ (x + \pi)^2 & \text{if } x > 0. \end{cases}$$

Find the indicated limits.

61. $\lim\limits_{x \to -1-} f(x)$

62. $\lim\limits_{x \to -1+} f(x)$

63. $\lim\limits_{x \to 0+} f(x)$

64. $\lim\limits_{x \to 0-} f(x)$

65. Suppose $\lim\limits_{x \to 4} f(x) = 2$ and $\lim\limits_{x \to 4} g(x) = -3$. Find:

(a) $\lim\limits_{x \to 4} \left(g(x) + 3 \right)$

(b) $\lim\limits_{x \to 4} x f(x)$

(c) $\lim\limits_{x \to 4} \left(g(x) \right)^2$

(d) $\lim\limits_{x \to 4} \dfrac{g(x)}{f(x) - 1}.$

66. Suppose $\lim\limits_{x \to a} f(x) = 4$ and $\lim\limits_{x \to a} g(x) = -2$. Find:

(a) $\lim\limits_{x \to a} \left(f(x) + g(x) \right)$

(b) $\lim\limits_{x \to a} f(x) \cdot g(x)$

(c) $\lim\limits_{x \to a} 4g(x)$

(d) $\lim\limits_{x \to a} f(x)/g(x).$

67. If $\lim\limits_{x \to 2} \dfrac{f(x) - 5}{x - 2} = 3$, find $\lim\limits_{x \to 2} f(x).$

68. If $\lim\limits_{x \to 0} \dfrac{f(x)}{x^2} = -2$, find $\lim\limits_{x \to 0} f(x)$ and $\lim\limits_{x \to 0} \dfrac{f(x)}{x}.$

Using Graphing Utilities to Find Limits

Graphing calculators or computer software can be used to evaluate limits at least approximately. Simply "zoom" the plot window to show smaller and smaller parts of the graph near the point where the limit is to be found. Find the following limits by graphical techniques. Where you think it justified, give an exact answer. Otherwise, give the answer correct to 4 decimal places. Remember to ensure that your calculator or software is set for radian mode when using trigonometric functions.

69. $\lim\limits_{x \to 0} \dfrac{\sin x}{x}$

70. $\lim\limits_{x \to 0} \dfrac{\sin(2\pi x)}{\sin(3\pi x)}$

71. $\lim\limits_{x \to 1-} \dfrac{\sin \sqrt{1 - x}}{\sqrt{1 - x^2}}$

72. $\lim\limits_{x \to 0+} \dfrac{x - \sqrt{x}}{\sqrt{\sin x}}$

73. On the same graph, plot the three functions $y = x \sin(1/x)$, $y = x$, and $y = -x$ for $-0.2 \le x \le 0.2, -0.2 \le y \le 0.2$. Describe the behaviour of $f(x) = x \sin(1/x)$ near $x = 0$. Does $\lim\limits_{x \to 0} f(x)$ exist, and if so, what is its value? Could you have predicted this before drawing the graph? Why?

Using the Squeeze Theorem

74. If $\sqrt{5 - 2x^2} \le f(x) \le \sqrt{5 - x^2}$ for $-1 \le x \le 1$, find $\lim\limits_{x \to 0} f(x).$

75. If $2 - x^2 \le g(x) \le 2 \cos x$ for all x, find $\lim\limits_{x \to 0} g(x).$

76. (a) Sketch the curves $y = x^2$ and $y = x^4$ on the same graph. Where do they intersect?

(b) The function $f(x)$ satisfies:

$$\begin{cases} x^2 \le f(x) \le x^4 & \text{if } x < -1 \text{ or } x > 1 \\ x^4 \le f(x) \le x^2 & \text{if } -1 \le x \le 1 \end{cases}$$

Find (i) $\lim\limits_{x \to -1} f(x)$, (ii) $\lim\limits_{x \to 0} f(x)$, (iii) $\lim\limits_{x \to 1} f(x).$

77. On what intervals is $x^{1/3} < x^3$? On what intervals is $x^{1/3} > x^3$? If the graph of $y = h(x)$ always lies between the graphs of $y = x^{1/3}$ and $y = x^3$, for what real numbers a can you determine the value of $\lim\limits_{x \to a} h(x)$? Find the limit for each of these values of a.

78. What is the domain of $x \sin \dfrac{1}{x}$? Evaluate $\lim\limits_{x \to 0} x \sin \dfrac{1}{x}.$

79. Suppose $|f(x)| \le g(x)$ for all x. What can you conclude about $\lim\limits_{x \to a} f(x)$ if $\lim\limits_{x \to a} g(x) = 0$? What if $\lim\limits_{x \to a} g(x) = 3$?

1.3 Limits at Infinity and Infinite Limits

In this section we will extend the concept of limit to allow for two situations not covered by the definitions of limit and one-sided limit in the previous section:

(i) limits at infinity, where x becomes arbitrarily large, positive or negative;

(ii) infinite limits, which are not really limits at all but provide useful symbolism for describing the behaviour of functions whose values become arbitrarily large, positive or negative.

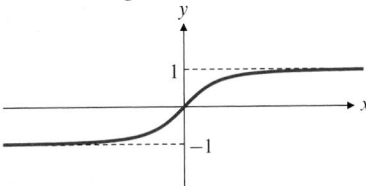

Figure 1.15   The graph of $x/\sqrt{x^2 + 1}$

Limits at Infinity

Consider the function

$$f(x) = \frac{x}{\sqrt{x^2 + 1}}$$

whose graph is shown in Figure 1.15 and for which some values (rounded to 7 decimal places) are given in Table 5. The values of $f(x)$ seem to approach 1 as x takes on larger and larger positive values, and -1 as x takes on negative values that get larger and larger in absolute value. (See Example 2 below for confirmation.) We express this behaviour by writing

$$\lim_{x \to \infty} f(x) = 1 \qquad \text{``} f(x) \text{ approaches 1 as } x \text{ approaches infinity.''}$$

$$\lim_{x \to -\infty} f(x) = -1 \qquad \text{``} f(x) \text{ approaches } -1 \text{ as } x \text{ approaches negative infinity.''}$$

Table 5.

x	$f(x) = x/\sqrt{x^2 + 1}$
$-1{,}000$	-0.9999995
-100	-0.9999500
-10	-0.9950372
-1	-0.7071068
0	0.0000000
1	0.7071068
10	0.9950372
100	0.9999500
$1{,}000$	0.9999995

The graph of f conveys this limiting behaviour by approaching the horizontal lines $y = 1$ as x moves far to the right and $y = -1$ as x moves far to the left. These lines are called **horizontal asymptotes** of the graph. In general, if a curve approaches a straight line as it recedes very far away from the origin, that line is called an **asymptote** of the curve.

DEFINITION

3

Limits at infinity and negative infinity (informal definition)

If the function f is defined on an interval (a, ∞) and if we can ensure that $f(x)$ is as close as we want to the number L by taking x large enough, then we say that $f(x)$ **approaches the limit L as x approaches infinity**, and we write

$$\lim_{x \to \infty} f(x) = L.$$

If f is defined on an interval $(-\infty, b)$ and if we can ensure that $f(x)$ is as close as we want to the number M by taking x negative and large enough in absolute value, then we say that $f(x)$ **approaches the limit M as x approaches negative infinity**, and we write

$$\lim_{x \to -\infty} f(x) = M.$$

Recall that the symbol ∞, called **infinity**, does *not* represent a real number. We cannot use ∞ in arithmetic in the usual way, but we can use the phrase "approaches ∞" to mean "becomes arbitrarily large positive" and the phrase "approaches $-\infty$" to mean "becomes arbitrarily large negative."

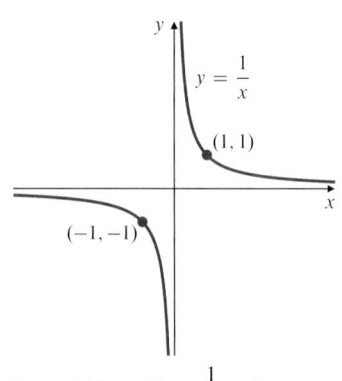

Figure 1.16 $\lim\limits_{x\to\pm\infty}\dfrac{1}{x}=0$

EXAMPLE 1   In Figure 1.16, we can see that $\lim_{x\to\infty} 1/x = \lim_{x\to-\infty} 1/x = 0$. The x-axis is a horizontal asymptote of the graph $y = 1/x$.

The theorems of Section 1.2 have suitable counterparts for limits at infinity or negative infinity. In particular, it follows from the example above and from the Product Rule for limits that $\lim_{x\to\pm\infty} 1/x^n = 0$ for any positive integer n. We will use this fact in the following examples. Example 2 shows how to obtain the limits at $\pm\infty$ for the function $x/\sqrt{x^2 + 1}$ by algebraic means, without resorting to making a table of values or drawing a graph, as we did above.

EXAMPLE 2   Evaluate $\lim\limits_{x\to\infty} f(x)$ and $\lim\limits_{x\to-\infty} f(x)$ for $f(x) = \dfrac{x}{\sqrt{x^2 + 1}}$.

Solution   Rewrite the expression for $f(x)$ as follows:

$$f(x) = \frac{x}{\sqrt{x^2\left(1 + \dfrac{1}{x^2}\right)}} = \frac{x}{\sqrt{x^2}\sqrt{1 + \dfrac{1}{x^2}}} \qquad \text{Remember } \sqrt{x^2} = |x|.$$

$$= \frac{x}{|x|\sqrt{1 + \dfrac{1}{x^2}}}$$

$$= \frac{\operatorname{sgn} x}{\sqrt{1 + \dfrac{1}{x^2}}}, \qquad \text{where } \operatorname{sgn} x = \frac{x}{|x|} = \begin{cases} 1 & \text{if } x > 0 \\ -1 & \text{if } x < 0. \end{cases}$$

The factor $\sqrt{1 + (1/x^2)}$ approaches 1 as x approaches ∞ or $-\infty$, so $f(x)$ must have the same limits as $x \to \pm\infty$ as does sgn (x). Therefore (see Figure 1.15),

$$\lim_{x\to\infty} f(x) = 1 \qquad \text{and} \qquad \lim_{x\to-\infty} f(x) = -1.$$

Limits at Infinity for Rational Functions

The only polynomials that have limits at $\pm\infty$ are constant ones, $P(x) = c$. The situation is more interesting for rational functions. Recall that a rational function is a quotient of two polynomials. The following examples show how to render such a function in a form where its limits at infinity and negative infinity (if they exist) are apparent. The way to do this is to *divide the numerator and denominator by the highest power of x appearing in the denominator*. The limits of a rational function at infinity and negative infinity either both fail to exist or both exist and are equal.

EXAMPLE 3   **(Numerator and denominator of the same degree)** Evaluate $\lim_{x\to\pm\infty} \dfrac{2x^2 - x + 3}{3x^2 + 5}$.

Solution   Divide the numerator and the denominator by x^2, the highest power of x appearing in the denominator:

$$\lim_{x\to\pm\infty} \frac{2x^2 - x + 3}{3x^2 + 5} = \lim_{x\to\pm\infty} \frac{2 - (1/x) + (3/x^2)}{3 + (5/x^2)} = \frac{2 - 0 + 0}{3 + 0} = \frac{2}{3}.$$

EXAMPLE 4   **(Degree of numerator less than degree of denominator)** Evaluate $\lim_{x\to\pm\infty} \dfrac{5x + 2}{2x^3 - 1}$.

Solution Divide the numerator and the denominator by the largest power of x in the denominator, namely, x^3:

$$\lim_{x \to \pm\infty} \frac{5x + 2}{2x^3 - 1} = \lim_{x \to \pm\infty} \frac{(5/x^2) + (2/x^3)}{2 - (1/x^3)} = \frac{0 + 0}{2 - 0} = 0.$$

The limiting behaviour of rational functions at infinity and negative infinity is summarized at the left.

The technique used in the previous examples can also be applied to more general kinds of functions. The function in the following example is not rational, and the limit seems to produce a meaningless $\infty - \infty$ until we resolve matters by rationalizing the numerator.

Summary of limits at $\pm\infty$ for rational functions

Let $P_m(x) = a_m x^m + \cdots + a_0$ and $Q_n(x) = b_n x^n + \cdots + b_0$ be polynomials of degree m and n, respectively, so that $a_m \neq 0$ and $b_n \neq 0$. Then

$$\lim_{x \to \pm\infty} \frac{P_m(x)}{Q_n(x)}$$

(a) equals zero if $m < n$,

(b) equals $\dfrac{a_m}{b_n}$ if $m = n$,

(c) does not exist if $m > n$.

EXAMPLE 5 Find $\lim_{x \to \infty} \left(\sqrt{x^2 + x} - x \right)$.

Solution We are trying to find the limit of the difference of two functions, each of which becomes arbitrarily large as x increases to infinity. We rationalize the expression by multiplying the numerator and the denominator (which is 1) by the conjugate expression $\sqrt{x^2 + x} + x$:

$$
\begin{aligned}
\lim_{x \to \infty} \left(\sqrt{x^2 + x} - x \right) &= \lim_{x \to \infty} \frac{\left(\sqrt{x^2 + x} - x \right) \left(\sqrt{x^2 + x} + x \right)}{\sqrt{x^2 + x} + x} \\
&= \lim_{x \to \infty} \frac{x^2 + x - x^2}{\sqrt{x^2 \left(1 + \dfrac{1}{x} \right)} + x} \\
&= \lim_{x \to \infty} \frac{x}{x \sqrt{1 + \dfrac{1}{x}} + x} = \lim_{x \to \infty} \frac{1}{\sqrt{1 + \dfrac{1}{x}} + 1} = \frac{1}{2}.
\end{aligned}
$$

(Here, $\sqrt{x^2} = x$ because $x > 0$ as $x \to \infty$.)

Remark The limit $\lim_{x \to -\infty} (\sqrt{x^2 + x} - x)$ is not nearly so subtle. Since $-x > 0$ as $x \to -\infty$, we have $\sqrt{x^2 + x} - x > \sqrt{x^2 + x}$, which grows arbitrarily large as $x \to -\infty$. The limit does not exist.

Infinite Limits

A function whose values grow arbitrarily large can sometimes be said to have an infinite limit. Since infinity is not a number, infinite limits are not really limits at all, but they provide a way of describing the behaviour of functions that grow arbitrarily large positive or negative. A few examples will make the terminology clear.

EXAMPLE 6 **(A two-sided infinite limit)** Describe the behaviour of the function $f(x) = 1/x^2$ near $x = 0$.

Solution As x approaches 0 from either side, the values of $f(x)$ are positive and grow larger and larger (see Figure 1.17), so the limit of $f(x)$ as x approaches 0 *does not exist*. It is nevertheless convenient to describe the behaviour of f near 0 by saying that $f(x)$ *approaches* ∞ as x approaches zero. We write

$$\lim_{x \to 0} f(x) = \lim_{x \to 0} \frac{1}{x^2} = \infty.$$

Note that in writing this we are *not* saying that $\lim_{x \to 0} 1/x^2$ exists. Rather, we are saying that that limit *does not exist because $1/x^2$ becomes arbitrarily large near $x = 0$. Observe how the graph of f approaches the y-axis as x approaches 0. The y-axis is a **vertical asymptote** of the graph.

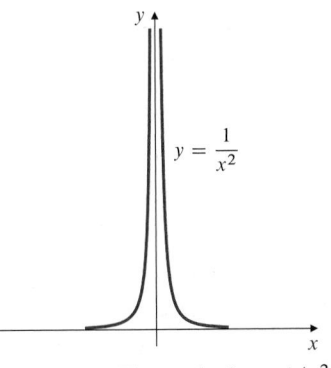

Figure 1.17   The graph of $y = 1/x^2$ (not to scale)

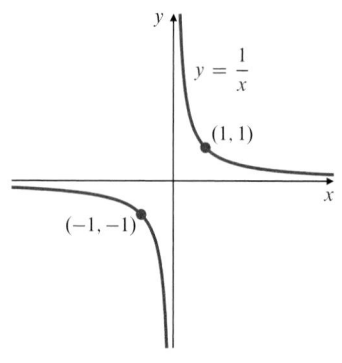

Figure 1.18 $\lim_{x\to 0-} 1/x = -\infty$, $\lim_{x\to 0+} 1/x = \infty$

EXAMPLE 7   **(One-sided infinite limits)** Describe the behaviour of the function $f(x) = 1/x$ near $x = 0$. (See Figure 1.18.)

Solution   As x approaches 0 from the right, the values of $f(x)$ become larger and larger positive numbers, and we say that f has right-hand limit infinity at $x = 0$:

$$\lim_{x\to 0+} f(x) = \infty.$$

Similarly, the values of $f(x)$ become larger and larger negative numbers as x approaches 0 from the left, so f has left-hand limit $-\infty$ at $x = 0$:

$$\lim_{x\to 0-} f(x) = -\infty.$$

These statements do not say that the one-sided limits *exist*; they do not exist because ∞ and $-\infty$ are not numbers. Since the one-sided limits are not equal even as infinite symbols, all we can say about the two-sided $\lim_{x\to 0} f(x)$ is that it does not exist.

EXAMPLE 8   **(Polynomial behaviour at infinity)**

(a) $\lim_{x\to\infty} (3x^3 - x^2 + 2) = \infty$ (b) $\lim_{x\to-\infty} (3x^3 - x^2 + 2) = -\infty$

(c) $\lim_{x\to\infty} (x^4 - 5x^3 - x) = \infty$ (d) $\lim_{x\to-\infty} (x^4 - 5x^3 - x) = \infty$

The highest-degree term of a polynomial dominates the other terms as $|x|$ grows large, so the limits of this term at ∞ and $-\infty$ determine the limits of the whole polynomial. For the polynomial in parts (a) and (b) we have

$$3x^3 - x^2 + 2 = 3x^3 \left(1 - \frac{1}{3x} + \frac{2}{3x^3}\right).$$

The factor in the large parentheses approaches 1 as x approaches $\pm\infty$, so the behaviour of the polynomial is just that of its highest-degree term $3x^3$.

We can now say a bit more about the limits at infinity and negative infinity of a rational function whose numerator has higher degree than the denominator. Earlier in this section we said that such a limit *does not exist*. This is true, but we can assign ∞ or $-\infty$ to such limits, as the following example shows.

EXAMPLE 9   **(Rational functions with numerator of higher degree)** Evaluate $\lim_{x\to\infty} \dfrac{x^3 + 1}{x^2 + 1}$.

Solution   Divide the numerator and the denominator by x^2, the largest power of x in the denominator:

$$\lim_{x\to\infty} \frac{x^3 + 1}{x^2 + 1} = \lim_{x\to\infty} \frac{x + \dfrac{1}{x^2}}{1 + \dfrac{1}{x^2}} = \frac{\lim_{x\to\infty}\left(x + \dfrac{1}{x^2}\right)}{1} = \infty.$$

A polynomial $Q(x)$ of degree $n > 0$ can have at most n *zeros*; that is, there are at most n different real numbers r for which $Q(r) = 0$. If $Q(x)$ is the denominator of a rational function $R(x) = P(x)/Q(x)$, that function will be defined for all x except those finitely many zeros of Q. At each of those zeros, $R(x)$ may have limits, infinite limits, or one-sided infinite limits. Here are some examples.

EXAMPLE 10

(a) $\displaystyle\lim_{x\to 2}\frac{(x-2)^2}{x^2-4} = \lim_{x\to 2}\frac{(x-2)^2}{(x-2)(x+2)} = \lim_{x\to 2}\frac{x-2}{x+2} = 0.$

(b) $\displaystyle\lim_{x\to 2}\frac{x-2}{x^2-4} = \lim_{x\to 2}\frac{x-2}{(x-2)(x+2)} = \lim_{x\to 2}\frac{1}{x+2} = \frac{1}{4}.$

(c) $\displaystyle\lim_{x\to 2+}\frac{x-3}{x^2-4} = \lim_{x\to 2+}\frac{x-3}{(x-2)(x+2)} = -\infty.$ (The values are negative for $x > 2$, x near 2.)

(d) $\displaystyle\lim_{x\to 2-}\frac{x-3}{x^2-4} = \lim_{x\to 2-}\frac{x-3}{(x-2)(x+2)} = \infty.$ (The values are positive for $x < 2$, x near 2.)

(e) $\displaystyle\lim_{x\to 2}\frac{x-3}{x^2-4} = \lim_{x\to 2}\frac{x-3}{(x-2)(x+2)}$ does not exist.

(f) $\displaystyle\lim_{x\to 2}\frac{2-x}{(x-2)^3} = \lim_{x\to 2}\frac{-(x-2)}{(x-2)^3} = \lim_{x\to 2}\frac{-1}{(x-2)^2} = -\infty.$

In parts (a) and (b) the effect of the zero in the denominator at $x = 2$ is cancelled because the numerator is zero there also. Thus a finite limit exists. This is not true in part (f) because the numerator only vanishes once at $x = 2$, while the denominator vanishes three times there.

Using Maple to Calculate Limits

Maple's `limit` procedure can be easily used to calculate limits, one-sided limits, limits at infinity, and infinite limits. Here is the syntax for calculating

$$\lim_{x\to 2}\frac{x^2-4}{x^2-5x+6}, \quad \lim_{x\to 0}\frac{x\sin x}{1-\cos x}, \quad \lim_{x\to -\infty}\frac{x}{\sqrt{x^2+1}}, \quad \lim_{x\to \infty}\frac{x}{\sqrt{x^2+1}},$$

$$\lim_{x\to 0}\frac{1}{x}, \quad \lim_{x\to 0-}\frac{1}{x}, \quad \lim_{x\to a-}\frac{x^2-a^2}{|x-a|}, \quad \text{and} \quad \lim_{x\to a+}\frac{x^2-a^2}{|x-a|}.$$

```
> limit((x^2-4)/(x^2-5*x+6),x=2);
```
$$-4$$

```
> limit(x*sin(x)/(1-cos(x)),x=0);
```
$$2$$

```
> limit(x/sqrt(x^2+1),x=-infinity);
```
$$-1$$

```
> limit(x/sqrt(x^2+1),x=infinity);
```
$$1$$

```
> limit(1/x,x=0); limit(1/x,x=0,left);
```
$$undefined$$

$$-\infty$$

```
> limit((x^2-a^2)/(abs(x-a)),x=a,left);
```
$$-2a$$

```
>  limit((x^2-a^2)/(abs(x-a)),x=a,right);
```
$$2a$$

Finally, we use Maple to confirm the limit discussed in Example 2 in Section 1.2.

```
>  limit((1+x^2)^(1/x^2), x=0); evalf(%);
```
$$e$$

$$2.718281828$$

We will learn a great deal about this very important number in Chapter 3.

EXERCISES 1.3

Find the limits in Exercises 1–10.

1. $\displaystyle\lim_{x\to\infty}\frac{x}{2x-3}$

2. $\displaystyle\lim_{x\to\infty}\frac{x}{x^2-4}$

3. $\displaystyle\lim_{x\to\infty}\frac{3x^3-5x^2+7}{8+2x-5x^3}$

4. $\displaystyle\lim_{x\to-\infty}\frac{x^2-2}{x-x^2}$

5. $\displaystyle\lim_{x\to-\infty}\frac{x^2+3}{x^3+2}$

6. $\displaystyle\lim_{x\to\infty}\frac{x^2+\sin x}{x^2+\cos x}$

7. $\displaystyle\lim_{x\to\infty}\frac{3x+2\sqrt{x}}{1-x}$

8. $\displaystyle\lim_{x\to\infty}\frac{2x-1}{\sqrt{3x^2+x+1}}$

9. $\displaystyle\lim_{x\to-\infty}\frac{2x-1}{\sqrt{3x^2+x+1}}$

10. $\displaystyle\lim_{x\to-\infty}\frac{2x-5}{|3x+2|}$

In Exercises 11–32 evaluate the indicated limit. If it does not exist, is the limit ∞, $-\infty$, or neither?

11. $\displaystyle\lim_{x\to3}\frac{1}{3-x}$

12. $\displaystyle\lim_{x\to3}\frac{1}{(3-x)^2}$

13. $\displaystyle\lim_{x\to3-}\frac{1}{3-x}$

14. $\displaystyle\lim_{x\to3+}\frac{1}{3-x}$

15. $\displaystyle\lim_{x\to-5/2}\frac{2x+5}{5x+2}$

16. $\displaystyle\lim_{x\to-2/5}\frac{2x+5}{5x+2}$

17. $\displaystyle\lim_{x\to-(2/5)-}\frac{2x+5}{5x+2}$

18. $\displaystyle\lim_{x\to-(2/5)+}\frac{2x+5}{5x+2}$

19. $\displaystyle\lim_{x\to2+}\frac{x}{(2-x)^3}$

20. $\displaystyle\lim_{x\to1-}\frac{x}{\sqrt{1-x^2}}$

21. $\displaystyle\lim_{x\to1+}\frac{1}{|x-1|}$

22. $\displaystyle\lim_{x\to1-}\frac{1}{|x-1|}$

23. $\displaystyle\lim_{x\to2}\frac{x-3}{x^2-4x+4}$

24. $\displaystyle\lim_{x\to1+}\frac{\sqrt{x^2-x}}{x-x^2}$

25. $\displaystyle\lim_{x\to\infty}\frac{x+x^3+x^5}{1+x^2+x^3}$

26. $\displaystyle\lim_{x\to\infty}\frac{x^3+3}{x^2+2}$

❗ 27. $\displaystyle\lim_{x\to\infty}\frac{x\sqrt{x+1}\left(1-\sqrt{2x+3}\right)}{7-6x+4x^2}$

28. $\displaystyle\lim_{x\to\infty}\left(\frac{x^2}{x+1}-\frac{x^2}{x-1}\right)$

❗ 29. $\displaystyle\lim_{x\to-\infty}\left(\sqrt{x^2+2x}-\sqrt{x^2-2x}\right)$

❗ 30. $\displaystyle\lim_{x\to\infty}\left(\sqrt{x^2+2x}-\sqrt{x^2-2x}\right)$

31. $\displaystyle\lim_{x\to\infty}\frac{1}{\sqrt{x^2-2x}-x}$

32. $\displaystyle\lim_{x\to-\infty}\frac{1}{\sqrt{x^2+2x}-x}$

33. What are the horizontal asymptotes of $y=\dfrac{1}{\sqrt{x^2-2x}-x}$? What are its vertical asymptotes?

34. What are the horizontal and vertical asymptotes of
$$y=\frac{2x-5}{|3x+2|}?$$

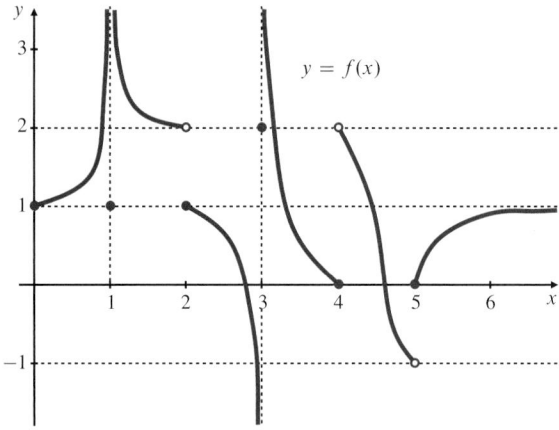

Figure 1.19

The function f whose graph is shown in Figure 1.19 has domain $[0,\infty)$. Find the limits of f indicated in Exercises 35–45.

35. $\displaystyle\lim_{x\to0+}f(x)$

36. $\displaystyle\lim_{x\to1}f(x)$

37. $\displaystyle\lim_{x\to2+}f(x)$

38. $\displaystyle\lim_{x\to2-}f(x)$

39. $\displaystyle\lim_{x\to3-}f(x)$

40. $\displaystyle\lim_{x\to3+}f(x)$

41. $\displaystyle\lim_{x\to4+}f(x)$

42. $\displaystyle\lim_{x\to4-}f(x)$

43. $\displaystyle\lim_{x\to5-}f(x)$

44. $\displaystyle\lim_{x\to5+}f(x)$

45. $\displaystyle\lim_{x\to\infty}f(x)$

46. What asymptotes does the graph in Figure 1.19 have?

Exercises 47–52 refer to the **greatest integer function** $\lfloor x\rfloor$ graphed in Figure 1.20. Find the indicated limit or explain why it does not exist.

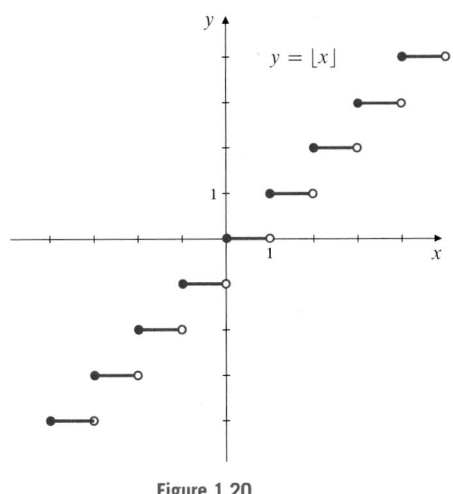

Figure 1.20

47. $\lim_{x \to 3+} \lfloor x \rfloor$

48. $\lim_{x \to 3-} \lfloor x \rfloor$

49. $\lim_{x \to 3} \lfloor x \rfloor$

50. $\lim_{x \to 2.5} \lfloor x \rfloor$

51. $\lim_{x \to 0+} \lfloor 2 - x \rfloor$

52. $\lim_{x \to -3-} \lfloor x \rfloor$

53. Parking in a certain parking lot costs $1.50 for each hour or part of an hour. Sketch the graph of the function $C(t)$ representing the cost of parking for t hours. At what values of t does $C(t)$ have a limit? Evaluate $\lim_{t \to t_0-} C(t)$ and $\lim_{t \to t_0+} C(t)$ for an arbitrary number $t_0 > 0$.

54. If $\lim_{x \to 0+} f(x) = L$, find $\lim_{x \to 0-} f(x)$ if (a) f is even, (b) f is odd.

55. If $\lim_{x \to 0+} f(x) = A$ and $\lim_{x \to 0-} f(x) = B$, find

(a) $\lim_{x \to 0+} f(x^3 - x)$

(b) $\lim_{x \to 0-} f(x^3 - x)$

(c) $\lim_{x \to 0-} f(x^2 - x^4)$

(d) $\lim_{x \to 0+} f(x^2 - x^4)$.

1.4   Continuity

When a car is driven along a highway, its distance from its starting point depends on time in a *continuous* way, changing by small amounts over short intervals of time. But not all quantities change in this way. When the car is parked in a parking lot where the rate is quoted as "$2.00 per hour or portion," the parking charges remain at $2.00 for the first hour and then suddenly jump to $4.00 as soon as the first hour has passed. The function relating parking charges to parking time will be called *discontinuous* at each hour. In this section we will define continuity and show how to tell whether a function is continuous. We will also examine some important properties possessed by continuous functions.

Continuity at a Point

Most functions that we encounter have domains that are intervals, or unions of separate intervals. A point P in the domain of such a function is called an **interior point** of the domain if it belongs to some *open* interval contained in the domain. If it is not an interior point, then P is called an **endpoint** of the domain. For example, the domain of the function $f(x) = \sqrt{4 - x^2}$ is the closed interval $[-2, 2]$, which consists of interior points in the interval $(-2, 2)$, a left endpoint -2, and a right endpoint 2. The domain of the function $g(x) = 1/x$ is the union of open intervals $(-\infty, 0) \cup (0, \infty)$ and consists entirely of interior points. Note that although 0 is an endpoint of each of those intervals, it does not belong to the domain of g and so is not an endpoint of that domain.

DEFINITION

4

Continuity at an interior point

We say that a function f is **continuous** at an interior point c of its domain if

$$\lim_{x \to c} f(x) = f(c).$$

If either $\lim_{x \to c} f(x)$ fails to exist or it exists but is not equal to $f(c)$, then we will say that f is **discontinuous** at c.

In graphical terms, f is continuous at an interior point c of its domain if its graph has no break in it at the point $(c, f(c))$; in other words, if you can draw the graph through that point without lifting your pen from the paper. Consider Figure 1.21. In (a), f is continuous at c. In (b), f is discontinuous at c because $\lim_{x \to c} f(x) \neq f(c)$. In (c),

f is discontinuous at c because $\lim_{x \to c} f(x)$ does not exist. In both (b) and (c) the graph of f has a break at $x = c$.

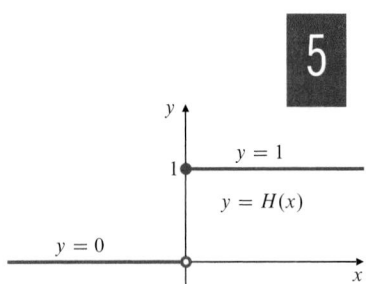

Figure 1.21

(a) f is continuous at c

(b) $\lim_{x \to c} f(x) \neq f(c)$

(c) $\lim_{x \to c} f(x)$ does not exist

Although a function cannot have a limit at an endpoint of its domain, it can still have a one-sided limit there. We extend the definition of continuity to provide for such situations.

DEFINITION

5

Right and left continuity

We say that f is **right continuous** at c if $\lim_{x \to c+} f(x) = f(c)$.

We say that f is **left continuous** at c if $\lim_{x \to c-} f(x) = f(c)$.

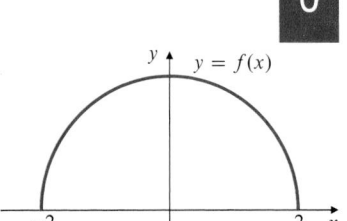

Figure 1.22 The Heaviside function

EXAMPLE 1 The Heaviside function $H(x)$, whose graph is shown in Figure 1.22, is continuous at every number x except 0. It is right continuous at 0 but is not left continuous or continuous there.

The relationship between continuity and one-sided continuity is summarized in the following theorem.

THEOREM

5

Function f is continuous at c if and only if it is both right continuous and left continuous at c.

DEFINITION

6

Continuity at an endpoint

We say that f is continuous at a left endpoint c of its domain if it is right continuous there.

We say that f is continuous at a right endpoint c of its domain if it is left continuous there.

Figure 1.23 $f(x) = \sqrt{4 - x^2}$ is continuous at every point of its domain

EXAMPLE 2 The function $f(x) = \sqrt{4 - x^2}$ has domain $[-2, 2]$. It is continuous at the right endpoint 2 because it is left continuous there, that is, because $\lim_{x \to 2-} f(x) = 0 = f(2)$. It is continuous at the left endpoint -2 because it is right continuous there: $\lim_{x \to -2+} f(x) = 0 = f(-2)$. Of course, f is also continuous at every interior point of its domain. If $-2 < c < 2$, then $\lim_{x \to c} f(x) = \sqrt{4 - c^2} = f(c)$. (See Figure 1.23.)

Continuity on an Interval

We have defined the concept of continuity at a point. Of greater importance is the concept of continuity on an interval.

DEFINITION

Continuity on an interval

We say that function f is **continuous on the interval** I if it is continuous at each point of I. In particular, we will say that f is a **continuous function** if f is continuous at every point of its domain.

EXAMPLE 3 The function $f(x) = \sqrt{x}$ is a continuous function. Its domain is $[0, \infty)$. It is continuous at the left endpoint 0 because it is right continuous there. Also, f is continuous at every number $c > 0$ since $\lim_{x \to c} \sqrt{x} = \sqrt{c}$.

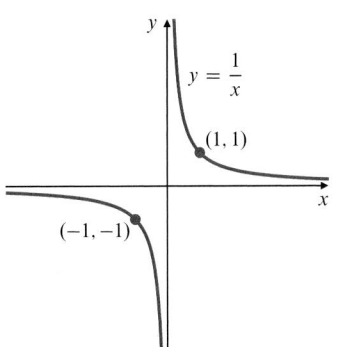

Figure 1.24 $1/x$ is continuous on its domain

EXAMPLE 4 The function $g(x) = 1/x$ is also a continuous function. This may seem wrong to you at first glance because its graph is broken at $x = 0$. (See Figure 1.24.) However, the number 0 is not in the domain of g, so we will prefer to say that g is undefined rather than discontinuous there. (Some authors would say that g is discontinuous at $x = 0$.) If we were to define $g(0)$ to be some number, say 0, then we would say that $g(x)$ is discontinuous at 0. There is no way of defining $g(0)$ so that g becomes continuous at 0.

EXAMPLE 5 The greatest integer function $\lfloor x \rfloor$ (see Figure 1.20) is continuous on every interval $[n, n + 1)$, where n is an integer. It is right continuous at each integer n but is not left continuous there, so it is discontinuous at the integers.

$$\lim_{x \to n+} \lfloor x \rfloor = n = \lfloor n \rfloor, \qquad \lim_{x \to n-} \lfloor x \rfloor = n - 1 \neq n = \lfloor n \rfloor.$$

There Are Lots of Continuous Functions

The following functions are continuous wherever they are defined:

(a) all polynomials;

(b) all rational functions;

(c) all rational powers $x^{m/n} = \sqrt[n]{x^m}$;

(d) the sine, cosine, tangent, secant, cosecant, and cotangent functions defined in Section P.7; and

(e) the absolute value function $|x|$.

Theorem 3 of Section 1.2 assures us that every polynomial is continuous everywhere on the real line, and every rational function is continuous everywhere on its domain (which consists of all real numbers except the finitely many where its denominator is zero). If m and n are integers and $n \neq 0$, the rational power function $x^{m/n}$ is defined for all positive numbers x, and also for all negative numbers x if n is odd. The domain includes 0 if and only if $m/n \geq 0$.

The following theorems show that if we combine continuous functions in various ways, the results will be continuous.

THEOREM

6

Combining continuous functions

If the functions f and g are both defined on an interval containing c and both are continuous at c, then the following functions are also continuous at c:

1. the sum $f + g$ and the difference $f - g$;
2. the product fg;
3. the constant multiple kf, where k is any number;
4. the quotient f/g (provided $g(c) \neq 0$); and
5. the nth root $(f(x))^{1/n}$, provided $f(c) > 0$ if n is even.

The proof involves using the various limit rules in Theorem 2 of Section 1.2. For example,

$$\lim_{x \to c} \left(f(x) + g(x) \right) = \lim_{x \to c} f(x) + \lim_{x \to c} g(x) = f(c) + g(c),$$

so $f + g$ is continuous.

THEOREM

7

Composites of continuous functions are continuous

If $f(g(x))$ is defined on an interval containing c, and if f is continuous at L and $\lim_{x \to c} g(x) = L$, then

$$\lim_{x \to c} f(g(x)) = f(L) = f\left(\lim_{x \to c} g(x) \right).$$

In particular, if g is continuous at c (so $L = g(c)$), then the composition $f \circ g$ is continuous at c:

$$\lim_{x \to c} f(g(x)) = f(g(c)).$$

(See Exercise 37 in Section 1.5.)

EXAMPLE 6 The following functions are continuous everywhere on their respective domains:

(a) $3x^2 - 2x$

(b) $\dfrac{x - 2}{x^2 - 4}$

(c) $|x^2 - 1|$

(d) $\sqrt{x}$

(e) $\sqrt{x^2 - 2x - 5}$

(f) $\dfrac{|x|}{\sqrt{|x + 2|}}$.

Continuous Extensions and Removable Discontinuities

As we have seen in Section 1.2, a rational function may have a limit even at a point where its denominator is zero. If $f(c)$ is not defined, but $\lim_{x \to c} f(x) = L$ exists, we can define a new function $F(x)$ by

$$F(x) = \begin{cases} f(x) & \text{if } x \text{ is in the domain of } f \\ L & \text{if } x = c. \end{cases}$$

$F(x)$ is continuous at $x = c$. It is called the **continuous extension** of $f(x)$ to $x = c$. For rational functions f, continuous extensions are usually found by cancelling common factors.

EXAMPLE 7 Show that $f(x) = \dfrac{x^2 - x}{x^2 - 1}$ has a continuous extension to $x = 1$, and find that extension.

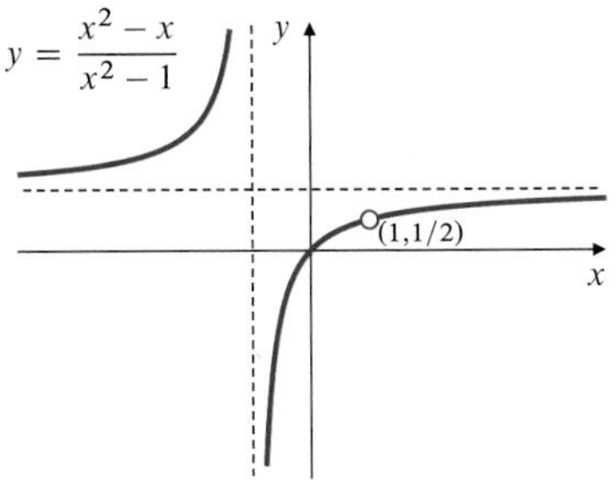

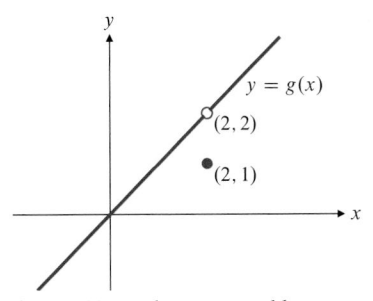

$y = \dfrac{x^2 - x}{x^2 - 1}$

Figure 1.25 This function has a continuous extension to $x = 1$

Figure 1.26 g has a removable discontinuity at 2

Solution Although $f(1)$ is not defined, if $x \neq 1$ we have

$$f(x) = \frac{x^2 - x}{x^2 - 1} = \frac{x(x - 1)}{(x + 1)(x - 1)} = \frac{x}{x + 1}.$$

The function

$$F(x) = \frac{x}{x + 1}$$

is equal to $f(x)$ for $x \neq 1$ but is also continuous at $x = 1$, having there the value $1/2$. The graph of f is shown in Figure 1.25. The continuous extension of $f(x)$ to $x = 1$ is $F(x)$. It has the same graph as $f(x)$ except with no hole at $(1, 1/2)$.

If a function f is undefined or discontinuous at a point a but can be (re)defined at that *single point* so that it becomes continuous there, then we say that f has a **removable discontinuity** at a. The function f in the above example has a removable discontinuity at $x = 1$. To remove it, define $f(1) = 1/2$.

EXAMPLE 8 The function $g(x) = \begin{cases} x & \text{if } x \neq 2 \\ 1 & \text{if } x = 2 \end{cases}$ has a removable discontinuity at $x = 2$. To remove it, redefine $g(2) = 2$. (See Figure 1.26.)

Continuous Functions on Closed, Finite Intervals

Continuous functions that are defined on *closed, finite intervals* have special properties that make them particularly useful in mathematics and its applications. We will discuss two of these properties here. Although they may appear obvious, these properties are much more subtle than the results about limits stated earlier in this chapter; their proofs (see Appendix III) require a careful study of the implications of the completeness property of the real numbers.

The first of the properties states that a function $f(x)$ that is continuous on a closed, finite interval $[a, b]$ must have an **absolute maximum value** and an **absolute minimum value**. This means that the values of $f(x)$ at all points of the interval lie between the values of $f(x)$ at two particular points in the interval; the graph of f has a highest point and a lowest point.

THEOREM

8

The Max-Min Theorem

If $f(x)$ is continuous on the closed, finite interval $[a, b]$, then there exist numbers p and q in $[a, b]$ such that for all x in $[a, b]$,

$$f(p) \leq f(x) \leq f(q).$$

Thus, f has the absolute minimum value $m = f(p)$, taken on at the point p, and the absolute maximum value $M = f(q)$, taken on at the point q.

Many important problems in mathematics and its applications come down to having to find maximum and minimum values of functions. Calculus provides some very useful tools for solving such problems. Observe, however, that the theorem above merely asserts that minimum and maximum values *exist;* it doesn't tell us how to find them. In Chapter 4 we will develop techniques for calculating maximum and minimum values of functions. For now, we can solve some simple maximum and minimum value problems involving quadratic functions by completing the square without using any calculus.

EXAMPLE 9 What is the largest possible area of a rectangular field that can be enclosed by 200 m of fencing?

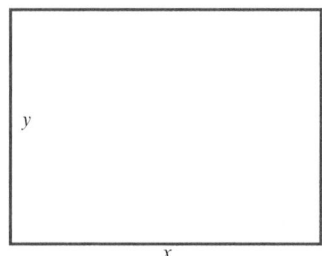

Figure 1.27   Rectangular field: perimeter $= 2x + 2y$, area $= xy$

Solution   If the sides of the field are x m and y m (Figure 1.27), then its perimeter is $P = 2x + 2y$ m, and its area is $A = xy$ m^2. We are given that $P = 200$, so $x + y = 100$, and $y = 100 - x$. Neither side can be negative, so x must belong to the closed interval $[0, 100]$. The area of the field can be expressed as a function of x by substituting $100 - x$ for y:

$$A = x(100 - x) = 100x - x^2.$$

We want to find the maximum value of the quadratic function $A(x) = 100x - x^2$ on the interval $[0, 100]$. Theorem 8 assures us that such a maximum exists.

To find the maximum, we complete the square of the function $A(x)$. Note that $x^2 - 100x$ are the first two terms of the square $(x - 50)^2 = x^2 - 100x + 2,500$. Thus,

$$A(x) = 2,500 - (x - 50)^2.$$

Observe that $A(50) = 2,500$ and $A(x) < 2,500$ if $x \neq 50$, because we are subtracting a positive number $(x - 50)^2$ from 2,500 in this case. Therefore, the maximum value of $A(x)$ is 2,500. The largest field has area 2,500 m^2 and is actually a square with dimensions $x = y = 50$ m.

Theorem 8 implies that a function that is continuous on a closed, finite interval is **bounded**. This means that it cannot take on arbitrarily large positive or negative values; there must exist a number K such that

$$|f(x)| \leq K; \qquad \text{that is,} \qquad -K \leq f(x) \leq K.$$

In fact, for K we can use the larger of the numbers $|f(p)|$ and $|f(q)|$ in the theorem.

The conclusions of Theorem 8 may fail if the function f is not continuous or if the interval is not closed. See Figures 1.28–1.31 for examples of how such failure can occur.

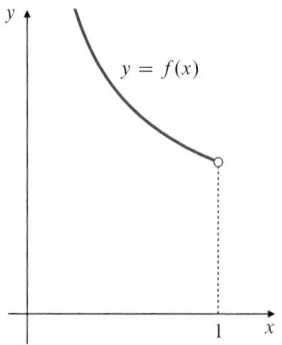

Figure 1.28 $f(x) = 1/x$ is continuous on the open interval $(0, 1)$. It is not bounded and has neither a maximum nor a minimum value

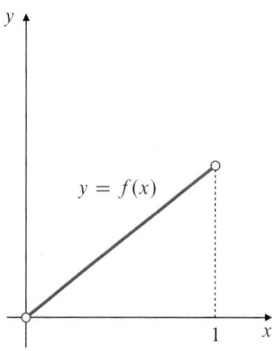

Figure 1.29 $f(x) = x$ is continuous on the open interval $(0, 1)$. It is bounded but has neither a maximum nor a minimum value

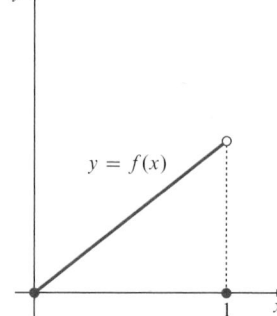

Figure 1.30   This function is defined on the closed interval $[0, 1]$ but is discontinuous at the endpoint $x = 1$. It has a minimum value but no maximum value

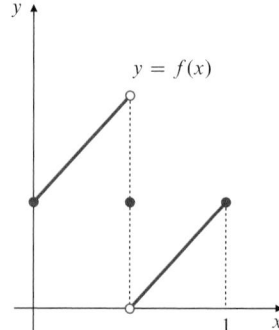

Figure 1.31   This function is discontinuous at an interior point of its domain, the closed interval $[0, 1]$. It is bounded but has neither maximum nor minimum values

The second property of a continuous function defined on a closed, finite interval is that the function takes on all real values between any two of its values. This property is called the **intermediate-value property**.

THEOREM

9

The Intermediate-Value Theorem

If $f(x)$ is continuous on the interval $[a, b]$ and if s is a number between $f(a)$ and $f(b)$, then there exists a number c in $[a, b]$ such that $f(c) = s$.

In particular, a continuous function defined on a closed interval takes on all values between its minimum value m and its maximum value M, so its range is also a closed interval, $[m, M]$.

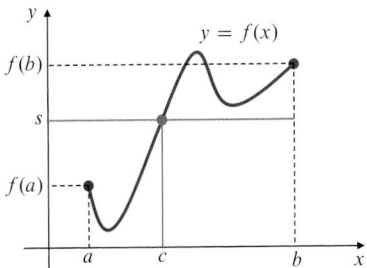

Figure 1.32 The continuous function f takes on the value s at some point c between a and b

Figure 1.32 shows a typical situation. The points $(a, f(a))$ and $(b, f(b))$ are on opposite sides of the horizontal line $y = s$. Being unbroken, the graph $y = f(x)$ must cross this line in order to go from one point to the other. In the figure, it crosses the line only once, at $x = c$. If the line $y = s$ were somewhat higher, there might have been three crossings and three possible values for c.

Theorem 9 is the reason why the graph of a function that is continuous on an interval I cannot have any breaks. It must be **connected**, a single, unbroken curve with no jumps.

EXAMPLE 10 Determine the intervals on which $f(x) = x^3 - 4x$ is positive and negative.

Solution Since $f(x) = x(x^2 - 4) = x(x - 2)(x + 2)$, $f(x) = 0$ only at $x = 0, 2$, and -2. Because f is continuous on the whole real line, it must have constant sign on each of the intervals $(-\infty, -2)$, $(-2, 0)$, $(0, 2)$, and $(2, \infty)$. (If there were points a and b in one of those intervals, say in $(0, 2)$, such that $f(a) < 0$ and $f(b) > 0$, then by the Intermediate-Value Theorem there would exist c between a and b, and therefore between 0 and 2, such that $f(c) = 0$. But we know f has no such zero in $(0, 2)$.)

To find whether $f(x)$ is positive or negative throughout each interval, pick a point in the interval and evaluate f at that point:

Since $f(-3) = -15 < 0$, $f(x)$ is negative on $(-\infty, -2)$.
Since $f(-1) = 3 > 0$, $f(x)$ is positive on $(-2, 0)$.
Since $f(1) = -3 < 0$, $f(x)$ is negative on $(0, 2)$.
Since $f(3) = 15 > 0$, $f(x)$ is positive on $(2, \infty)$.

Finding Roots of Equations

Among the many useful tools that calculus will provide are ones that enable us to calculate solutions to equations of the form $f(x) = 0$ to any desired degree of accuracy. Such a solution is called a **root** of the equation, or a **zero** of the function f. Using these tools usually requires previous knowledge that the equation has a solution in some interval. The Intermediate-Value Theorem can provide this information.

EXAMPLE 11 Show that the equation $x^3 - x - 1 = 0$ has a solution in the interval $[1, 2]$.

Solution The function $f(x) = x^3 - x - 1$ is a polynomial and is therefore continuous everywhere. Now $f(1) = -1$ and $f(2) = 5$. Since 0 lies between -1 and 5, the Intermediate-Value Theorem assures us that there must be a number c in $[1, 2]$ such that $f(c) = 0$.

One method for finding a zero of a function that is continuous and changes sign on an interval involves bisecting the interval many times, each time determining which half of the previous interval must contain the root, because the function has opposite signs at the two ends of that half. This method is slow. For example, if the original interval has length 1, it will take 11 bisections to cut down to an interval of length less than 0.0005 (because $2^{11} > 2,000 = 1/(0.0005)$), and thus to ensure that we have found the root correct to 3 decimal places.

EXAMPLE 12 (**The Bisection Method**) Solve the equation $x^3 - x - 1 = 0$ of Example 11 correct to 3 decimal places by successive bisections.

Solution We start out knowing that there is a root in $[1, 2]$. Table 6 shows the results of the bisections.

Table 6. The Bisection Method for $f(x) = x^3 - x - 1 = 0$

Bisection Number	x	$f(x)$	Root in Interval	Midpoint
	1	-1		
	2	5	$[1, 2]$	1.5
1	1.5	0.8750	$[1, 1.5]$	1.25
2	1.25	-0.2969	$[1.25, 1.5]$	1.375
3	1.375	0.2246	$[1.25, 1.375]$	1.3125
4	1.3125	-0.0515	$[1.3125, 1.375]$	1.3438
5	1.3438	0.0826	$[1.3125, 1.3438]$	1.3282
6	1.3282	0.0147	$[1.3125, 1.3282]$	1.3204
7	1.3204	-0.0186	$[1.3204, 1.3282]$	1.3243
8	1.3243	-0.0018	$[1.3243, 1.3282]$	1.3263
9	1.3263	0.0065	$[1.3243, 1.3263]$	1.3253
10	1.3253	0.0025	$[1.3243, 1.3253]$	1.3248
11	1.3248	0.0003	$[1.3243, 1.3248]$	1.3246
12	1.3246	-0.0007	$[1.3246, 1.3248]$	

The root is 1.325, rounded to 3 decimal places.

In Section 4.2, calculus will provide us with much faster methods of solving equations such as the one in the example above. Many programmable calculators and computer algebra software packages have built-in routines for solving equations. For example, Maple's `fsolve` routine can be used to find the real solution of $x^3 - x - 1 = 0$ in $[1, 2]$ in Example 11:

```
>   fsolve(x^3-x-1=0,x=1..2);
```

$$1.324717957$$

Remark The Max-Min Theorem and the Intermediate-Value Theorem are examples of what mathematicians call **existence theorems**. Such theorems assert that something exists without telling you how to find it. Students sometimes complain that mathematicians worry too much about proving that a problem has a solution and not enough about how to find that solution. They argue: "If I can calculate a solution to a problem, then surely I do not need to worry about whether a solution exists." This is, however, false logic. Suppose we pose the problem: "Find the largest positive integer." Of course, this problem has no solution; there is no largest positive integer because we can add 1 to any integer and get a larger integer. Suppose, however, that we forget this and try to calculate a solution. We could proceed as follows:

Let N be the largest positive integer.
Since 1 is a positive integer, we must have $N \geq 1$.
Since N^2 is a positive integer, it cannot exceed the largest positive integer.
Therefore, $N^2 \leq N$ and so $N^2 - N \leq 0$.
Thus, $N(N - 1) \leq 0$ and we must have $N - 1 \leq 0$.
Therefore, $N \leq 1$. Since also $N \geq 1$, we have $N = 1$.
Therefore, 1 is the largest positive integer.

The only error we have made here is in the assumption (in the first line) that the problem has a solution. It is partly to avoid logical pitfalls like this that mathematicians prove existence theorems.

EXERCISES 1.4

Exercises 1–3 refer to the function g defined on $[-2, 2]$, whose graph is shown in Figure 1.33.

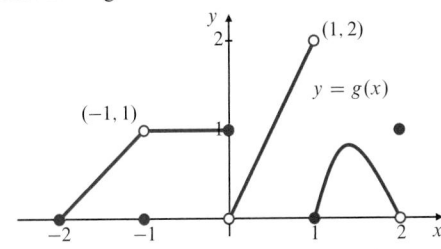

Figure 1.33

1. State whether g is (a) continuous, (b) left continuous, (c) right continuous, and (d) discontinuous at each of the points -2, -1, 0, 1, and 2.

2. At what points in its domain does g have a removable discontinuity, and how should g be redefined at each of those points so as to be continuous there?

3. Does g have an absolute maximum value on $[-2, 2]$? an absolute minimum value?

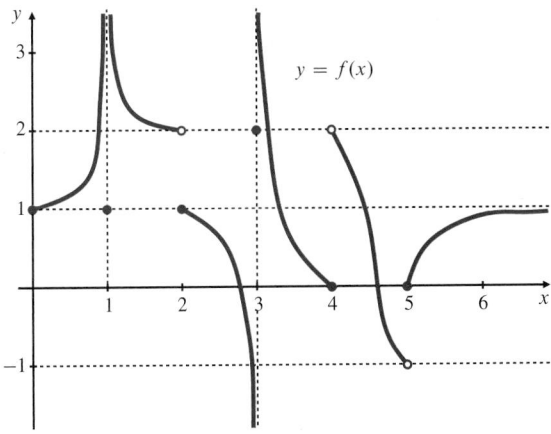

Figure 1.34

4. At what points is the function f, whose graph is shown in Figure 1.34, discontinuous? At which of those points is it left continuous? right continuous?

5. Can the function f graphed in Figure 1.34 be redefined at the single point $x = 1$ so that it becomes continuous there?

6. The function $\operatorname{sgn}(x) = x/|x|$ is neither continuous nor discontinuous at $x = 0$. How is this possible?

In Exercises 7–12, state where in its domain the given function is continuous, where it is left or right continuous, and where it is just discontinuous.

7. $f(x) = \begin{cases} x & \text{if } x < 0 \\ x^2 & \text{if } x \geq 0 \end{cases}$

8. $f(x) = \begin{cases} x & \text{if } x < -1 \\ x^2 & \text{if } x \geq -1 \end{cases}$

9. $f(x) = \begin{cases} 1/x^2 & \text{if } x \neq 0 \\ 0 & \text{if } x = 0 \end{cases}$

10. $f(x) = \begin{cases} x^2 & \text{if } x \leq 1 \\ 0.987 & \text{if } x > 1 \end{cases}$

11. The least integer function $\lceil x \rceil$ of Example 11 in Section P.5.

12. The cost function $C(t)$ of Exercise 53 in Section 1.3.

In Exercises 13–16, how should the given function be defined at the given point to be continuous there? Give a formula for the continuous extension to that point.

13. $\dfrac{x^2 - 4}{x - 2}$ at $x = 2$

14. $\dfrac{1 + t^3}{1 - t^2}$ at $t = -1$

15. $\dfrac{t^2 - 5t + 6}{t^2 - t - 6}$ at 3

16. $\dfrac{x^2 - 2}{x^4 - 4}$ at $\sqrt{2}$

17. Find k so that $f(x) = \begin{cases} x^2 & \text{if } x \leq 2 \\ k - x^2 & \text{if } x > 2 \end{cases}$ is a continuous function.

18. Find m so that $g(x) = \begin{cases} x - m & \text{if } x < 3 \\ 1 - mx & \text{if } x \geq 3 \end{cases}$ is continuous for all x.

19. Does the function x^2 have a maximum value on the open interval $-1 < x < 1$? a minimum value? Explain.

20. The Heaviside function of Example 1 has both absolute maximum and minimum values on the interval $[-1, 1]$, but it is not continuous on that interval. Does this violate the Max-Min Theorem? Why?

Exercises 21–24 ask for maximum and minimum values of functions. They can all be done by the method of Example 9.

21. The sum of two nonnegative numbers is 8. What is the largest possible value of their product?

22. The sum of two nonnegative numbers is 8. What is (a) the smallest and (b) the largest possible value for the sum of their squares?

23. A software company estimates that if it assigns x programmers to work on the project, it can develop a new product in T days, where

$$T = 100 - 30x + 3x^2.$$

How many programmers should the company assign in order to complete the development as quickly as possible?

24. It costs a desk manufacturer $\$(245x - 30x^2 + x^3)$ to send a shipment of x desks to its warehouse. How many desks should it include in each shipment to minimize the average shipping cost per desk?

Find the intervals on which the functions $f(x)$ in Exercises 25–28 are positive and negative.

25. $f(x) = \dfrac{x^2 - 1}{x}$

26. $f(x) = x^2 + 4x + 3$

27. $f(x) = \dfrac{x^2 - 1}{x^2 - 4}$

28. $f(x) = \dfrac{x^2 + x - 2}{x^3}$

29. Show that $f(x) = x^3 + x - 1$ has a zero between $x = 0$ and $x = 1$.

30. Show that the equation $x^3 - 15x + 1 = 0$ has three solutions in the interval $[-4, 4]$.

31. Show that the function $F(x) = (x - a)^2(x - b)^2 + x$ has the value $(a + b)/2$ at some point x.

32. (**A fixed-point theorem**) Suppose that f is continuous on the closed interval $[0, 1]$ and that $0 \le f(x) \le 1$ for every x in $[0, 1]$. Show that there must exist a number c in $[0, 1]$ such that $f(c) = c$. (c is called a fixed point of the function f.) *Hint:* If $f(0) = 0$ or $f(1) = 1$, you are done. If not, apply the Intermediate-Value Theorem to $g(x) = f(x) - x$.

33. If an even function f is right continuous at $x = 0$, show that it is continuous at $x = 0$.

34. If an odd function f is right continuous at $x = 0$, show that it is continuous at $x = 0$ and that it satisfies $f(0) = 0$.

Use a graphing utility to find maximum and minimum values of the functions in Exercises 35–38 and the points x where they occur. Obtain 3-decimal-place accuracy for all answers.

35. $f(x) = \dfrac{x^2 - 2x}{x^4 + 1}$ on $[-5, 5]$

36. $f(x) = \dfrac{\sin x}{6 + x}$ on $[-\pi, \pi]$

37. $f(x) = x^2 + \dfrac{4}{x}$ on $[1, 3]$

38. $f(x) = \sin(\pi x) + x(\cos(\pi x) + 1)$ on $[0, 1]$

Use a graphing utility or a programmable calculator and the Bisection Method to solve the equations in Exercises 39–40 to 3 decimal places. As a first step, try to guess a small interval that you can be sure contains a root.

39. $x^3 + x - 1 = 0$ **40.** $\cos x - x = 0$

Use Maple's `fsolve` routine to solve the equations in Exercises 41–42.

41. $\sin x + 1 - x^2 = 0$ (two roots)

42. $x^4 - x - 1 = 0$ (two roots)

43. Investigate the difference between the Maple routines `fsolve(f,x)`, `solve(f,x)`, and `evalf(solve(f,x))`, where `f := x^3-x-1=0`. Note that no interval is specified for x here.

1.5 The Formal Definition of Limit

The material in this section is optional.

The *informal* definition of limit given in Section 1.2 is not precise enough to enable us to prove results about limits such as those given in Theorems 2–4 of Section 1.2. A more precise *formal* definition is based on the idea of controlling the input x of a function f so that the output $f(x)$ will lie in a specific interval.

EXAMPLE 1 The area of a circular disk of radius r cm is $A = \pi r^2$ cm^2. A machinist is required to manufacture a circular metal disk having area 400π cm^2 within an error tolerance of ± 5 cm^2. How close to 20 cm must the machinist control the radius of the disk to achieve this?

Solution The machinist wants $|\pi r^2 - 400\pi| < 5$, that is,

$$400\pi - 5 < \pi r^2 < 400\pi + 5,$$

or, equivalently,

$$\sqrt{400 - (5/\pi)} < r < \sqrt{400 + (5/\pi)}$$
$$19.96017 < r < 20.03975.$$

Thus, the machinist needs $|r - 20| < 0.03975$; she must ensure that the radius of the disk differs from 20 cm by less than 0.4 mm so that the area of the disk will lie within the required error tolerance.

When we say that $f(x)$ has limit L as x approaches a, we are really saying that we can ensure that the *error* $|f(x) - L|$ will be less than *any* allowed tolerance, no matter how small, by taking x *close enough* to a (but not equal to a). It is traditional to use ϵ, the Greek letter "epsilon," for the size of the allowable *error* and δ, the Greek letter "delta," for the *difference* $x - a$ that measures how close x must be to a to ensure that the error is within that tolerance. These are the letters that Cauchy and Weierstrass used in their pioneering work on limits and continuity in the nineteenth century.

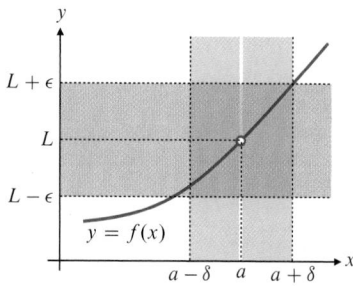

Figure 1.35 If $x \ne a$ and $|x - a| < \delta$, then $|f(x) - L| < \epsilon$

If ϵ is any positive number, *no matter how small*, we must be able to ensure that $|f(x) - L| < \epsilon$ by restricting x to be *close enough to* (but not equal to) a. How close is close enough? It is sufficient that the distance $|x - a|$ from x to a be less than a positive number δ that depends on ϵ. (See Figure 1.35.) If we can find such a δ for any positive ϵ, we are entitled to conclude that $\lim_{x \to a} f(x) = L$.

DEFINITION

8

> **A formal definition of limit**
>
> We say that $f(x)$ **approaches the limit** L as x **approaches** a, and we write
>
> $$\lim_{x \to a} f(x) = L \qquad \text{or} \qquad \lim_{x \to a} f(x) = L,$$
>
> if the following condition is satisfied:
> for every number $\epsilon > 0$ there exists a number $\delta > 0$, possibly depending on ϵ, such that if $0 < |x - a| < \delta$, then x belongs to the domain of f and
>
> $$|f(x) - L| < \epsilon.$$

Though precise, the above definition is more restrictive than it needs to be. It requires that the domain of f must contain open intervals with right and left endpoints at a. In Section 12.2 of Chapter 12 we will give a new, more general definition of limit for functions of any number of variables. For functions of one variable, it replaces the requirement that f be defined on open intervals with right and left endpoints at a with the weaker requirement that every open interval containing a must contain a point of the domain of f different from a. For now, we prefer the simpler but more restrictive definition given above.

The formal definition of limit does not tell you how to find the limit of a function, but it does enable you to verify that a suspected limit is correct. The following examples show how it can be used to verify limit statements for specific functions. The first of these gives a formal verification of the two limits found in Example 3 of Section 1.2.

EXAMPLE 2   (**Two important limits**) Verify that:
(a) $\lim_{x \to a} x = a$ and   (b) $\lim_{x \to a} k = k$ (k = constant).

Solution

(a) Let $\epsilon > 0$ be given. We must find $\delta > 0$ so that

$$0 < |x - a| < \delta \qquad \text{implies} \qquad |x - a| < \epsilon.$$

Clearly, we can take $\delta = \epsilon$ and the implication above will be true. This proves that $\lim_{x \to a} x = a$.

(b) Let $\epsilon > 0$ be given. We must find $\delta > 0$ so that

$$0 < |x - a| < \delta \qquad \text{implies} \qquad |k - k| < \epsilon.$$

Since $k - k = 0$, we can use any positive number for δ and the implication above will be true. This proves that $\lim_{x \to a} k = k$.

EXAMPLE 3   Verify that $\lim_{x \to 2} x^2 = 4$.

Solution   Here $a = 2$ and $L = 4$. Let ϵ be a given positive number. We want to find $\delta > 0$ so that if $0 < |x - 2| < \delta$, then $|f(x) - 4| < \epsilon$. Now

$$|f(x) - 4| = |x^2 - 4| = |(x + 2)(x - 2)| = |x + 2||x - 2|.$$

We want the expression above to be less than ϵ. We can make the factor $|x - 2|$ as small as we wish by choosing δ properly, but we need to control the factor $|x + 2|$ so that it does not become too large. If we first assume $\delta \leq 1$ and require that $|x - 2| < \delta$, then we have

$$|x - 2| < 1 \quad \Rightarrow \quad 1 < x < 3 \quad \Rightarrow \quad 3 < x + 2 < 5$$
$$\Rightarrow \quad |x + 2| < 5.$$

Hence,

$$|f(x) - 4| < 5|x - 2| \quad \text{if} \quad |x - 2| < \delta \le 1.$$

But $5|x-2| < \epsilon$ if $|x-2| < \epsilon/5$. Therefore, if we take $\delta = \min\{1, \epsilon/5\}$, the *minimum* (the smaller) of the two numbers 1 and $\epsilon/5$, then

$$|f(x) - 4| < 5|x - 2| < 5 \times \frac{\epsilon}{5} = \epsilon \quad \text{if} \quad |x - 2| < \delta.$$

This proves that $\lim_{x \to 2} f(x) = 4$.

Using the Definition of Limit to Prove Theorems

We do not usually rely on the formal definition of limit to verify specific limits such as those in the two examples above. Rather, we appeal to general theorems about limits, in particular Theorems 2–4 of Section 1.2. The definition is used to prove these theorems. As an example, we prove part 1 of Theorem 2, the *Sum Rule*.

EXAMPLE 4   **(Proving the rule for the limit of a sum)** If $\lim_{x \to a} f(x) = L$ and $\lim_{x \to a} g(x) = M$, prove that $\lim_{x \to a} \big(f(x) + g(x)\big) = L + M$.

Solution   Let $\epsilon > 0$ be given. We want to find a positive number δ such that

$$0 < |x - a| < \delta \quad \Rightarrow \quad \big|\big(f(x) + g(x)\big) - (L + M)\big| < \epsilon.$$

Observe that

$$\begin{aligned}
\big|\big(f(x) + g(x)\big) - (L + M)\big| & \quad \text{Regroup terms.}\\
= \big|\big(f(x) - L\big) + \big(g(x) - M\big)\big| & \quad \text{(Use the triangle inequality:}\\
& \quad |a + b| \le |a| + |b|).\\
\le |f(x) - L| + |g(x) - M|.
\end{aligned}$$

Since $\lim_{x \to a} f(x) = L$ and $\epsilon/2$ is a positive number, there exists a number $\delta_1 > 0$ such that

$$0 < |x - a| < \delta_1 \quad \Rightarrow \quad |f(x) - L| < \epsilon/2.$$

Similarly, since $\lim_{x \to a} g(x) = M$, there exists a number $\delta_2 > 0$ such that

$$0 < |x - a| < \delta_2 \quad \Rightarrow \quad |g(x) - M| < \epsilon/2.$$

Let $\delta = \min\{\delta_1, \delta_2\}$, the smaller of δ_1 and δ_2. If $0 < |x - a| < \delta$, then $|x - a| < \delta_1$, so $|f(x) - L| < \epsilon/2$, and $|x - a| < \delta_2$, so $|g(x) - M| < \epsilon/2$. Therefore,

$$\big|\big(f(x) + g(x)\big) - (L + M)\big| < \frac{\epsilon}{2} + \frac{\epsilon}{2} = \epsilon.$$

This shows that $\lim_{x \to a} \big(f(x) + g(x)\big) = L + M$.

Other Kinds of Limits

The formal definition of limit can be modified to give precise definitions of one-sided limits, limits at infinity, and infinite limits. We give some of the definitions here and leave you to supply the others.

DEFINITION

9

Right limits

We say that $f(x)$ has **right limit** L at a, and we write

$$\lim_{x \to a+} f(x) = L,$$

if the following condition is satisfied:
for every number $\epsilon > 0$ there exists a number $\delta > 0$, possibly depending on ϵ, such that if $a < x < a + \delta$, then x belongs to the domain of f and

$$|f(x) - L| < \epsilon.$$

Notice how the condition $0 < |x - a| < \delta$ in the definition of limit becomes $a < x < a + \delta$ in the right limit case (Figure 1.36). The definition for a left limit is formulated in a similar way.

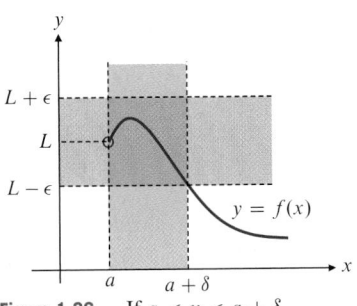

Figure 1.36   If $a < x < a + \delta$, then $|f(x) - L| < \epsilon$

EXAMPLE 5   Show that $\lim_{x \to 0+} \sqrt{x} = 0$.

Solution   Let $\epsilon > 0$ be given. If $x > 0$, then $|\sqrt{x} - 0| = \sqrt{x}$. We can ensure that $\sqrt{x} < \epsilon$ by requiring $x < \epsilon^2$. Thus, we can take $\delta = \epsilon^2$ and the condition of the definition will be satisfied:

$$0 < x < \delta = \epsilon^2 \quad \text{implies} \quad |\sqrt{x} - 0| < \epsilon.$$

Therefore, $\lim_{x \to 0+} \sqrt{x} = 0$.

To claim that a function f has a limit L at infinity, we must be able to ensure that the error $|f(x) - L|$ is less than any given positive number ϵ by restricting x to be *sufficiently large*, that is, by requiring $x > R$ for some positive number R depending on ϵ.

DEFINITION

10

Limit at infinity

We say that $f(x)$ **approaches the limit** L **as** x **approaches infinity**, and we write

$$\lim_{x \to \infty} f(x) = L,$$

if the following condition is satisfied:
for every number $\epsilon > 0$ there exists a number R, possibly depending on ϵ, such that if $x > R$, then x belongs to the domain of f and

$$|f(x) - L| < \epsilon.$$

You are invited to formulate a version of the definition of a limit at negative infinity.

EXAMPLE 6   Show that $\lim_{x \to \infty} \dfrac{1}{x} = 0$.

Solution   Let ϵ be a given positive number. For $x > 0$ we have

$$\left| \frac{1}{x} - 0 \right| = \frac{1}{|x|} = \frac{1}{x} < \epsilon \quad \text{provided} \quad x > \frac{1}{\epsilon}.$$

Therefore, the condition of the definition is satisfied with $R = 1/\epsilon$. We have shown that $\lim_{x \to \infty} 1/x = 0$.

To show that $f(x)$ has an infinite limit at a, we must ensure that $f(x)$ is larger than any given positive number (say B) by restricting x to a sufficiently small interval centred at a, and requiring that $x \neq a$.

DEFINITION

11

Infinite limits

We say that $f(x)$ approaches infinity as x approaches a and write

$$\lim_{x \to a} f(x) = \infty,$$

if for every positive number B we can find a positive number δ, possibly depending on B, such that if $0 < |x - a| < \delta$, then x belongs to the domain of f and $f(x) > B$.

Try to formulate the corresponding definition for the concept $\lim_{x \to a} f(x) = -\infty$. Then try to modify both definitions to cover the case of infinite one-sided limits and infinite limits at infinity.

EXAMPLE 7 Verify that $\lim_{x \to 0} \dfrac{1}{x^2} = \infty$.

Solution Let B be any positive number. We have

$$\frac{1}{x^2} > B \qquad \text{provided that} \quad x^2 < \frac{1}{B}.$$

If $\delta = 1/\sqrt{B}$, then

$$0 < |x| < \delta \quad \Rightarrow \quad x^2 < \delta^2 = \frac{1}{B} \quad \Rightarrow \quad \frac{1}{x^2} > B.$$

Therefore, $\lim_{x \to 0} 1/x^2 = \infty$.

EXERCISES 1.5

1. The length L of a metal rod is given in terms of the temperature T (°C) by $L = 39.6 + 0.025T$ cm. Within what range of temperature must the rod be kept if its length must be maintained within ± 1 mm of 40 cm?

2. What is the largest tolerable error in the 20 cm edge length of a cubical cardboard box if the volume of the box must be within $\pm 1.2\%$ of 8,000 cm^3?

In Exercises 3–6, in what interval must x be confined if $f(x)$ must be within the given distance ϵ of the number L?

3. $f(x) = 2x - 1$, $\quad L = 3$, $\quad \epsilon = 0.02$
4. $f(x) = x^2$, $\quad\quad\quad L = 4$, $\quad \epsilon = 0.1$
5. $f(x) = \sqrt{x}$, $\quad\quad L = 1$, $\quad \epsilon = 0.1$
6. $f(x) = 1/x$, $\quad\quad L = -2$, $\epsilon = 0.01$

In Exercises 7–10, find a number $\delta > 0$ such that if $|x - a| < \delta$, then $|f(x) - L|$ will be less than the given number ϵ.

7. $f(x) = 3x + 1$, $\quad a = 2$, $\quad L = 7$, $\quad \epsilon = 0.03$
8. $f(x) = \sqrt{2x + 3}$, $a = 3$, $\quad L = 3$, $\quad \epsilon = 0.01$
9. $f(x) = x^3$, $\quad\quad\quad a = 2$, $\quad L = 8$, $\quad \epsilon = 0.2$
10. $f(x) = 1/(x + 1)$, $a = 0$, $\quad L = 1$, $\quad \epsilon = 0.05$

In Exercises 11–20, use the formal definition of limit to verify the indicated limit.

11. $\lim\limits_{x \to 1} (3x + 1) = 4$

12. $\lim\limits_{x \to 2} (5 - 2x) = 1$

13. $\lim\limits_{x \to 0} x^2 = 0$

14. $\lim\limits_{x \to 2} \dfrac{x - 2}{1 + x^2} = 0$

15. $\lim\limits_{x \to 1/2} \dfrac{1 - 4x^2}{1 - 2x} = 2$

16. $\lim\limits_{x \to -2} \dfrac{x^2 + 2x}{x + 2} = -2$

17. $\lim\limits_{x \to 1} \dfrac{1}{x + 1} = \dfrac{1}{2}$

18. $\lim\limits_{x \to -1} \dfrac{x + 1}{x^2 - 1} = -\dfrac{1}{2}$

19. $\lim\limits_{x \to 1} \sqrt{x} = 1$

20. $\lim\limits_{x \to 2} x^3 = 8$

Give formal definitions of the limit statements in Exercises 21–26.

21. $\lim\limits_{x \to a-} f(x) = L$

22. $\lim\limits_{x \to -\infty} f(x) = L$

23. $\lim\limits_{x \to a} f(x) = -\infty$

24. $\lim\limits_{x \to \infty} f(x) = \infty$

25. $\lim\limits_{x \to a+} f(x) = -\infty$

26. $\lim\limits_{x \to a-} f(x) = \infty$

Use formal definitions of the various kinds of limits to prove the statements in Exercises 27–30.

27. $\lim\limits_{x \to 1+} \dfrac{1}{x-1} = \infty$ **28.** $\lim\limits_{x \to 1-} \dfrac{1}{x-1} = -\infty$

29. $\lim\limits_{x \to \infty} \dfrac{1}{\sqrt{x^2+1}} = 0$ **30.** $\lim\limits_{x \to \infty} \sqrt{x} = \infty$

Proving Theorems with the Definition of Limit

❗ 31. Prove that limits are unique; that is, if $\lim_{x \to a} f(x) = L$ and $\lim_{x \to a} f(x) = M$, prove that $L = M$. *Hint:* Suppose $L \neq M$ and let $\epsilon = |L - M|/3$.

❷ 32. If $\lim_{x \to a} g(x) = M$, show that there exists a number $\delta > 0$ such that

$$0 < |x - a| < \delta \quad \Rightarrow \quad |g(x)| < 1 + |M|.$$

(*Hint:* Take $\epsilon = 1$ in the definition of limit.) This says that the values of $g(x)$ are **bounded** near a point where g has a limit.

❗ 33. If $\lim_{x \to a} f(x) = L$ and $\lim_{x \to a} g(x) = M$, prove that $\lim_{x \to a} f(x)g(x) = LM$ (the Product Rule part of Theorem 2). *Hint:* Reread Example 4. Let $\epsilon > 0$ and write

$$
\begin{aligned}
|f(x)g(x) - LM| &= |f(x)g(x) - Lg(x) + Lg(x) - LM| \\
&= |(f(x) - L)g(x) + L(g(x) - M)| \\
&\leq |(f(x) - L)g(x)| + |L(g(x) - M)| \\
&= |g(x)||f(x) - L| + |L||g(x) - M|
\end{aligned}
$$

Now try to make each term in the last line less than $\epsilon/2$ by taking x close enough to a. You will need the result of Exercise 32.

❷ 34. If $\lim_{x \to a} g(x) = M$, where $M \neq 0$, show that there exists a number $\delta > 0$ such that

$$0 < |x - a| < \delta \quad \Rightarrow \quad |g(x)| > |M|/2.$$

❷ 35. If $\lim_{x \to a} g(x) = M$, where $M \neq 0$, show that

$$\lim\limits_{x \to a} \dfrac{1}{g(x)} = \dfrac{1}{M}.$$

Hint: You will need the result of Exercise 34.

❷ 36. Use the facts proved in Exercises 33 and 35 to prove the Quotient Rule (part 5 of Theorem 2): if $\lim_{x \to a} f(x) = L$ and $\lim_{x \to a} g(x) = M$, where $M \neq 0$, then

$$\lim\limits_{x \to a} \dfrac{f(x)}{g(x)} = \dfrac{L}{M}.$$

❗ 37. Use the definition of limit twice to prove Theorem 7 of Section 1.4; that is, if f is continuous at L and if $\lim_{x \to c} g(x) = L$, then

$$\lim\limits_{x \to c} f(g(x)) = f(L) = f\left(\lim\limits_{x \to c} g(x)\right).$$

❗ 38. Prove the Squeeze Theorem (Theorem 4 in Section 1.2). *Hint:* If $f(x) \leq g(x) \leq h(x)$, then

$$
\begin{aligned}
|g(x) - L| &= |g(x) - f(x) + f(x) - L| \\
&\leq |g(x) - f(x)| + |f(x) - L| \\
&\leq |h(x) - f(x)| + |f(x) - L| \\
&= |h(x) - L - (f(x) - L)| + |f(x) - L| \\
&\leq |h(x) - L| + |f(x) - L| + |f(x) - L|
\end{aligned}
$$

Now you can make each term in the last expression less than $\epsilon/3$ and so complete the proof.

CHAPTER REVIEW

Key Ideas

- **What do the following statements and phrases mean?**
  ◇ the average rate of change of $f(x)$ on $[a, b]$
  ◇ the instantaneous rate of change of $f(x)$ at $x = a$
  ◇ $\lim_{x \to a} f(x) = L$
  ◇ $\lim_{x \to a+} f(x) = L$, $\lim_{x \to a-} f(x) = L$
  ◇ $\lim_{x \to \infty} f(x) = L$, $\lim_{x \to -\infty} f(x) = L$
  ◇ $\lim_{x \to a} f(x) = \infty$, $\lim_{x \to a+} f(x) = -\infty$
  ◇ f is continuous at c.
  ◇ f is left (or right) continuous at c.
  ◇ f has a continuous extension to c.
  ◇ f is a continuous function.
  ◇ f takes on maximum and minimum values on interval I.
  ◇ f is bounded on interval I.
  ◇ f has the intermediate-value property on interval I.

- **State as many "laws of limits" as you can.**

- **What properties must a function have if it is continuous and its domain is a closed, finite interval?**

- **How can you find zeros (roots) of a continuous function?**

Review Exercises

1. Find the average rate of change of x^3 over $[1, 3]$.

2. Find the average rate of change of $1/x$ over $[-2, -1]$.

3. Find the rate of change of x^3 at $x = 2$.

4. Find the rate of change of $1/x$ at $x = -3/2$.

Evaluate the limits in Exercises 5–30 or explain why they do not exist.

5. $\lim\limits_{x \to 1} (x^2 - 4x + 7)$ **6.** $\lim\limits_{x \to 2} \dfrac{x^2}{1 - x^2}$

7. $\lim\limits_{x \to 1} \dfrac{x^2}{1 - x^2}$ **8.** $\lim\limits_{x \to 2} \dfrac{x^2 - 4}{x^2 - 5x + 6}$

9. $\lim\limits_{x \to 2} \dfrac{x^2 - 4}{x^2 - 4x + 4}$ **10.** $\lim\limits_{x \to 2-} \dfrac{x^2 - 4}{x^2 - 4x + 4}$

11. $\lim_{x \to -2+} \dfrac{x^2 - 4}{x^2 + 4x + 4}$

12. $\lim_{x \to 4} \dfrac{2 - \sqrt{x}}{x - 4}$

13. $\lim_{x \to 3} \dfrac{x^2 - 9}{\sqrt{x} - \sqrt{3}}$

14. $\lim_{h \to 0} \dfrac{h}{\sqrt{x + 3h} - \sqrt{x}}$

15. $\lim_{x \to 0+} \sqrt{x - x^2}$

16. $\lim_{x \to 0} \sqrt{x - x^2}$

17. $\lim_{x \to 1} \sqrt{x - x^2}$

18. $\lim_{x \to 1-} \sqrt{x - x^2}$

19. $\lim_{x \to \infty} \dfrac{1 - x^2}{3x^2 - x - 1}$

20. $\lim_{x \to -\infty} \dfrac{2x + 100}{x^2 + 3}$

21. $\lim_{x \to -\infty} \dfrac{x^3 - 1}{x^2 + 4}$

22. $\lim_{x \to \infty} \dfrac{x^4}{x^2 - 4}$

23. $\lim_{x \to 0+} \dfrac{1}{\sqrt{x - x^2}}$

24. $\lim_{x \to 1/2} \dfrac{1}{\sqrt{x - x^2}}$

25. $\lim_{x \to \infty} \sin x$

26. $\lim_{x \to \infty} \dfrac{\cos x}{x}$

27. $\lim_{x \to 0} x \sin \dfrac{1}{x}$

28. $\lim_{x \to 0} \sin \dfrac{1}{x^2}$

29. $\lim_{x \to -\infty} [x + \sqrt{x^2 - 4x + 1}]$

30. $\lim_{x \to \infty} [x + \sqrt{x^2 - 4x + 1}]$

At what, if any, points in its domain is the function f in Exercises 31–38 discontinuous? Is f left or right continuous at these points? In Exercises 35 and 36, H refers to the Heaviside function: $H(x) = 1$ if $x \geq 0$ and $H(x) = 0$ if $x < 0$.

31. $f(x) = x^3 - 4x^2 + 1$

32. $f(x) = \dfrac{x}{x + 1}$

33. $f(x) = \begin{cases} x^2 & \text{if } x > 2 \\ x & \text{if } x \leq 2 \end{cases}$

34. $f(x) = \begin{cases} x^2 & \text{if } x > 1 \\ x & \text{if } x \leq 1 \end{cases}$

35. $f(x) = H(x - 1)$

36. $f(x) = H(9 - x^2)$

37. $f(x) = |x| + |x + 1|$

38. $f(x) = \begin{cases} |x|/|x + 1| & \text{if } x \neq -1 \\ 1 & \text{if } x = -1 \end{cases}$

Challenging Problems

1. Show that the average rate of change of the function x^3 over the interval $[a, b]$, where $0 < a < b$, is equal to the instantaneous rate of change of x^3 at $x = \sqrt{(a^2 + ab + b^2)/3}$. Is this point to the left or to the right of the midpoint $(a + b)/2$ of the interval $[a, b]$?

2. Evaluate $\lim_{x \to 0} \dfrac{x}{|x - 1| - |x + 1|}$.

3. Evaluate $\lim_{x \to 3} \dfrac{|5 - 2x| - |x - 2|}{|x - 5| - |3x - 7|}$.

4. Evaluate $\lim_{x \to 64} \dfrac{x^{1/3} - 4}{x^{1/2} - 8}$.

5. Evaluate $\lim_{x \to 1} \dfrac{\sqrt{3 + x} - 2}{\sqrt[3]{7 + x} - 2}$.

6. The equation $ax^2 + 2x - 1 = 0$, where a is a constant, has two roots if $a > -1$ and $a \neq 0$:

$$r_+(a) = \frac{-1 + \sqrt{1 + a}}{a} \text{ and } r_-(a) = \frac{-1 - \sqrt{1 + a}}{a}.$$

(a) What happens to the root $r_-(a)$ when $a \to 0$?

(b) Investigate numerically what happens to the root $r_+(a)$ when $a \to 0$ by trying the values $a = 1, \pm 0.1, \pm 0.01, \dots$. For values such as $a = 10^{-8}$, the limited precision of your calculator may produce some interesting results. What happens, and why?

(c) Evaluate $\lim_{a \to 0} r_+(a)$ mathematically by using the identity

$$\sqrt{A} - \sqrt{B} = \frac{A - B}{\sqrt{A} + \sqrt{B}}.$$

❷ 7. TRUE or FALSE? If TRUE, give reasons; if FALSE, give a counterexample.

(a) If $\lim_{x \to a} f(x)$ exists but $\lim_{x \to a} g(x)$ does not exist, then $\lim_{x \to a} (f(x) + g(x))$ does not exist.

(b) If neither $\lim_{x \to a} f(x)$ nor $\lim_{x \to a} g(x)$ exists, then $\lim_{x \to a} (f(x) + g(x))$ does not exist.

(c) If f is continuous at a, then so is $|f|$.

(d) If $|f|$ is continuous at a, then so is f.

(e) If $f(x) < g(x)$ for all x in an interval around a, and if $\lim_{x \to a} f(x)$ and $\lim_{x \to a} g(x)$ both exist, then $\lim_{x \to a} f(x) < \lim_{x \to a} g(x)$.

❷ 8. (a) If f is a continuous function defined on a closed interval $[a, b]$, show that $R(f)$ is a closed interval.

(b) What are the possibilities for $R(f)$ if $D(f)$ is an open interval (a, b)?

9. Consider the function $f(x) = \dfrac{x^2 - 1}{|x^2 - 1|}$. Find all points where f is not continuous. Does f have one-sided limits at those points, and if so, what are they?

❷ 10. Find the minimum value of $f(x) = 1/(x - x^2)$ on the interval $(0, 1)$. Explain how you know such a minimum value must exist.

❗ 11. (a) Suppose f is a continuous function on the interval $[0, 1]$, and $f(0) = f(1)$. Show that $f(a) = f\left(a + \dfrac{1}{2}\right)$ for some $a \in \left[0, \dfrac{1}{2}\right]$.

Hint: Let $g(x) = f\left(x + \dfrac{1}{2}\right) - f(x)$, and use the Intermediate-Value Theorem.

(b) If n is an integer larger than 2, show that $f(a) = f\left(a + \dfrac{1}{n}\right)$ for some $a \in \left[0, 1 - \dfrac{1}{n}\right]$.

CHAPTER 2

Differentiation

❝ 'All right,' said Deep Thought. 'The Answer to the Great Question ...'
'Yes ...!'
'Of Life, the Universe and Everything ...' said Deep Thought.
'Yes ...!'
'Is ...' said Deep Thought, and paused.
'Yes ...! ...?'
'Forty-two,' said Deep Thought, with infinite majesty and calm.
...
'Forty-two!' yelled Loonquawl. 'Is that all you've got to show for seven and a half million years' work?'
'I checked it very thoroughly,' said the computer, 'and that quite definitely is the answer. I think the problem, to be quite honest with you, is that you've never actually known what the question is.' ❞

Douglas Adams   1952–2001
from *The Hitchhiker's Guide to the Galaxy*

Introduction

Two fundamental problems are considered in calculus. The **problem of slopes** is concerned with finding the slope of (the tangent line to) a given curve at a given point on the curve. The **problem of areas** is concerned with finding the area of a plane region bounded by curves and straight lines. The solution of the problem of slopes is the subject of **differential calculus**. As we will see, it has many applications in mathematics and other disciplines. The problem of areas is the subject of **integral calculus**, which we begin in Chapter 5.

2.1 Tangent Lines and Their Slopes

This section deals with the problem of finding a straight line L that is tangent to a curve C at a point P. As is often the case in mathematics, the most important step in the solution of such a fundamental problem is making a suitable definition.

For simplicity, and to avoid certain problems best postponed until later, we will not deal with the most general kinds of curves now, but only with those that are the *graphs of continuous functions*. Let C be the graph of $y = f(x)$ and let P be the point (x_0, y_0) on C, so that $y_0 = f(x_0)$. We assume that P is not an endpoint of C. Therefore, C extends some distance on both sides of P. (See Figure 2.1.)

What do we mean when we say that the line L is tangent to C at P? Past experience with tangent lines to circles does not help us to define tangency for more general curves. A tangent line to a circle at P has the following properties (see Figure 2.2):

(i) It meets the circle at only the one point P.

(ii) The circle lies on only one side of the line.

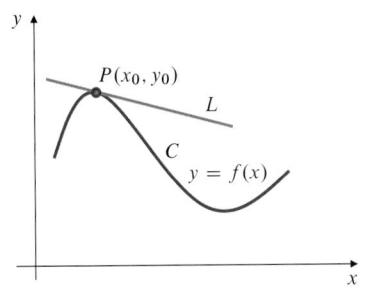

Figure 2.1 L is tangent to C at P

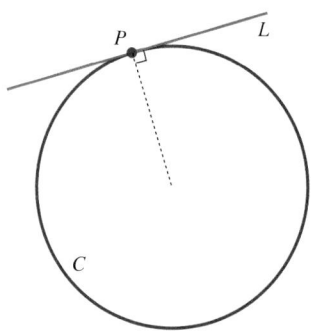

Figure 2.2   *L* is tangent to *C* at *P*

Figure 2.3

(a) *L* meets *C* only at *P* but is not tangent to *C*

(b) *L* meets *C* at several points but is tangent to *C* at *P*

(c) *L* is tangent to *C* at *P* but crosses *C* at *P*

(d) Many lines meet *C* only at *P* but none of them is tangent to *C* at *P*

(iii) The tangent is perpendicular to the line joining the centre of the circle to *P*.

Most curves do not have obvious *centres*, so (iii) is useless for characterizing tangents to them. The curves in Figure 2.3 show that (i) and (ii) cannot be used to define tangency either. In particular, the curve in Figure 2.3(d) is not "smooth" at *P*, so that curve should not have any tangent line there. A tangent line should have the "same direction" as the curve does at the point of tangency.

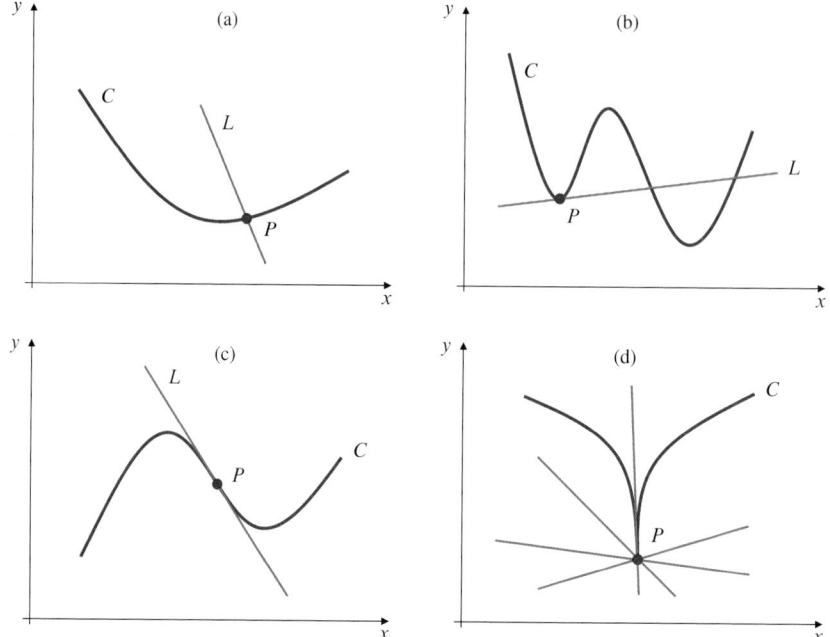

A reasonable definition of tangency can be stated in terms of limits. If *Q* is a point on *C* different from *P*, then the line through *P* and *Q* is called a **secant line** to the curve. This line rotates around *P* as *Q* moves along the curve. If *L* is a line through *P* whose slope is the limit of the slopes of these secant lines *PQ* as *Q* approaches *P* along *C* (Figure 2.4), then we will say that *L* is tangent to *C* at *P*.

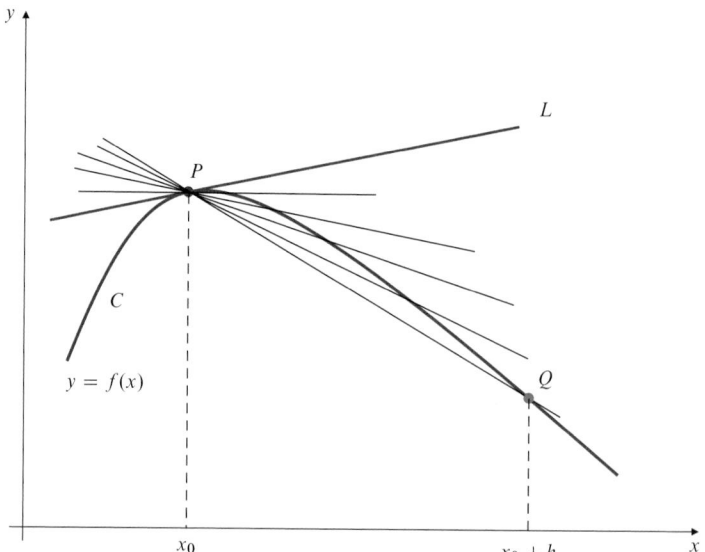

Figure 2.4   Secant lines *PQ* approach tangent line *L* as *Q* approaches *P* along the curve *C*

Since *C* is the graph of the *function* $y = f(x)$, then vertical lines can meet *C* only once. Since $P = (x_0, f(x_0))$, a different point *Q* on the graph must have a different *x*-coordinate, say $x_0 + h$, where $h \neq 0$. Thus $Q = (x_0 + h, f(x_0 + h))$, and the slope of the line *PQ* is

$$\frac{f(x_0 + h) - f(x_0)}{h}.$$

This expression is called the **Newton quotient** or **difference quotient** for f at x_0. Note that h can be positive or negative, depending on whether Q is to the right or left of P.

DEFINITION

1

Nonvertical tangent lines

Suppose that the function f is continuous at $x = x_0$ and that

$$\lim_{h \to 0} \frac{f(x_0 + h) - f(x_0)}{h} = m$$

exists. Then the straight line having slope m and passing through the point $P = (x_0, f(x_0))$ is called the **tangent line** (or simply the **tangent**) to the graph of $y = f(x)$ at P. An equation of this tangent is

$$y = m(x - x_0) + y_0.$$

EXAMPLE 1 Find an equation of the tangent line to the curve $y = x^2$ at the point $(1, 1)$.

Solution Here $f(x) = x^2$, $x_0 = 1$, and $y_0 = f(1) = 1$. The slope of the required tangent is

$$m = \lim_{h \to 0} \frac{f(1 + h) - f(1)}{h} = \lim_{h \to 0} \frac{(1 + h)^2 - 1}{h}$$
$$= \lim_{h \to 0} \frac{1 + 2h + h^2 - 1}{h}$$
$$= \lim_{h \to 0} \frac{2h + h^2}{h} = \lim_{h \to 0} (2 + h) = 2.$$

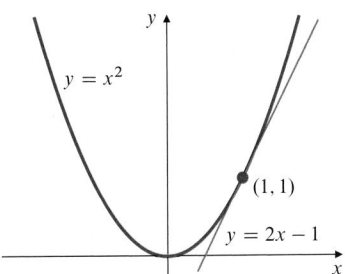

Figure 2.5 The tangent to $y = x^2$ at $(1, 1)$

Accordingly, the equation of the tangent line at $(1, 1)$ is $y = 2(x-1)+1$, or $y = 2x-1$. See Figure 2.5.

Definition 1 deals only with tangents that have finite slopes and are, therefore, not vertical. It is also possible for the graph of a continuous function to have a *vertical* tangent line.

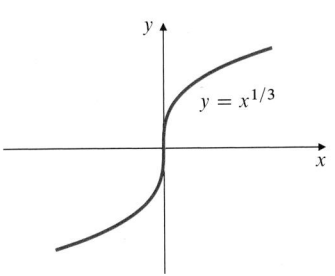

Figure 2.6 The y-axis is tangent to $y = x^{1/3}$ at the origin

EXAMPLE 2 Consider the graph of the function $f(x) = \sqrt[3]{x} = x^{1/3}$, which is shown in Figure 2.6. The graph is a smooth curve, and it seems evident that the y-axis is tangent to this curve at the origin. Let us try to calculate the limit of the Newton quotient for f at $x = 0$:

$$\lim_{h \to 0} \frac{f(0 + h) - f(0)}{h} = \lim_{h \to 0} \frac{h^{1/3}}{h} = \lim_{h \to 0} \frac{1}{h^{2/3}} = \infty.$$

Although the limit does not exist, the slope of the secant line joining the origin to another point Q on the curve approaches infinity as Q approaches the origin from either side.

EXAMPLE 3 On the other hand, the function $f(x) = x^{2/3}$, whose graph is shown in Figure 2.7, does not have a tangent line at the origin because it is not "smooth" there. In this case the Newton quotient is

$$\frac{f(0 + h) - f(0)}{h} = \frac{h^{2/3}}{h} = \frac{1}{h^{1/3}},$$

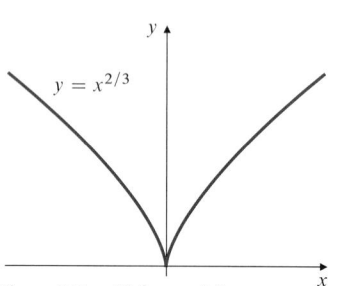

$y = x^{2/3}$

Figure 2.7 This graph has no tangent at the origin

which has no limit as h approaches zero. (The right limit is ∞; the left limit is $-\infty$.) We say this curve has a **cusp** at the origin. A cusp is an infinitely sharp point; if you were travelling along the curve, you would have to stop and turn $180°$ at the origin.

In the light of the two preceding examples, we extend the definition of tangent line to allow for vertical tangents as follows:

DEFINITION

2

Vertical tangents

If f is continuous at $P = (x_0, y_0)$, where $y_0 = f(x_0)$, and if either

$$\lim_{h \to 0} \frac{f(x_0 + h) - f(x_0)}{h} = \infty \quad \text{or} \quad \lim_{h \to 0} \frac{f(x_0 + h) - f(x_0)}{h} = -\infty,$$

then the vertical line $x = x_0$ is tangent to the graph $y = f(x)$ at P. If the limit of the Newton quotient fails to exist in any other way than by being ∞ or $-\infty$, the graph $y = f(x)$ has no tangent line at P.

EXAMPLE 4 Does the graph of $y = |x|$ have a tangent line at $x = 0$?

Solution The Newton quotient here is

$$\frac{|0 + h| - |0|}{h} = \frac{|h|}{h} = \operatorname{sgn} h = \begin{cases} 1, & \text{if } h > 0 \\ -1, & \text{if } h < 0. \end{cases}$$

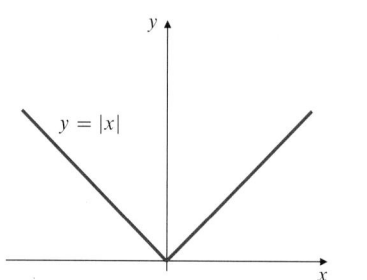

$y = |x|$

Figure 2.8 $y = |x|$ has no tangent at the origin

Since sgn h has different right and left limits at 0 (namely, 1 and -1), the Newton quotient has no limit as $h \to 0$, so $y = |x|$ has no tangent line at $(0,0)$. (See Figure 2.8.) The graph does not have a cusp at the origin, but it is kinked at that point; *it suddenly changes direction and is not smooth.* Curves have tangents only at points where they are smooth. The graphs of $y = x^{2/3}$ and $y = |x|$ have tangent lines everywhere except at the origin, where they are not smooth.

DEFINITION

3

The slope of a curve

The **slope** of a curve C at a point P is the slope of the tangent line to C at P if such a tangent line exists. In particular, the slope of the graph of $y = f(x)$ at the point x_0 is

$$\lim_{h \to 0} \frac{f(x_0 + h) - f(x_0)}{h}.$$

EXAMPLE 5   Find the slope of the curve $y = x/(3x + 2)$ at the point $x = -2$.

Solution   If $x = -2$, then $y = 1/2$, so the required slope is

$$
\begin{aligned}
m &= \lim_{h \to 0} \frac{\dfrac{-2 + h}{3(-2 + h) + 2} - \dfrac{1}{2}}{h} \\
&= \lim_{h \to 0} \frac{-4 + 2h - (-6 + 3h + 2)}{2(-6 + 3h + 2)h} \\
&= \lim_{h \to 0} \frac{-h}{2h(-4 + 3h)} = \lim_{h \to 0} \frac{-1}{2(-4 + 3h)} = \frac{1}{8}.
\end{aligned}
$$

Normals

If a curve C has a tangent line L at point P, then the straight line N through P perpendicular to L is called the **normal** to C at P. If L is horizontal, then N is vertical; if L is vertical, then N is horizontal. If L is neither horizontal nor vertical, then, as shown in Section P.2, the slope of N is the negative reciprocal of the slope of L; that is,

$$
\text{slope of the normal} = \frac{-1}{\text{slope of the tangent}}.
$$

EXAMPLE 6   Find an equation of the normal to $y = x^2$ at $(1, 1)$.

Solution   By Example 1, the tangent to $y = x^2$ at $(1, 1)$ has slope 2. Hence, the normal has slope $-1/2$, and its equation is

$$
y = -\frac{1}{2}(x - 1) + 1 \qquad \text{or} \qquad y = -\frac{x}{2} + \frac{3}{2}.
$$

EXAMPLE 7   Find equations of the straight lines that are tangent and normal to the curve $y = \sqrt{x}$ at the point $(4, 2)$.

Solution   The slope of the tangent at $(4, 2)$ (Figure 2.9) is

$$
\begin{aligned}
m &= \lim_{h \to 0} \frac{\sqrt{4 + h} - 2}{h} = \lim_{h \to 0} \frac{(\sqrt{4 + h} - 2)(\sqrt{4 + h} + 2)}{h(\sqrt{4 + h} + 2)} \\
&= \lim_{h \to 0} \frac{4 + h - 4}{h(\sqrt{4 + h} + 2)} \\
&= \lim_{h \to 0} \frac{1}{\sqrt{4 + h} + 2} = \frac{1}{4}.
\end{aligned}
$$

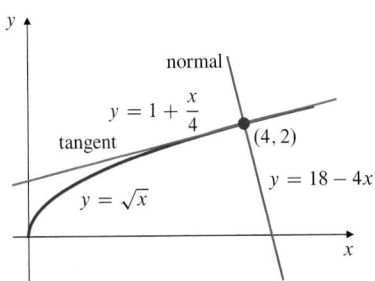

Figure 2.9   The tangent (blue) and normal (green) to $y = \sqrt{x}$ at $(4, 2)$

The tangent line has equation

$$
y = \frac{1}{4}(x - 4) + 2 \qquad \text{or} \qquad x - 4y + 4 = 0,
$$

and the normal has slope -4 and, therefore, equation

$$
y = -4(x - 4) + 2 \qquad \text{or} \qquad y = -4x + 18.
$$

EXERCISES 2.1

In Exercises 1–12, find an equation of the straight line tangent to the given curve at the point indicated.

1. $y = 3x - 1$ at $(1, 2)$

2. $y = x/2$ at $(a, a/2)$

3. $y = 2x^2 - 5$ at $(2, 3)$

4. $y = 6 - x - x^2$ at $x = -2$

5. $y = x^3 + 8$ at $x = -2$

6. $y = \dfrac{1}{x^2 + 1}$ at $(0, 1)$

7. $y = \sqrt{x + 1}$ at $x = 3$

8. $y = \dfrac{1}{\sqrt{x}}$ at $x = 9$

9. $y = \dfrac{2x}{x + 2}$ at $x = 2$

10. $y = \sqrt{5 - x^2}$ at $x = 1$

11. $y = x^2$ at $x = x_0$

12. $y = \dfrac{1}{x}$ at $\left(a, \dfrac{1}{a}\right)$

Do the graphs of the functions f in Exercises 13–17 have tangent lines at the given points? If yes, what is the tangent line?

13. $f(x) = \sqrt{|x|}$ at $x = 0$

14. $f(x) = (x - 1)^{4/3}$ at $x = 1$

15. $f(x) = (x + 2)^{3/5}$ at $x = -2$

16. $f(x) = |x^2 - 1|$ at $x = 1$

17. $f(x) = \begin{cases} \sqrt{x} & \text{if } x \geq 0 \\ -\sqrt{-x} & \text{if } x < 0 \end{cases}$ at $x = 0$

18. Find the slope of the curve $y = x^2 - 1$ at the point $x = x_0$. What is the equation of the tangent line to $y = x^2 - 1$ that has slope -3?

19. (a) Find the slope of $y = x^3$ at the point $x = a$.

   (b) Find the equations of the straight lines having slope 3 that are tangent to $y = x^3$.

20. Find all points on the curve $y = x^3 - 3x$ where the tangent line is parallel to the x-axis.

21. Find all points on the curve $y = x^3 - x + 1$ where the tangent line is parallel to the line $y = 2x + 5$.

22. Find all points on the curve $y = 1/x$ where the tangent line is perpendicular to the line $y = 4x - 3$.

23. For what value of the constant k is the line $x + y = k$ normal to the curve $y = x^2$?

24. For what value of the constant k do the curves $y = kx^2$ and $y = k(x - 2)^2$ intersect at right angles? *Hint:* Where do the curves intersect? What are their slopes there?

Use a graphics utility to plot the following curves. Where does the curve have a horizontal tangent? Does the curve fail to have a tangent line anywhere?

25. $y = x^3(5 - x)^2$

26. $y = 2x^3 - 3x^2 - 12x + 1$

27. $y = |x^2 - 1| - x$

28. $y = |x + 1| - |x - 1|$

29. $y = (x^2 - 1)^{1/3}$

30. $y = ((x^2 - 1)^2)^{1/3}$

31. If line L is tangent to curve C at point P, then the smaller angle between L and the secant line PQ joining P to another point Q on C approaches 0 as Q approaches P along C. Is the converse true: if the angle between PQ and line L (which passes through P) approaches 0, must L be tangent to C?

32. Let $P(x)$ be a polynomial. If a is a real number, then $P(x)$ can be expressed in the form

$$P(x) = a_0 + a_1(x - a) + a_2(x - a)^2 + \cdots + a_n(x - a)^n$$

for some $n \geq 0$. If $\ell(x) = m(x - a) + b$, show that the straight line $y = \ell(x)$ is tangent to the graph of $y = P(x)$ at $x = a$ provided $P(x) - \ell(x) = (x - a)^2 Q(x)$, where $Q(x)$ is a polynomial.

2.2   The Derivative

A straight line has the property that its slope is the same at all points. For any other graph, however, the slope may vary from point to point. Thus, the slope of the graph of $y = f(x)$ at the point x is itself a function of x. At any point x where the graph has a finite slope, we say that f is differentiable, and we call the slope the derivative of f. The derivative is therefore the limit of the Newton quotient.

DEFINITION

4

The **derivative** of a function f is another function f' defined by

$$f'(x) = \lim_{h \to 0} \frac{f(x + h) - f(x)}{h}$$

at all points x for which the limit exists (i.e., is a finite real number). If $f'(x)$ exists, we say that f is **differentiable** at x.

The domain of the derivative f' (read "f prime") is the set of numbers x in the domain of f where the graph of f has a *nonvertical* tangent line, and the value $f'(x_0)$ of f' at such a point x_0 is the slope of the tangent line to $y = f(x)$ there. Thus, the equation of the tangent line to $y = f(x)$ at $(x_0, f(x_0))$ is

$$y = f(x_0) + f'(x_0)(x - x_0).$$

The domain $\mathcal{D}(f')$ of f' may be smaller than the domain $\mathcal{D}(f)$ of f because it contains only those points in $\mathcal{D}(f)$ at which f is differentiable. Values of x in $\mathcal{D}(f)$ where f is not differentiable and that are not endpoints of $\mathcal{D}(f)$ are **singular points** of f.

Remark The value of the derivative of f at a particular point x_0 can be expressed as a limit in either of two ways:

$$f'(x_0) = \lim_{h \to 0} \frac{f(x_0 + h) - f(x_0)}{h} = \lim_{x \to x_0} \frac{f(x) - f(x_0)}{x - x_0}.$$

In the second limit $x_0 + h$ is replaced by x, so that $h = x - x_0$ and $h \to 0$ is equivalent to $x \to x_0$.

The process of calculating the derivative f' of a given function f is called **differentiation**. The graph of f' can often be sketched directly from that of f by visualizing slopes, a procedure called **graphical differentiation**. In Figure 2.10 the graphs of f' and g' were obtained by measuring the slopes at the corresponding points in the graphs of f and g lying above them. The height of the graph $y = f'(x)$ at x is the slope of the graph of $y = f(x)$ at x. Note that -1 and 1 are singular points of f. Although $f(-1)$ and $f(1)$ are defined, $f'(-1)$ and $f'(1)$ are not defined; the graph of f has no tangent at -1 or at 1.

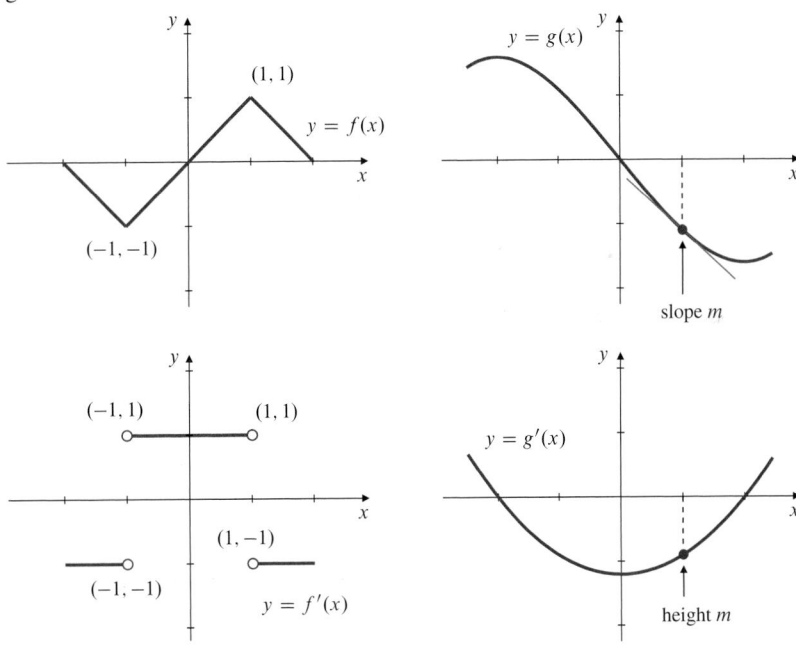

Figure 2.10   Graphical differentiation

A function is differentiable on a set S if it is differentiable at every point x in S. Typically, the functions we encounter are defined on intervals or unions of intervals. If f is defined on a closed interval $[a, b]$, Definition 4 does not allow for the existence of a derivative at the endpoints $x = a$ or $x = b$. (Why?) As we did for continuity in Section 1.4, we extend the definition to allow for a **right derivative** at $x = a$ and a **left derivative** at $x = b$:

$$f'_+(a) = \lim_{h \to 0+} \frac{f(a + h) - f(a)}{h}, \qquad f'_-(b) = \lim_{h \to 0-} \frac{f(b + h) - f(b)}{h}.$$

We now say that f is **differentiable** on $[a, b]$ if $f'(x)$ exists for all x in (a, b) and $f'_+(a)$ and $f'_-(b)$ both exist.

Some Important Derivatives

We now give several examples of the calculation of derivatives algebraically from the definition of derivative. Some of these are the basic building blocks from which more complicated derivatives can be calculated later. They are collected in Table 1 later in this section and should be memorized.

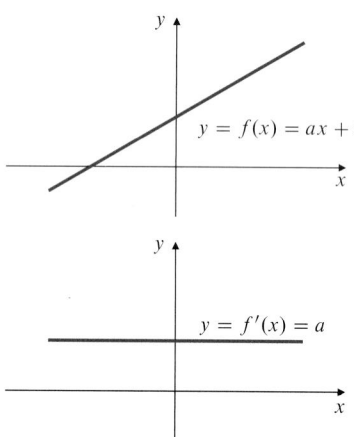

Figure 2.11 The derivative of the linear function $f(x) = ax + b$ is the constant function $f'(x) = a$

EXAMPLE 1 **(The derivative of a linear function)** Show that if $f(x) = ax + b$, then $f'(x) = a$.

Solution The result is apparent from the graph of f (Figure 2.11), but we will do the calculation using the definition:

$$
\begin{aligned}
f'(x) &= \lim_{h \to 0} \frac{f(x+h) - f(x)}{h} \\
&= \lim_{h \to 0} \frac{a(x+h) + b - (ax+b)}{h} \\
&= \lim_{h \to 0} \frac{ah}{h} = a.
\end{aligned}
$$

An important special case of Example 1 says that the derivative of a constant function is the zero function:

> If $g(x) = c$ (constant), then $g'(x) = 0$.

EXAMPLE 2 Use the definition of the derivative to calculate the derivatives of the functions

$$\text{(a) } f(x) = x^2, \quad \text{(b) } g(x) = \frac{1}{x}, \quad \text{and} \quad \text{(c) } k(x) = \sqrt{x}.$$

Solution Figures 2.12–2.14 show the graphs of these functions and their derivatives.

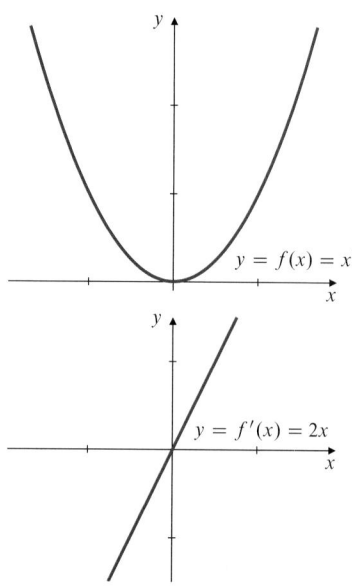

Figure 2.12 The derivative of $f(x) = x^2$ is $f'(x) = 2x$

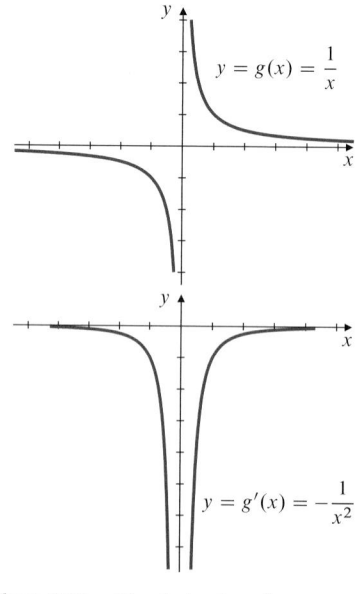

Figure 2.13 The derivative of $g(x) = 1/x$ is $g'(x) = -1/x^2$

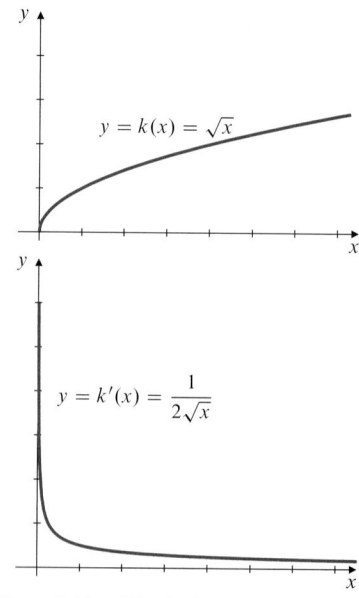

Figure 2.14 The derivative of $k(x) = \sqrt{x}$ is $k'(x) = 1/(2\sqrt{x})$

(a) $f'(x) = \lim_{h \to 0} \dfrac{f(x+h) - f(x)}{h}$

$= \lim_{h \to 0} \dfrac{(x+h)^2 - x^2}{h}$

$= \lim_{h \to 0} \dfrac{2hx + h^2}{h} = \lim_{h \to 0} (2x + h) = 2x.$

(b) $g'(x) = \lim_{h \to 0} \dfrac{g(x+h) - g(x)}{h}$

$= \lim_{h \to 0} \dfrac{\dfrac{1}{x+h} - \dfrac{1}{x}}{h}$

$= \lim_{h \to 0} \dfrac{x - (x+h)}{h(x+h)x} = \lim_{h \to 0} -\dfrac{1}{(x+h)x} = -\dfrac{1}{x^2}.$

(c) $k'(x) = \lim_{h \to 0} \dfrac{k(x+h) - k(x)}{h}$

$= \lim_{h \to 0} \dfrac{\sqrt{x+h} - \sqrt{x}}{h}$

$= \lim_{h \to 0} \dfrac{\sqrt{x+h} - \sqrt{x}}{h} \times \dfrac{\sqrt{x+h} + \sqrt{x}}{\sqrt{x+h} + \sqrt{x}}$

$= \lim_{h \to 0} \dfrac{x+h-x}{h(\sqrt{x+h} + \sqrt{x})} = \lim_{h \to 0} \dfrac{1}{\sqrt{x+h} + \sqrt{x}} = \dfrac{1}{2\sqrt{x}}.$

Note that k is not differentiable at the endpoint $x = 0$.

The three derivative formulas calculated in Example 2 are special cases of the following **General Power Rule**:

If $f(x) = x^r$, then $f'(x) = r\, x^{r-1}$.

This formula, which we will verify in Section 3.3, is valid for *all values of r and x for which x^{r-1} makes sense as a real number.*

EXAMPLE 3 **(Differentiating powers)**

If $f(x) = x^{5/3}$, then $f'(x) = \dfrac{5}{3}x^{(5/3)-1} = \dfrac{5}{3}x^{2/3}$ for all real x.

If $g(t) = \dfrac{1}{\sqrt{t}} = t^{-1/2}$, then $g'(t) = -\dfrac{1}{2}t^{-(1/2)-1} = -\dfrac{1}{2}t^{-3/2}$ for $t > 0$.

Eventually, we will prove all appropriate cases of the General Power Rule. For the time being, here is a proof of the case $r = n$, a positive integer, based on the *factoring of a difference of nth powers*:

$a^n - b^n = (a-b)(a^{n-1} + a^{n-2}b + a^{n-3}b^2 + \cdots + ab^{n-2} + b^{n-1}).$

(Check that this formula is correct by multiplying the two factors on the right-hand side.) If $f(x) = x^n$, $a = x + h$, and $b = x$, then $a - b = h$ and

$f'(x) = \lim_{h \to 0} \dfrac{(x+h)^n - x^n}{h}$

$= \lim_{h \to 0} \dfrac{h\,\overbrace{[(x+h)^{n-1} + (x+h)^{n-2}x + (x+h)^{n-3}x^2 + \cdots + x^{n-1}]}^{n \text{ terms}}}{h}$

$= nx^{n-1}.$

An alternative proof based on the product rule and mathematical induction will be given in Section 2.3. The factorization method used above can also be used to demonstrate the General Power Rule for negative integers, $r = -n$, and reciprocals of integers, $r = 1/n$. (See Exercises 52 and 54 at the end of this section.)

EXAMPLE 4 **(Differentiating the absolute value function)** Verify that:

$$\text{If } f(x) = |x|, \quad \text{then} \quad f'(x) = \frac{x}{|x|} = \text{sgn}\, x.$$

Solution We have

$$f(x) = \begin{cases} x, & \text{if } x \geq 0 \\ -x, & \text{if } x < 0 \end{cases}.$$

Thus, from Example 1 above, $f'(x) = 1$ if $x > 0$ and $f'(x) = -1$ if $x < 0$. Also, Example 4 of Section 2.1 shows that f is not differentiable at $x = 0$, which is a singular point of f. Therefore (see Figure 2.15),

$$f'(x) = \begin{cases} 1, & \text{if } x > 0 \\ -1, & \text{if } x < 0 \end{cases} = \frac{x}{|x|} = \text{sgn}\, x.$$

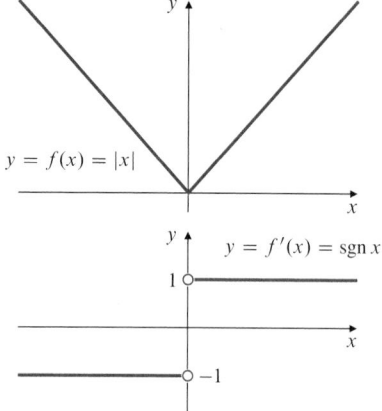

$y = f(x) = |x|$

$y = f'(x) = \text{sgn}\, x$

Figure 2.15 The derivative of $|x|$ is $\text{sgn}\, x = x/|x|$

Table 1 lists the elementary derivatives calculated above. Beginning in Section 2.3 we will develop general rules for calculating the derivatives of functions obtained by combining simpler functions. Thereafter, we will seldom have to revert to the definition of the derivative and to the calculation of limits to evaluate derivatives. It is important, therefore, to remember the derivatives of some elementary functions. Memorize those in Table 1.

Table 1. Some elementary functions and their derivatives

$f(x)$	$f'(x)$
c (constant)	0
x	1
x^2	$2x$
$\dfrac{1}{x}$	$-\dfrac{1}{x^2}$ $(x \neq 0)$
$\sqrt{x}$	$\dfrac{1}{2\sqrt{x}}$ $(x > 0)$
x^r	$r\,x^{r-1}$ $(x^{r-1}$ real$)$
$|x|$	$\dfrac{x}{|x|} = \text{sgn}\, x$

Leibniz Notation

Because functions can be written in different ways, it is useful to have more than one notation for derivatives. If $y = f(x)$, we can use the dependent variable y to represent the function, and we can denote the derivative of the function with respect to x in any of the following ways:

$$D_x y = y' = \frac{dy}{dx} = \frac{d}{dx} f(x) = f'(x) = D_x f(x) = Df(x).$$

(In the forms using "D_x," we can omit the subscript x if the variable of differentiation is obvious.) Often the most convenient way of referring to the derivative of a function given explicitly as an expression in the variable x is to write $\frac{d}{dx}$ in front of that expression. The symbol $\frac{d}{dx}$ is a *differential operator* and should be read "the derivative with respect to x of ..." For example,

$$\frac{d}{dx} x^2 = 2x \quad \text{(the derivative with respect to } x \text{ of } x^2 \text{ is } 2x)$$

$$\frac{d}{dx}\sqrt{x} = \frac{1}{2\sqrt{x}}$$

$$\frac{d}{dt} t^{100} = 100\, t^{99}$$

if $y = u^3$, then $\dfrac{dy}{du} = 3u^2$.

The value of the derivative of a function at a particular number x_0 in its domain can also be expressed in several ways:

$$D_x y\Big|_{x=x_0} = y'\Big|_{x=x_0} = \frac{dy}{dx}\Big|_{x=x_0} = \frac{d}{dx} f(x)\Big|_{x=x_0} = f'(x_0) = D_x f(x_0).$$

Do not confuse the expressions

$$\frac{d}{dx} f(x) \text{ and } \frac{d}{dx} f(x)\Big|_{x=x_0} .$$

The first expression represents a *function*, $f'(x)$. The second represents a *number*, $f'(x_0)$.

The symbol $\Big|_{x=x_0}$ is called an **evaluation symbol**. It signifies that the expression preceding it should be evaluated at $x = x_0$. Thus,

$$\frac{d}{dx} x^4\Big|_{x=-1} = 4x^3\Big|_{x=-1} = 4(-1)^3 = -4.$$

Here is another example in which a derivative is computed from the definition, this time for a somewhat more complicated function.

EXAMPLE 5   Use the definition of derivative to calculate $\dfrac{d}{dx}\left(\dfrac{x}{x^2+1}\right)\Big|_{x=2}$.

Solution   We could calculate $\dfrac{d}{dx}\left(\dfrac{x}{x^2+1}\right)$ and then substitute $x = 2$, but it is easier to put $x = 2$ in the expression for the Newton quotient before taking the limit:

$$\frac{d}{dx}\left(\frac{x}{x^2+1}\right)\Big|_{x=2} = \lim_{h \to 0} \frac{\dfrac{2+h}{(2+h)^2+1} - \dfrac{2}{2^2+1}}{h}$$

$$= \lim_{h \to 0} \frac{\dfrac{2+h}{5+4h+h^2} - \dfrac{2}{5}}{h}$$

$$= \lim_{h \to 0} \frac{5(2+h) - 2(5+4h+h^2)}{5(5+4h+h^2)h}$$

$$= \lim_{h \to 0} \frac{-3h - 2h^2}{5(5+4h+h^2)h}$$

$$= \lim_{h \to 0} \frac{-3 - 2h}{5(5+4h+h^2)} = -\frac{3}{25}.$$

The notations dy/dx and $\frac{d}{dx} f(x)$ are called **Leibniz notations** for the derivative, after Gottfried Wilhelm Leibniz (1646–1716), one of the creators of calculus, who used such notations. The main ideas of calculus were developed independently by Leibniz and Isaac Newton (1642–1727); Newton used notations similar to the prime (y') notations we use here.

The Leibniz notation is suggested by the definition of derivative. The Newton quotient $[f(x + h) - f(x)]/h$, whose limit we take to find the derivative dy/dx, can be written in the form $\Delta y/\Delta x$, where $\Delta y = f(x + h) - f(x)$ is the increment in y, and $\Delta x = (x + h) - x = h$ is the corresponding increment in x as we pass from the point $(x, f(x))$ to the point $(x + h, f(x + h))$ on the graph of f. (See Figure 2.16.) Δ is the uppercase Greek letter Delta. Using symbols:

$$\frac{dy}{dx} = \lim_{\Delta x \to 0} \frac{\Delta y}{\Delta x}.$$

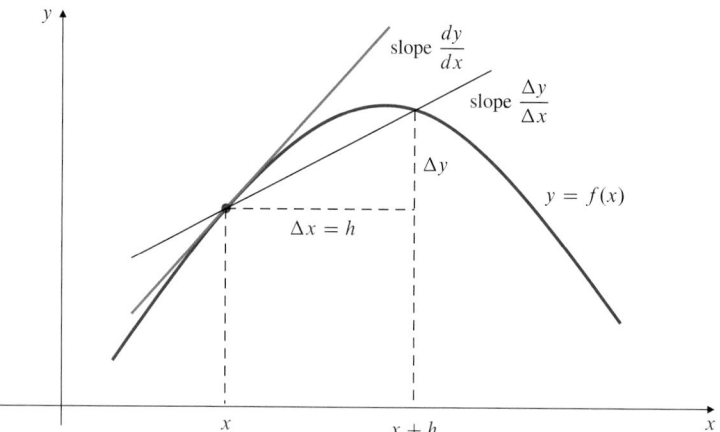

Figure 2.16 $\dfrac{dy}{dx} = \lim\limits_{\Delta x \to 0} \dfrac{\Delta y}{\Delta x}$

Differentials

The Newton quotient $\Delta y/\Delta x$ is actually the quotient of two quantities, Δy and Δx. It is not at all clear, however, that the derivative dy/dx, the limit of $\Delta y/\Delta x$ as Δx approaches zero, can be regarded as a quotient. If y is a continuous function of x, then Δy approaches zero when Δx approaches zero, so dy/dx appears to be the meaningless quantity $0/0$. Nevertheless, it is sometimes useful to be able to refer to quantities dy and dx in such a way that their quotient is the derivative dy/dx. We can justify this by regarding dx as a new *independent* variable (called **the differential of x**) and defining a new *dependent* variable dy (**the differential of y**) as a function of x and dx by

$$dy = \frac{dy}{dx}\, dx = f'(x)\, dx.$$

For example, if $y = x^2$, we can write $dy = 2x\, dx$ to mean the same thing as $dy/dx = 2x$. Similarly, if $f(x) = 1/x$, we can write $df(x) = -(1/x^2)\, dx$ as the equivalent differential form of the assertion that $(d/dx)f(x) = f'(x) = -1/x^2$. This *differential notation* is useful in applications (see Sections 2.7 and 12.6), and especially for the interpretation and manipulation of integrals beginning in Chapter 5.

Note that, defined as above, differentials are merely variables that may or may not be small in absolute value. The differentials dy and dx were originally regarded (by Leibniz and his successors) as "infinitesimals" (infinitely small but nonzero) quantities whose quotient dy/dx gave the slope of the tangent line (a secant line meeting the graph of $y = f(x)$ at two points infinitely close together). It can be shown that such "infinitesimal" quantities cannot exist (as real numbers). It is possible to extend the number system to contain infinitesimals and use these to develop calculus, but we will not consider this approach here.

Derivatives Have the Intermediate-Value Property

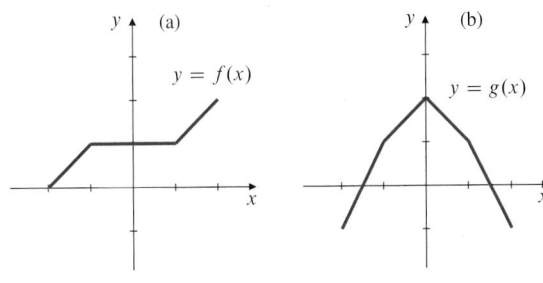

Is a function f defined on an interval I necessarily the derivative of some other function defined on I? The answer is no; some functions are derivatives and some are not. Although a derivative need not be a continuous function (see Exercise 28 in Section 2.8), it must, like a continuous function, have the intermediate-value property: on an interval $[a, b]$, a derivative $f'(x)$ takes on every value between $f'(a)$ and $f'(b)$. (See Exercise 29 in Section 2.8 for a proof of this fact.) An everywhere-defined step function such as the Heaviside function $H(x)$ considered in Example 1 in Section 1.4 (see Figure 2.17) does not have this property on, say, the interval $[-1, 1]$, so cannot be the derivative of a function on that interval. This argument does not apply to the signum function, which is the derivative of the absolute value function on any interval where it is defined. (See Example 4.) Such an interval cannot contain the origin, as sgn (x) is not defined at $x = 0$.

If $g(x)$ is continuous on an interval I, then $g(x) = f'(x)$ for some function f that is differentiable on I. We will discuss this fact further in Chapter 5 and prove it in Appendix IV.

Figure 2.17 This function is not a derivative on $[-1, 1]$; it does not have the intermediate-value property.

EXERCISES 2.2

Make rough sketches of the graphs of the derivatives of the functions in Exercises 1–4.

1. The function f graphed in Figure 2.18(a).

2. The function g graphed in Figure 2.18(b).

3. The function h graphed in Figure 2.18(c).

4. The function k graphed in Figure 2.18(d).

5. Where is the function f graphed in Figure 2.18(a) differentiable?

6. Where is the function g graphed in Figure 2.18(b) differentiable?

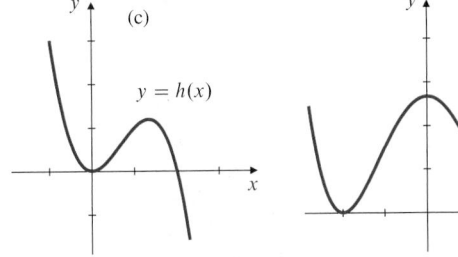

Figure 2.18

Use a graphics utility with differentiation capabilities to plot the graphs of the following functions and their derivatives. Observe the relationships between the graph of y and that of y' in each case. What features of the graph of y can you infer from the graph of y'?

7. $y = 3x - x^2 - 1$

8. $y = x^3 - 3x^2 + 2x + 1$

9. $y = |x^3 - x|$

10. $y = |x^2 - 1| - |x^2 - 4|$

In Exercises 11–24, (a) calculate the derivative of the given function directly from the definition of derivative, and (b) express the result of (a) using differentials.

11. $y = x^2 - 3x$

12. $f(x) = 1 + 4x - 5x^2$

13. $f(x) = x^3$

14. $s = \dfrac{1}{3 + 4t}$

15. $g(x) = \dfrac{2 - x}{2 + x}$

16. $y = \dfrac{1}{3}x^3 - x$

17. $F(t) = \sqrt{2t + 1}$

18. $f(x) = \dfrac{3}{4}\sqrt{2 - x}$

19. $y = x + \dfrac{1}{x}$

20. $z = \dfrac{s}{1 + s}$

21. $F(x) = \dfrac{1}{\sqrt{1 + x^2}}$

22. $y = \dfrac{1}{x^2}$

23. $y = \dfrac{1}{\sqrt{1 + x}}$

24. $f(t) = \dfrac{t^2 - 3}{t^2 + 3}$

25. How should the function $f(x) = x\,\text{sgn}\,x$ be defined at $x = 0$ so that it is continuous there? Is it then differentiable there?

26. How should the function $g(x) = x^2\text{sgn}\,x$ be defined at $x = 0$ so that it is continuous there? Is it then differentiable there?

27. Where does $h(x) = |x^2 + 3x + 2|$ fail to be differentiable?

28. Using a calculator, find the slope of the secant line to $y = x^3 - 2x$ passing through the points corresponding to $x = 1$ and $x = 1 + \Delta x$, for several values of Δx of decreasing size, say $\Delta x = \pm 0.1, \pm 0.01, \pm 0.001, \pm 0.0001$. (Make a table.) Also, calculate $\dfrac{d}{dx}\left(x^3 - 2x\right)\Big|_{x=1}$ using the definition of derivative.

29. Repeat Exercise 28 for the function $f(x) = \dfrac{1}{x}$ and the points $x = 2$ and $x = 2 + \Delta x$.

Using the definition of derivative, find equations for the tangent lines to the curves in Exercises 30–33 at the points indicated.

30. $y = 5 + 4x - x^2$ at the point where $x = 2$

31. $y = \sqrt{x + 6}$ at the point $(3, 3)$

32. $y = \dfrac{t}{t^2 - 2}$ at the point where $t = -2$

33. $y = \dfrac{2}{t^2 + t}$ at the point where $t = a$

Calculate the derivatives of the functions in Exercises 34–39 using the General Power Rule. Where is each derivative valid?

34. $f(x) = x^{-17}$

35. $g(t) = t^{22}$

36. $y = x^{1/3}$

37. $y = x^{-1/3}$

38. $t^{-2.25}$

39. $s^{119/4}$

In Exercises 40–50, you may use the formulas for derivatives established in this section.

40. Calculate $\dfrac{d}{ds}\sqrt{s}\,\Big|_{s=9}$.

41. Find $F'(\frac{1}{4})$ if $F(x) = \dfrac{1}{x}$.

42. Find $f'(8)$ if $f(x) = x^{-2/3}$.

43. Find $dy/dt\,\Big|_{t=4}$ if $y = t^{1/4}$.

44. Find an equation of the straight line tangent to the curve $y = \sqrt{x}$ at $x = x_0$.

45. Find an equation of the straight line normal to the curve $y = 1/x$ at the point where $x = a$.

46. Show that the curve $y = x^2$ and the straight line $x + 4y = 18$ intersect at right angles at one of their two intersection points. *Hint:* Find the product of their slopes at their intersection points.

47. There are two distinct straight lines that pass through the point $(1, -3)$ and are tangent to the curve $y = x^2$. Find their equations. *Hint:* Draw a sketch. The points of tangency are not given; let them be denoted (a, a^2).

48. Find equations of two straight lines that have slope -2 and are tangent to the graph of $y = 1/x$.

49. Find the slope of a straight line that passes through the point $(-2, 0)$ and is tangent to the curve $y = \sqrt{x}$.

❷ 50. Show that there are two distinct tangent lines to the curve $y = x^2$ passing through the point (a, b) provided $b < a^2$. How many tangent lines to $y = x^2$ pass through (a, b) if $b = a^2$? if $b > a^2$?

❷ 51. Show that the derivative of an odd differentiable function is even and that the derivative of an even differentiable function is odd.

❗ 52. Prove the case $r = -n$ (n is a positive integer) of the General Power Rule; that is, prove that

$$\frac{d}{dx}x^{-n} = -n\,x^{-n-1}.$$

Use the factorization of a difference of nth powers given in this section.

❗ 53. Use the factoring of a difference of cubes:

$$a^3 - b^3 = (a - b)(a^2 + ab + b^2),$$

to help you calculate the derivative of $f(x) = x^{1/3}$ directly from the definition of derivative.

❗ 54. Prove the General Power Rule for $\frac{d}{dx}x^r$, where $r = 1/n$, n being a positive integer. (*Hint:*

$$\frac{d}{dx}x^{1/n} = \lim_{h \to 0}\frac{(x+h)^{1/n} - x^{1/n}}{h}$$

$$= \lim_{h \to 0}\frac{(x+h)^{1/n} - x^{1/n}}{((x+h)^{1/n})^n - (x^{1/n})^n}.$$

Apply the factorization of the difference of nth powers to the denominator of the latter quotient.)

55. Give a proof of the power rule $\frac{d}{dx}x^n = nx^{n-1}$ for positive integers n using the Binomial Theorem:

$$(x+h)^n = x^n + \frac{n}{1}x^{n-1}h + \frac{n(n-1)}{1\times 2}x^{n-2}h^2$$

$$+ \frac{n(n-1)(n-2)}{1\times 2 \times 3}x^{n-3}h^3 + \cdots + h^n.$$

❗ 56. Use right and left derivatives, $f'_+(a)$ and $f'_-(a)$, to define the concept of a half-line starting at $(a, f(a))$ being a right or left tangent to the graph of f at $x = a$. Show that the graph has a tangent line at $x = a$ if and only if it has right and left tangents that are opposite halves of the same straight line. What are the left and right tangents to the graphs of $y = x^{1/3}$, $y = x^{2/3}$, and $y = |x|$ at $x = 0$?

2.3   Differentiation Rules

If every derivative had to be calculated directly from the definition of derivative as in the examples of Section 2.2, calculus would indeed be a painful subject. Fortunately, there is an easier way. We will develop several general *differentiation rules* that enable us to calculate the derivatives of complicated combinations of functions easily if we already know the derivatives of the elementary functions from which they are constructed. For instance, we will be able to find the derivative of $\dfrac{x^2}{\sqrt{x^2 + 1}}$ if we know the derivatives of x^2 and $\sqrt{x}$. The rules we develop in this section tell us how to differentiate sums, constant multiples, products, and quotients of functions whose derivatives we already know. In Section 2.4 we will learn how to differentiate composite functions.

Before developing these differentiation rules we need to establish one obvious but very important theorem which states, roughly, that the graph of a function cannot possibly have a break at a point where it is smooth.

THEOREM

1

Differentiability implies continuity

If f is differentiable at x, then f is continuous at x.

PROOF Since f is differentiable at x, we know that

$$\lim_{h \to 0} \frac{f(x+h) - f(x)}{h} = f'(x)$$

exists. Using the limit rules (Theorem 2 of Section 1.2), we have

$$\lim_{h \to 0} \left(f(x+h) - f(x) \right) = \lim_{h \to 0} \left(\frac{f(x+h) - f(x)}{h} \right) (h) = \left(f'(x) \right)(0) = 0.$$

This is equivalent to $\lim_{h \to 0} f(x+h) = f(x)$, which says that f is continuous at x.

Sums and Constant Multiples

The derivative of a sum (or difference) of functions is the sum (or difference) of the derivatives of those functions. The derivative of a constant multiple of a function is the same constant multiple of the derivative of the function.

THEOREM

2

Differentiation rules for sums, differences, and constant multiples

If functions f and g are differentiable at x, and if C is a constant, then the functions $f + g$, $f - g$, and Cf are all differentiable at x and

$$(f + g)'(x) = f'(x) + g'(x),$$
$$(f - g)'(x) = f'(x) - g'(x),$$
$$(Cf)'(x) = Cf'(x).$$

PROOF The proofs of all three assertions are straightforward, using the corresponding limit rules from Theorem 2 of Section 1.2. For the sum, we have

$$(f + g)'(x) = \lim_{h \to 0} \frac{(f+g)(x+h) - (f+g)(x)}{h}$$
$$= \lim_{h \to 0} \frac{(f(x+h) + g(x+h)) - (f(x) + g(x))}{h}$$
$$= \lim_{h \to 0} \left(\frac{f(x+h) - f(x)}{h} + \frac{g(x+h) - g(x)}{h} \right)$$
$$= f'(x) + g'(x),$$

because the limit of a sum is the sum of the limits. The proof for the difference $f - g$ is similar. For the constant multiple, we have

$$(Cf)'(x) = \lim_{h \to 0} \frac{Cf(x+h) - Cf(x)}{h}$$
$$= C \lim_{h \to 0} \frac{f(x+h) - f(x)}{h} = Cf'(x).$$

The rule for differentiating sums extends to sums of any finite number of terms

$$(f_1 + f_2 + \cdots + f_n)' = f_1' + f_2' + \cdots + f_n'. \qquad (*)$$

To see this we can use a technique called **mathematical induction**. (See the note in the margin.) Theorem 2 shows that the case $n = 2$ is true; this is STEP 1. For STEP 2, we must show that *if* the formula $(*)$ holds for some integer $n = k \geq 2$, *then* it must also hold for $n = k + 1$. Therefore, *assume* that

$$(f_1 + f_2 + \cdots + f_k)' = f_1' + f_2' + \cdots + f_k'.$$

Then we have

$$(f_1 + f_2 + \cdots + f_k + f_{k+1})'$$
$$= \Big(\underbrace{(f_1 + f_2 + \cdots + f_k)}_{\text{Let this function be } f} + f_{k+1}\Big)'$$
$$= (f + f_{k+1})' \qquad \text{(Now use the known case } n = 2.)$$
$$= f' + f_{k+1}'$$
$$= f_1' + f_2' + \cdots + f_k' + f_{k+1}'.$$

With both steps verified, we can claim that $(*)$ holds for any $n \geq 2$ *by induction*. In particular, therefore, the derivative of any polynomial is the sum of the derivatives of its terms.

Mathematical Induction

Mathematical induction is a technique for proving that a statement about an integer n is true for every integer n greater than or equal to some starting integer n_0. The proof requires us to carry out two steps:

STEP 1. Prove that the statement is true for $n = n_0$.

STEP 2. Prove that if the statement is true for some integer $n = k$, where $k \geq n_0$, then it is also true for the next larger integer, $n = k + 1$.

Step 2 prevents there from being a smallest integer greater than n_0 for which the statement is false. Being true for n_0, the statement must therefore be true for all larger integers.

EXAMPLE 1   Calculate the derivatives of the functions

(a) $2x^3 - 5x^2 + 4x + 7$,   (b) $f(x) = 5\sqrt{x} + \dfrac{3}{x} - 18$,   (c) $y = \dfrac{1}{7}t^4 - 3t^{7/3}$.

Solution   Each of these functions is a sum of constant multiples of functions that we already know how to differentiate.

(a) $\dfrac{d}{dx}(2x^3 - 5x^2 + 4x + 7) = 2(3x^2) - 5(2x) + 4(1) + 0 = 6x^2 - 10x + 4.$

(b) $f'(x) = 5\left(\dfrac{1}{2\sqrt{x}}\right) + 3\left(-\dfrac{1}{x^2}\right) - 0 = \dfrac{5}{2\sqrt{x}} - \dfrac{3}{x^2}.$

(c) $\dfrac{dy}{dt} = \dfrac{1}{7}(4t^3) - 3\left(\dfrac{7}{3}t^{4/3}\right) = \dfrac{4}{7}t^3 - 7t^{4/3}.$

EXAMPLE 2   Find an equation of the tangent to the curve $y = \dfrac{3x^3 - 4}{x}$ at the point on the curve where $x = -2$.

Solution   If $x = -2$, then $y = 14$. The slope of the curve at $(-2, 14)$ is

$$\left.\frac{dy}{dx}\right|_{x=-2} = \left.\frac{d}{dx}\left(3x^2 - \frac{4}{x}\right)\right|_{x=-2} = \left.\left(6x + \frac{4}{x^2}\right)\right|_{x=-2} = -11.$$

An equation of the tangent line is $y = 14 - 11(x + 2)$, or $y = -11x - 8$.

The Product Rule

The rule for differentiating a product of functions is a little more complicated than that for sums. It is *not* true that the derivative of a product is the product of the derivatives.

THEOREM 3

The Product Rule

If functions f and g are differentiable at x, then their product fg is also differentiable at x, and

$$(fg)'(x) = f'(x)g(x) + f(x)g'(x).$$

PROOF We set up the Newton quotient for fg and then add 0 to the numerator in a way that enables us to involve the Newton quotients for f and g separately:

$$(fg)'(x) = \lim_{h \to 0} \frac{f(x+h)g(x+h) - f(x)g(x)}{h}$$

$$= \lim_{h \to 0} \frac{f(x+h)g(x+h) - f(x)g(x+h) + f(x)g(x+h) - f(x)g(x)}{h}$$

$$= \lim_{h \to 0} \left(\frac{f(x+h) - f(x)}{h} g(x+h) + f(x) \frac{g(x+h) - g(x)}{h} \right)$$

$$= f'(x)g(x) + f(x)g'(x).$$

To get the last line, we have used the fact that f and g are differentiable and the fact that g is therefore continuous (Theorem 1), as well as limit rules from Theorem 2 of Section 1.2. A graphical proof of the Product Rule is suggested by Figure 2.19.

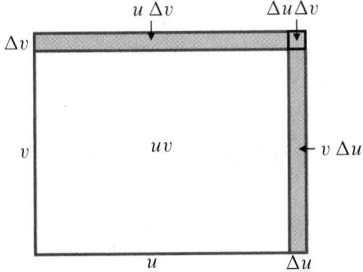

Figure 2.19

A graphical proof of the Product Rule

Here $u = f(x)$ and $v = g(x)$, so that the rectangular area uv represents $f(x)g(x)$. If x changes by an amount Δx, the corresponding increments in u and v are Δu and Δv. The change in the area of the rectangle is

$$\Delta(uv)$$

$$= (u + \Delta u)(v + \Delta v) - uv$$

$$= (\Delta u)v + u(\Delta v) + (\Delta u)(\Delta v),$$

the sum of the three shaded areas. Dividing by Δx and taking the limit as $\Delta x \to 0$, we get

$$\frac{d}{dx}(uv) = \left(\frac{du}{dx} \right) v + u \left(\frac{dv}{dx} \right),$$

since

$$\lim_{\Delta x \to 0} \frac{\Delta u}{\Delta x} \Delta v = \frac{du}{dx} \times 0 = 0.$$

EXAMPLE 3 Find the derivative of $(x^2 + 1)(x^3 + 4)$ using and without using the Product Rule.

Solution Using the Product Rule with $f(x) = x^2 + 1$ and $g(x) = x^3 + 4$, we calculate

$$\frac{d}{dx}\left((x^2 + 1)(x^3 + 4)\right) = 2x(x^3 + 4) + (x^2 + 1)(3x^2) = 5x^4 + 3x^2 + 8x.$$

On the other hand, we can calculate the derivative by first multiplying the two binomials and then differentiating the resulting polynomial:

$$\frac{d}{dx}\left((x^2 + 1)(x^3 + 4)\right) = \frac{d}{dx}(x^5 + x^3 + 4x^2 + 4) = 5x^4 + 3x^2 + 8x.$$

EXAMPLE 4 Find $\dfrac{dy}{dx}$ if $y = \left(2\sqrt{x} + \dfrac{3}{x}\right)\left(3\sqrt{x} - \dfrac{2}{x}\right)$.

Solution Applying the Product Rule with f and g being the two functions enclosed in the large parentheses, we obtain

$$\frac{dy}{dx} = \left(\frac{1}{\sqrt{x}} - \frac{3}{x^2} \right)\left(3\sqrt{x} - \frac{2}{x} \right) + \left(2\sqrt{x} + \frac{3}{x} \right)\left(\frac{3}{2\sqrt{x}} + \frac{2}{x^2} \right)$$

$$= 6 - \frac{5}{2x^{3/2}} + \frac{12}{x^3}.$$

EXAMPLE 5 Let $y = uv$ be the product of the functions u and v. Find $y'(2)$ if $u(2) = 2$, $u'(2) = -5$, $v(2) = 1$, and $v'(2) = 3$.

Solution From the Product Rule we have

$$y' = (uv)' = u'v + uv'.$$

Therefore,

$$y'(2) = u'(2)v(2) + u(2)v'(2) = (-5)(1) + (2)(3) = -5 + 6 = 1.$$

EXAMPLE 6 Use mathematical induction to verify the formula $\dfrac{d}{dx}x^n = n x^{n-1}$ for all positive integers n.

Solution For $n = 1$ the formula says that $\frac{d}{dx}x^1 = 1 = 1x^0$, so the formula is true in this case. We must show that if the formula is true for $n = k \geq 1$, then it is also true for $n = k + 1$. Therefore, assume that

$$\frac{d}{dx}x^k = kx^{k-1}.$$

Using the Product Rule we calculate

$$\frac{d}{dx}x^{k+1} = \frac{d}{dx}\left(x^k x\right) = (kx^{k-1})(x)+(x^k)(1) = (k+1)x^k = (k+1)x^{(k+1)-1}.$$

Thus, the formula is true for $n = k + 1$ also. The formula is true for all integers $n \geq 1$ *by induction.*

The Product Rule can be extended to products of any number of factors; for instance,

$$(fgh)'(x) = f'(x)(gh)(x) + f(x)(gh)'(x)$$
$$= f'(x)g(x)h(x) + f(x)g'(x)h(x) + f(x)g(x)h'(x).$$

In general, the derivative of a product of n functions will have n terms; each term will be the same product but with one of the factors replaced by its derivative:

$$(f_1 f_2 f_3 \cdots f_n)' = f_1' f_2 f_3 \cdots f_n + f_1 f_2' f_3 \cdots f_n + \cdots + f_1 f_2 f_3 \cdots f_n'.$$

This can be proved by mathematical induction. See Exercise 54 at the end of this section.

The Reciprocal Rule

THEOREM

4

The Reciprocal Rule

If f is differentiable at x and $f(x) \neq 0$, then $1/f$ is differentiable at x, and

$$\left(\frac{1}{f}\right)'(x) = \frac{-f'(x)}{(f(x))^2}.$$

PROOF Using the definition of the derivative, we calculate

$$\frac{d}{dx}\frac{1}{f(x)} = \lim_{h \to 0} \frac{\dfrac{1}{f(x + h)} - \dfrac{1}{f(x)}}{h}$$

$$= \lim_{h \to 0} \frac{f(x) - f(x + h)}{hf(x + h)f(x)}$$

$$= \lim_{h \to 0} \left(\frac{-1}{f(x + h)f(x)}\right)\frac{f(x + h) - f(x)}{h}$$

$$= \frac{-1}{(f(x))^2} f'(x).$$

Again we have to use the continuity of f (from Theorem 1) and the limit rules from Section 1.2.

EXAMPLE 7 Differentiate the functions

(a) $\dfrac{1}{x^2+1}$ and (b) $f(t) = \dfrac{1}{t+\dfrac{1}{t}}$.

Solution Using the Reciprocal Rule:

(a) $\dfrac{d}{dx}\left(\dfrac{1}{x^2+1}\right) = \dfrac{-2x}{(x^2+1)^2}$.

(b) $f'(t) = \dfrac{-1}{\left(t+\dfrac{1}{t}\right)^2}\left(1-\dfrac{1}{t^2}\right) = \dfrac{-t^2}{(t^2+1)^2}\dfrac{t^2-1}{t^2} = \dfrac{1-t^2}{(t^2+1)^2}$.

We can use the Reciprocal Rule to confirm the General Power Rule for negative integers:

$$\frac{d}{dx}x^{-n} = -n\,x^{-n-1},$$

since we have already proved the rule for positive integers. We have

$$\frac{d}{dx}x^{-n} = \frac{d}{dx}\frac{1}{x^n} = \frac{-n\,x^{n-1}}{(x^n)^2} = -n\,x^{-n-1}.$$

EXAMPLE 8 **(Differentiating sums of reciprocals)**

$$\frac{d}{dx}\left(\frac{x^2+x+1}{x^3}\right) = \frac{d}{dx}\left(\frac{1}{x}+\frac{1}{x^2}+\frac{1}{x^3}\right)$$
$$= \frac{d}{dx}(x^{-1}+x^{-2}+x^{-3})$$
$$= -x^{-2}-2x^{-3}-3x^{-4} = -\frac{1}{x^2}-\frac{2}{x^3}-\frac{3}{x^4}.$$

The Quotient Rule

The Product Rule and the Reciprocal Rule can be combined to provide a rule for differentiating a quotient of two functions. Observe that

$$\frac{d}{dx}\left(\frac{f(x)}{g(x)}\right) = \frac{d}{dx}\left(f(x)\frac{1}{g(x)}\right) = f'(x)\frac{1}{g(x)} + f(x)\left(-\frac{g'(x)}{(g(x))^2}\right)$$
$$= \frac{g(x)f'(x) - f(x)g'(x)}{(g(x))^2}.$$

Thus, we have proved the following Quotient Rule.

THEOREM

5

The Quotient Rule

If f and g are differentiable at x, and if $g(x) \neq 0$, then the quotient f/g is differentiable at x and

$$\left(\frac{f}{g}\right)'(x) = \frac{g(x)f'(x) - f(x)g'(x)}{(g(x))^2}.$$

Sometimes students have trouble remembering this rule. (Getting the order of the terms in the numerator wrong will reverse the sign.) Try to remember (and use) the Quotient Rule in the following form:

$$(\text{quotient})'$$
$$= \frac{(\text{denominator}) \times (\text{numerator})' - (\text{numerator}) \times (\text{denominator})'}{(\text{denominator})^2}$$

EXAMPLE 9   Find the derivatives of

(a) $y = \dfrac{1 - x^2}{1 + x^2}$,   (b) $\dfrac{\sqrt{t}}{3 - 5t}$,   and   (c) $f(\theta) = \dfrac{a + b\theta}{m + n\theta}$.

Solution   We use the Quotient Rule in each case.

(a) $\dfrac{dy}{dx} = \dfrac{(1 + x^2)(-2x) - (1 - x^2)(2x)}{(1 + x^2)^2} = -\dfrac{4x}{(1 + x^2)^2}.$

(b) $\dfrac{d}{dt}\left(\dfrac{\sqrt{t}}{3 - 5t}\right) = \dfrac{(3 - 5t)\dfrac{1}{2\sqrt{t}} - \sqrt{t}(-5)}{(3 - 5t)^2} = \dfrac{3 + 5t}{2\sqrt{t}(3 - 5t)^2}.$

(c) $f'(\theta) = \dfrac{(m + n\theta)(b) - (a + b\theta)(n)}{(m + n\theta)^2} = \dfrac{mb - na}{(m + n\theta)^2}.$

In all three parts of Example 9, the Quotient Rule yielded fractions with numerators that were complicated but could be simplified algebraically. It is advisable to attempt such simplifications when calculating derivatives; the usefulness of derivatives in applications of calculus often depends on such simplifications.

EXAMPLE 10   Find equations of any lines that pass through the point $(-1, 0)$ and are tangent to the curve $y = (x - 1)/(x + 1)$.

Solution   The point $(-1, 0)$ does not lie on the curve, so it is not the point of tangency. Suppose a line is tangent to the curve at $x = a$, so the point of tangency is $(a, (a - 1)/(a + 1))$. Note that a cannot be -1. The slope of the line must be

$$\left.\frac{dy}{dx}\right|_{x=a} = \left.\frac{(x + 1)(1) - (x - 1)(1)}{(x + 1)^2}\right|_{x=a} = \frac{2}{(a + 1)^2}.$$

If the line also passes through $(-1, 0)$, its slope must also be given by

$$\frac{\dfrac{a - 1}{a + 1} - 0}{a - (-1)} = \frac{a - 1}{(a + 1)^2}.$$

Equating these two expressions for the slope, we get an equation to solve for a:

$$\frac{a - 1}{(a + 1)^2} = \frac{2}{(a + 1)^2} \qquad \Longrightarrow \qquad a - 1 = 2.$$

Thus, $a = 3$, and the slope of the line is $2/4^2 = 1/8$. There is only one line through $(-1, 0)$ tangent to the given curve, and its equation is

$$y = 0 + \frac{1}{8}(x + 1) \qquad \text{or} \qquad x - 8y + 1 = 0.$$

Remark Derivatives of quotients of functions where the denominator is a monomial, such as in Example 8, are usually easier to do by breaking the quotient into a sum of several fractions (as was done in that example) rather than by using the Quotient Rule.

EXERCISES 2.3

In Exercises 1–32, calculate the derivatives of the given functions. Simplify your answers whenever possible.

1. $y = 3x^2 - 5x - 7$

2. $y = 4x^{1/2} - \dfrac{5}{x}$

3. $f(x) = Ax^2 + Bx + C$

4. $f(x) = \dfrac{6}{x^3} + \dfrac{2}{x^2} - 2$

5. $z = \dfrac{s^5 - s^3}{15}$

6. $y = x^{45} - x^{-45}$

7. $g(t) = t^{1/3} + 2t^{1/4} + 3t^{1/5}$

8. $y = 3\sqrt[3]{t^2} - \dfrac{2}{\sqrt{t^3}}$

9. $u = \dfrac{3}{5}x^{5/3} - \dfrac{5}{3}x^{-3/5}$

10. $F(x) = (3x - 2)(1 - 5x)$

11. $y = \sqrt{x}\left(5 - x - \dfrac{x^2}{3}\right)$

12. $g(t) = \dfrac{1}{2t - 3}$

13. $y = \dfrac{1}{x^2 + 5x}$

14. $y = \dfrac{4}{3 - x}$

15. $f(t) = \dfrac{\pi}{2 - \pi t}$

16. $g(y) = \dfrac{2}{1 - y^2}$

17. $f(x) = \dfrac{1 - 4x^2}{x^3}$

18. $g(u) = \dfrac{u\sqrt{u} - 3}{u^2}$

19. $y = \dfrac{2 + t + t^2}{\sqrt{t}}$

20. $z = \dfrac{x - 1}{x^{2/3}}$

21. $f(x) = \dfrac{3 - 4x}{3 + 4x}$

22. $z = \dfrac{t^2 + 2t}{t^2 - 1}$

23. $s = \dfrac{1 + \sqrt{t}}{1 - \sqrt{t}}$

24. $f(x) = \dfrac{x^3 - 4}{x + 1}$

25. $f(x) = \dfrac{ax + b}{cx + d}$

26. $F(t) = \dfrac{t^2 + 7t - 8}{t^2 - t + 1}$

27. $f(x) = (1 + x)(1 + 2x)(1 + 3x)(1 + 4x)$

28. $f(r) = (r^{-2} + r^{-3} - 4)(r^2 + r^3 + 1)$

29. $y = (x^2 + 4)(\sqrt{x} + 1)(5x^{2/3} - 2)$

30. $y = \dfrac{(x^2 + 1)(x^3 + 2)}{(x^2 + 2)(x^3 + 1)}$

❗ 31. $y = \dfrac{x}{2x + \dfrac{1}{3x + 1}}$

❗ 32. $f(x) = \dfrac{(\sqrt{x} - 1)(2 - x)(1 - x^2)}{\sqrt{x}(3 + 2x)}$

Calculate the derivatives in Exercises 33–36, given that $f(2) = 2$ and $f'(2) = 3$.

33. $\dfrac{d}{dx}\left(\dfrac{x^2}{f(x)}\right)\Big|_{x=2}$

34. $\dfrac{d}{dx}\left(\dfrac{f(x)}{x^2}\right)\Big|_{x=2}$

35. $\dfrac{d}{dx}\left(x^2 f(x)\right)\Big|_{x=2}$

36. $\dfrac{d}{dx}\left(\dfrac{f(x)}{x^2 + f(x)}\right)\Big|_{x=2}$

37. Find $\dfrac{d}{dx}\left(\dfrac{x^2 - 4}{x^2 + 4}\right)\Big|_{x=-2}$.

38. Find $\dfrac{d}{dt}\left(\dfrac{t(1 + \sqrt{t})}{5 - t}\right)\Big|_{t=4}$.

39. If $f(x) = \dfrac{\sqrt{x}}{x + 1}$, find $f'(2)$.

40. Find $\dfrac{d}{dt}\left((1 + t)(1 + 2t)(1 + 3t)(1 + 4t)\right)\Big|_{t=0}$.

41. Find an equation of the tangent line to $y = \dfrac{2}{3 - 4\sqrt{x}}$ at the point $(1, -2)$.

42. Find equations of the tangent and normal to $y = \dfrac{x + 1}{x - 1}$ at $x = 2$.

43. Find the points on the curve $y = x + 1/x$ where the tangent line is horizontal.

44. Find the equations of all horizontal lines that are tangent to the curve $y = x^2(4 - x^2)$.

45. Find the coordinates of all points where the curve $y = \dfrac{1}{x^2 + x + 1}$ has a horizontal tangent line.

46. Find the coordinates of points on the curve $y = \dfrac{x + 1}{x + 2}$ where the tangent line is parallel to the line $y = 4x$.

47. Find the equation of the straight line that passes through the point $(0, b)$ and is tangent to the curve $y = 1/x$. Assume $b \neq 0$.

❗ 48. Show that the curve $y = x^2$ intersects the curve $y = 1/\sqrt{x}$ at right angles.

49. Find two straight lines that are tangent to $y = x^3$ and pass through the point $(2, 8)$.

50. Find two straight lines that are tangent to $y = x^2/(x - 1)$ and pass through the point $(2, 0)$.

51. **(A Square Root Rule)** Show that if f is differentiable at x and $f(x) > 0$, then

$$\frac{d}{dx}\sqrt{f(x)} = \frac{f'(x)}{2\sqrt{f(x)}}.$$

Use this Square Root Rule to find the derivative of $\sqrt{x^2 + 1}$.

52. Show that $f(x) = |x^3|$ is differentiable at every real number x, and find its derivative.

Mathematical Induction

53. Use mathematical induction to prove that $\frac{d}{dx}x^{n/2} = \frac{n}{2}x^{(n/2)-1}$ for every positive integer n. Then use the Reciprocal Rule to get the same result for every negative integer n.

54. Use mathematical induction to prove the formula for the derivative of a product of n functions given earlier in this section.

2.4 The Chain Rule

Although we can differentiate $\sqrt{x}$ and $x^2 + 1$, we cannot yet differentiate $\sqrt{x^2 + 1}$. To do this, we need a rule that tells us how to differentiate *composites* of functions whose derivatives we already know. This rule is known as the Chain Rule and is the most often used of all the differentiation rules.

EXAMPLE 1 The function $\frac{1}{x^2 - 4}$ is the composite $f(g(x))$ of $f(u) = \frac{1}{u}$ and $g(x) = x^2 - 4$, which have derivatives

$$f'(u) = \frac{-1}{u^2} \quad \text{and} \quad g'(x) = 2x.$$

According to the Reciprocal Rule (which is a special case of the Chain Rule),

$$\frac{d}{dx}f(g(x)) = \frac{d}{dx}\left(\frac{1}{x^2 - 4}\right) = \frac{-2x}{(x^2 - 4)^2} = \frac{-1}{(x^2 - 4)^2}(2x)$$
$$= f'(g(x))g'(x).$$

This example suggests that the derivative of a composite function $f(g(x))$ is the derivative of f evaluated at $g(x)$ multiplied by the derivative of g evaluated at x. This is the Chain Rule:

$$\frac{d}{dx}f(g(x)) = f'(g(x))\,g'(x).$$

THEOREM

6

The Chain Rule

If $f(u)$ is differentiable at $u = g(x)$, and $g(x)$ is differentiable at x, then the composite function $f \circ g(x) = f(g(x))$ is differentiable at x, and

$$(f \circ g)'(x) = f'(g(x))g'(x).$$

In terms of Leibniz notation, if $y = f(u)$ where $u = g(x)$, then $y = f(g(x))$ and:

at u, y is changing $\dfrac{dy}{du}$ times as fast as u is changing;

at x, u is changing $\dfrac{du}{dx}$ times as fast as x is changing.

Therefore, at x, $y = f(u) = f(g(x))$ is changing $\dfrac{dy}{du} \times \dfrac{du}{dx}$ times as fast as x is changing. That is,

$$\frac{dy}{dx} = \frac{dy}{du}\frac{du}{dx}, \qquad \text{where } \frac{dy}{du} \text{ is evaluated at } u = g(x).$$

It appears as though the symbol du cancels from the numerator and denominator, but this is not meaningful because dy/du was not defined as the quotient of two quantities, but rather as a single quantity, the derivative of y with respect to u.

We would like to prove Theorem 6 by writing

$$\frac{\Delta y}{\Delta x} = \frac{\Delta y}{\Delta u}\frac{\Delta u}{\Delta x}$$

and taking the limit as $\Delta x \to 0$. Such a proof is valid for most composite functions but not all. (See Exercise 46 at the end of this section.) A correct proof will be given later in this section, but first we do more examples to get a better idea of how the Chain Rule works.

EXAMPLE 2 Find the derivative of $y = \sqrt{x^2 + 1}$.

Solution Here $y = f(g(x))$, where $f(u) = \sqrt{u}$ and $g(x) = x^2 + 1$. Since the derivatives of f and g are

$$f'(u) = \frac{1}{2\sqrt{u}} \qquad \text{and} \qquad g'(x) = 2x,$$

the Chain Rule gives

$$\frac{dy}{dx} = \frac{d}{dx}f(g(x)) = f'(g(x)) \cdot g'(x)$$

$$= \frac{1}{2\sqrt{g(x)}} \cdot g'(x) = \frac{1}{2\sqrt{x^2 + 1}} \cdot (2x) = \frac{x}{\sqrt{x^2 + 1}}.$$

Outside and Inside Functions

In the composite $f(g(x))$, the function f is "outside," and the function g is "inside." The Chain Rule says that the derivative of the composite is the derivative f' of the outside function evaluated at the inside function $g(x)$, multiplied by the derivative $g'(x)$ of the inside function:
$$\frac{d}{dx}f(g(x)) = f'(g(x)) \times g'(x).$$

Usually, when applying the Chain Rule, we do not introduce symbols to represent the functions being composed, but rather just proceed to calculate the derivative of the "outside" function and then multiply by the derivative of whatever is "inside." You can say to yourself: "the derivative of f of something is f' of that thing, multiplied by the derivative of that thing."

EXAMPLE 3 Find derivatives of the following functions:

(a) $(7x - 3)^{10}$, (b) $f(t) = |t^2 - 1|$, and (c) $\left(3x + \frac{1}{(2x+1)^3}\right)^{1/4}$.

Solution

(a) Here, the outside function is the 10th power; it must be differentiated first and the result multiplied by the derivative of the expression $7x - 3$:

$$\frac{d}{dx}(7x - 3)^{10} = 10(7x - 3)^9(7) = 70(7x - 3)^9.$$

(b) Here, we are differentiating the absolute value of something. The derivative is signum of that thing, multiplied by the derivative of that thing:

$$f'(t) = \left(\text{sgn}\,(t^2 - 1)\right)(2t) = \frac{2t(t^2 - 1)}{|t^2 - 1|} = \begin{cases} 2t & \text{if } t < -1 \text{ or } t > 1 \\ -2t & \text{if } -1 < t < 1 \\ \text{undefined} & \text{if } t = \pm 1. \end{cases}$$

(c) Here, we will need to use the Chain Rule twice. We begin by differentiating the 1/4 power of something, but the something involves the -3rd power of $2x + 1$, and the derivative of that will also require the Chain Rule:

$$\frac{d}{dx}\left(3x + \frac{1}{(2x+1)^3}\right)^{1/4} = \frac{1}{4}\left(3x + \frac{1}{(2x+1)^3}\right)^{-3/4}\frac{d}{dx}\left(3x + \frac{1}{(2x+1)^3}\right)$$

$$= \frac{1}{4}\left(3x + \frac{1}{(2x+1)^3}\right)^{-3/4}\left(3 - \frac{3}{(2x+1)^4}\frac{d}{dx}(2x+1)\right)$$

$$= \frac{3}{4}\left(1 - \frac{2}{(2x+1)^4}\right)\left(3x + \frac{1}{(2x+1)^3}\right)^{-3/4}.$$

When you start to feel comfortable with the Chain Rule, you may want to save a line or two by carrying out the whole differentiation in one step:

$$\frac{d}{dx}\left(3x + \frac{1}{(2x+1)^3}\right)^{1/4} = \frac{1}{4}\left(3x + \frac{1}{(2x+1)^3}\right)^{-3/4}\left(3 - \frac{3}{(2x+1)^4}(2)\right)$$

$$= \frac{3}{4}\left(1 - \frac{2}{(2x+1)^4}\right)\left(3x + \frac{1}{(2x+1)^3}\right)^{-3/4}.$$

Use of the Chain Rule produces products of factors that do not usually come out in the order you would naturally write them. Often you will want to rewrite the result with the factors in a different order. This is obvious in parts (a) and (c) of the example above. In monomials (expressions that are products of factors), it is common to write the factors in order of increasing complexity from left to right, with numerical factors coming first. One time when you would *not* waste time doing this, or trying to make any other simplification, is when you are going to evaluate the derivative at a particular number. In this case, substitute the number as soon as you have calculated the derivative, before doing any simplification:

$$\frac{d}{dx}(x^2 - 3)^{10}\Big|_{x=2} = 10(x^2-3)^9(2x)\Big|_{x=2} = (10)(1^9)(4) = 40.$$

EXAMPLE 4   Suppose that f is a differentiable function on the real line. In terms of the derivative f' of f, express the derivatives of:

(a) $f(3x)$,   (b) $f(x^2)$,   (c) $f(\pi f(x))$,   and   (d) $[f(3 - 2f(x))]^4$.

Solution

(a) $\dfrac{d}{dx}f(3x) = \big(f'(3x)\big)(3) = 3f'(3x)$.

(b) $\dfrac{d}{dx}f(x^2) = \big(f'(x^2)\big)(2x) = 2xf'(x^2)$.

(c) $\dfrac{d}{dx}f(\pi f(x)) = \big(f'(\pi f(x))\big)(\pi f'(x)) = \pi f'(x)f'(\pi f(x))$.

(d) $\dfrac{d}{dx}[f(3 - 2f(x))]^4 = 4[f(3 - 2f(x))]^3 f'(3 - 2f(x))(-2f'(x))$

$$= -8f'(x)f'(3 - 2f(x))[f(3 - 2f(x))]^3.$$

As a final example, we illustrate combinations of the Chain Rule with the Product and Quotient Rules.

EXAMPLE 5   Find and simplify the following derivatives:

(a) $f'(t)$ if $f(t) = \dfrac{t^2 + 1}{\sqrt{t^2 + 2}}$,   and   (b) $g'(-1)$ if $g(x) = (x^2 + 3x + 4)^5\sqrt{3 - 2x}$.

Solution

(a) $f'(t) = \dfrac{\sqrt{t^2+2}\,(2t) - (t^2+1)\,\dfrac{2t}{2\sqrt{t^2+2}}}{t^2+2}$

$= \dfrac{2t}{\sqrt{t^2+2}} - \dfrac{t^3+t}{\left(t^2+2\right)^{3/2}} = \dfrac{t^3+3t}{\left(t^2+2\right)^{3/2}}.$

(b) $g'(x) = 5\left(x^2+3x+4\right)^4(2x+3)\sqrt{3-2x} + \left(x^2+3x+4\right)^5 \dfrac{-2}{2\sqrt{3-2x}}$

$g'(-1) = (5)(2^4)(1)(\sqrt{5}) - \dfrac{2^5}{\sqrt{5}} = 80\sqrt{5} - \dfrac{32}{5}\sqrt{5} = \dfrac{368\sqrt{5}}{5}.$

Finding Derivatives with Maple

Computer algebra systems know the derivatives of elementary functions and can calculate the derivatives of combinations of these functions symbolically, using differentiation rules. Maple's D operator can be used to find the derivative function D(f) of a function f of one variable. Alternatively, you can use diff to differentiate an expression with respect to a variable and then use the substitution routine subs to evaluate the result at a particular number.

```
> f := x -> sqrt(1+2*x^2);
```

$$f := x \rightarrow \sqrt{1+2x^2}$$

```
> fprime := D(f);
```

$$fprime := x \rightarrow 2\frac{x}{\sqrt{1+2x^2}}$$

```
> fprime(2);
```

$$\frac{4}{3}$$

```
> diff(t^2*sin(3*t),t);
```

$$2t\,\sin(3t) + 3t^2\,\cos(3t)$$

```
> simplify(subs(t=Pi/12, %));
```

$$\frac{1}{12}\pi\sqrt{2} + \frac{1}{96}\pi^2\sqrt{2}$$

Building the Chain Rule into Differentiation Formulas

If u is a differentiable function of x and $y = u^n$, then the Chain Rule gives

$$\frac{d}{dx}u^n = \frac{dy}{dx} = \frac{dy}{du}\frac{du}{dx} = nu^{n-1}\frac{du}{dx}.$$

The formula

$$\frac{d}{dx}u^n = nu^{n-1}\frac{du}{dx}$$

is just the formula $\frac{d}{dx}x^n = nx^{n-1}$ with an application of the Chain Rule built in, so that it applies to functions of x rather than just to x. Some other differentiation rules with built-in Chain Rule applications are:

$$\frac{d}{dx}\left(\frac{1}{u}\right) = \frac{-1}{u^2}\frac{du}{dx} \qquad \text{(the Reciprocal Rule)}$$

$$\frac{d}{dx}\sqrt{u} = \frac{1}{2\sqrt{u}}\frac{du}{dx} \qquad \text{(the Square Root Rule)}$$

$$\frac{d}{dx}u^r = r\,u^{r-1}\frac{du}{dx} \qquad \text{(the General Power Rule)}$$

$$\frac{d}{dx}|u| = \operatorname{sgn} u\,\frac{du}{dx} = \frac{u}{|u|}\frac{du}{dx} \qquad \text{(the Absolute Value Rule)}$$

Proof of the Chain Rule (Theorem 6)

Suppose that f is differentiable at the point $u = g(x)$ and that g is differentiable at x. Let the function $E(k)$ be defined by

$$E(0) = 0,$$

$$E(k) = \frac{f(u+k)-f(u)}{k} - f'(u), \qquad \text{if } k \neq 0.$$

By the definition of derivative, $\lim_{k\to 0} E(k) = f'(u) - f'(u) = 0 = E(0)$, so $E(k)$ is continuous at $k = 0$. Also, whether $k = 0$ or not, we have

$$f(u+k) - f(u) = \big(f'(u) + E(k)\big)k.$$

Now put $u = g(x)$ and $k = g(x+h) - g(x)$, so that $u + k = g(x+h)$, and obtain

$$f(g(x+h)) - f(g(x)) = \big(f'(g(x)) + E(k)\big)(g(x+h) - g(x)).$$

Since g is differentiable at x, $\lim_{h\to 0}[g(x+h) - g(x)]/h = g'(x)$. Also, g is continuous at x by Theorem 1, so $\lim_{h\to 0} k = \lim_{h\to 0}(g(x+h) - g(x)) = 0$. Since E is continuous at 0, $\lim_{h\to 0} E(k) = \lim_{k\to 0} E(k) = E(0) = 0$. Hence,

$$\frac{d}{dx}f(g(x)) = \lim_{h\to 0}\frac{f(g(x+h)) - f(g(x))}{h}$$

$$= \lim_{h\to 0}\big(f'(g(x)) + E(k)\big)\frac{g(x+h) - g(x)}{h}$$

$$= \big(f'(g(x)) + 0\big)g'(x) = f'(g(x))g'(x),$$

which was to be proved.

EXERCISES 2.4

Find the derivatives of the functions in Exercises 1–16.

1. $y = (2x+3)^6$

2. $y = \left(1 - \frac{x}{3}\right)^{99}$

3. $f(x) = (4-x^2)^{10}$

4. $y = \sqrt{1-3x^2}$

5. $F(t) = \left(2 + \frac{3}{t}\right)^{-10}$

6. $(1 + x^{2/3})^{3/2}$

7. $\dfrac{3}{5-4x}$

8. $(1 - 2t^2)^{-3/2}$

9. $y = |1 - x^2|$

10. $f(t) = |2 + t^3|$

11. $y = 4x + |4x - 1|$

12. $y = (2 + |x|^3)^{1/3}$

13. $y = \dfrac{1}{2 + \sqrt{3x+4}}$

14. $f(x) = \left(1 + \sqrt{\dfrac{x-2}{3}}\right)^4$

15. $z = \left(u + \dfrac{1}{u-1}\right)^{-5/3}$

16. $y = \dfrac{x^5\sqrt{3+x^6}}{(4+x^2)^3}$

17. Sketch the graph of the function in Exercise 10.

18. Sketch the graph of the function in Exercise 11.

Verify that the General Power Rule holds for the functions in Exercises 19–21.

19. $x^{1/4} = \sqrt{\sqrt{x}}$ **20.** $x^{3/4} = \sqrt{x\sqrt{x}}$

21. $x^{3/2} = \sqrt{(x^3)}$

In Exercises 22–29, express the derivative of the given function in terms of the derivative f' of the differentiable function f.

22. $f(2t + 3)$ **23.** $f(5x - x^2)$

24. $\left[f\left(\dfrac{2}{x}\right)\right]^3$ **25.** $\sqrt{3 + 2f(x)}$

26. $f\left(\sqrt{3 + 2t}\right)$ **27.** $f\left(3 + 2\sqrt{x}\right)$

28. $f\left(2f(3f(x))\right)$ **29.** $f\left(2 - 3f(4 - 5t)\right)$

30. Find $\dfrac{d}{dx}\left(\dfrac{\sqrt{x^2 - 1}}{x^2 + 1}\right)\bigg|_{x=-2}$.

31. Find $\dfrac{d}{dt}\sqrt{3t - 7}\bigg|_{t=3}$.

32. If $f(x) = \dfrac{1}{\sqrt{2x + 1}}$, find $f'(4)$.

33. If $y = (x^3 + 9)^{17/2}$, find $y'\bigg|_{x=-2}$.

34. Find $F'(0)$ if $F(x) = (1 + x)(2 + x)^2(3 + x)^3(4 + x)^4$.

! 35. Calculate y' if $y = (x + ((3x)^5 - 2)^{-1/2})^{-6}$. Try to do it all in one step.

In Exercises 36–39, find an equation of the tangent line to the given curve at the given point.

36. $y = \sqrt{1 + 2x^2}$ at $x = 2$

37. $y = (1 + x^{2/3})^{3/2}$ at $x = -1$

38. $y = (ax + b)^8$ at $x = b/a$

39. $y = 1/(x^2 - x + 3)^{3/2}$ at $x = -2$

40. Show that the derivative of $f(x) = (x - a)^m(x - b)^n$ vanishes at some point between a and b if m and n are positive integers.

Use Maple or another computer algebra system to evaluate and simplify the derivatives of the functions in Exercises 41–44.

41. $y = \sqrt{x^2 + 1} + \dfrac{1}{(x^2 + 1)^{3/2}}$

42. $y = \dfrac{(x^2 - 1)(x^2 - 4)(x^2 - 9)}{x^6}$

43. $\dfrac{dy}{dt}\bigg|_{t=2}$ if $y = (t + 1)(t^2 + 2)(t^3 + 3)(t^4 + 4)(t^5 + 5)$

44. $f'(1)$ if $f(x) = \dfrac{(x^2 + 3)^{1/2}(x^3 + 7)^{1/3}}{(x^4 + 15)^{1/4}}$

45. Does the Chain Rule enable you to calculate the derivatives of $|x|^2$ and $|x^2|$ at $x = 0$? Do these functions have derivatives at $x = 0$? Why?

! 46. What is wrong with the following "proof" of the Chain Rule? Let $k = g(x + h) - g(x)$. Then $\lim_{h \to 0} k = 0$. Thus,

$$\lim_{h \to 0} \frac{f(g(x + h)) - f(g(x))}{h}$$

$$= \lim_{h \to 0} \frac{f(g(x + h)) - f(g(x))}{g(x + h) - g(x)} \cdot \frac{g(x + h) - g(x)}{h}$$

$$= \lim_{h \to 0} \frac{f(g(x) + k) - f(g(x))}{k} \cdot \frac{g(x + h) - g(x)}{h}$$

$$= f'(g(x)) \, g'(x).$$

2.5 Derivatives of Trigonometric Functions

The trigonometric functions, especially sine and cosine, play a very important role in the mathematical modelling of real-world phenomena. In particular, they arise whenever quantities fluctuate in a periodic way. Elastic motions, vibrations, and waves of all kinds naturally involve the trigonometric functions, and many physical and mechanical laws are formulated as differential equations having these functions as solutions.

In this section we will calculate the derivatives of the six trigonometric functions. We only have to work hard for one of them, sine; the others then follow from known identities and the differentiation rules of Section 2.3.

Some Special Limits

First, we have to establish some trigonometric limits that we will need to calculate the derivative of sine. It is assumed throughout that the arguments of the trigonometric functions are measured in radians.

THEOREM

7

The functions $\sin\theta$ and $\cos\theta$ are continuous at every value of θ. In particular, at $\theta = 0$ we have:

$$\lim_{\theta \to 0} \sin\theta = \sin 0 = 0 \qquad \text{and} \qquad \lim_{\theta \to 0} \cos\theta = \cos 0 = 1.$$

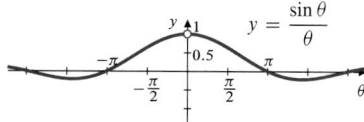

Figure 2.20 It appears that $\lim_{\theta \to 0} (\sin \theta)/\theta = 1$

This result is obvious from the graphs of sine and cosine, so we will not prove it here. A proof can be based on the Squeeze Theorem (Theorem 4 of Section 1.2). The method is suggested in Exercise 62 at the end of this section.

The graph of the function $y = (\sin \theta)/\theta$ is shown in Figure 2.20. Although it is not defined at $\theta = 0$, this function appears to have limit 1 as θ approaches 0.

THEOREM

8

An important trigonometric limit

$$\lim_{\theta \to 0} \frac{\sin \theta}{\theta} = 1 \qquad \text{(where } \theta \text{ is in radians)}.$$

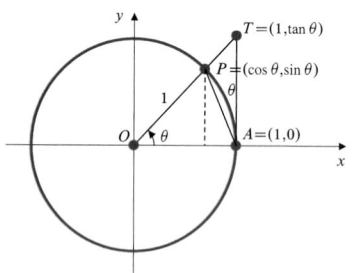

Figure 2.21 Area $\triangle OAP$
$< $ Area sector OAP
$< $ Area $\triangle OAT$

PROOF Let $0 < \theta < \pi/2$, and represent θ as shown in Figure 2.21. Points $A(1, 0)$ and $P(\cos \theta, \sin \theta)$ lie on the unit circle $x^2 + y^2 = 1$. The area of the circular sector OAP lies between the areas of triangles OAP and OAT:

$$\text{Area } \triangle OAP < \text{Area sector } OAP < \text{Area } \triangle OAT.$$

As shown in Section P.7, the area of a circular sector having central angle θ (radians) and radius 1 is $\theta/2$. The area of a triangle is $(1/2) \times$ base $\times$ height, so

$$\text{Area } \triangle OAP = \frac{1}{2} (1) (\sin \theta) = \frac{\sin \theta}{2},$$
$$\text{Area } \triangle OAT = \frac{1}{2} (1) (\tan \theta) = \frac{\sin \theta}{2 \cos \theta}.$$

Thus,

$$\frac{\sin \theta}{2} < \frac{\theta}{2} < \frac{\sin \theta}{2 \cos \theta},$$

or, upon multiplication by the positive number $2/\sin \theta$,

$$1 < \frac{\theta}{\sin \theta} < \frac{1}{\cos \theta}.$$

Now take reciprocals, thereby reversing the inequalities:

$$1 > \frac{\sin \theta}{\theta} > \cos \theta.$$

Since $\lim_{\theta \to 0+} \cos \theta = 1$ by Theorem 7, the Squeeze Theorem gives

$$\lim_{\theta \to 0+} \frac{\sin \theta}{\theta} = 1.$$

Finally, note that $\sin \theta$ and θ are *odd functions*. Therefore, $f(\theta) = (\sin \theta)/\theta$ is an *even function*: $f(-\theta) = f(\theta)$, as shown in Figure 2.20. This symmetry implies that the left limit at 0 must have the same value as the right limit:

$$\lim_{\theta \to 0-} \frac{\sin \theta}{\theta} = 1 = \lim_{\theta \to 0+} \frac{\sin \theta}{\theta},$$

so $\lim_{\theta \to 0} (\sin \theta)/\theta = 1$ by Theorem 1 of Section 1.2.

Theorem 8 can be combined with limit rules and known trigonometric identities to yield other trigonometric limits.

The full details of these internal political struggles will for ever escape us, but the kind of world we should be envisaging is clear. At this stage, as one commentator has evocatively called it, it was not so much a state as a 'glorified Hudson Bay company', composed of essentially independent trading operations located at various centres along the main river routes, loosely linked together by having to pay protection money to the most powerful among them. They acted in concert only in certain circumstances, such as when using their collective muscle to extract advantageous trade terms from the Byzantines, and no doubt also to engage in a little price-fixing. The Rus state began life, therefore, as a hierarchically organized umbrella organization for these merchants, no doubt established originally by force. Even so, the original merchant adventurers, or their descendants, were left with considerable powers and independence, and as late as 944 still ran their own localities.[25]

By the eleventh century, however, this stratum of independent non-Riurikid rulers in their own settlements had disappeared. By this date, the preferred solution to the dynastic mayhem which characteristically accompanied transfers of power between different Rurikid generations took the form of giving their own centre of power to each eligible contender. This was already happening by the year 1000, the *Chronicle* providing us with an exhaustive list of the twelve cities that were granted by Vladimir to his twelve sons, the products of five of his more official liaisons. How many other children he had generated from the 300 concubines he kept at Vyshgorod, the three hundred at Belgorod and the two hundred at Berestovoe is not recorded. At some point in the tenth century, then, the independent power of the descendants of the founding merchant princes had been curtailed, turning their formerly self-governing settlements into dynastic appanages. In fact, this was probably a steady process, which played itself out over a lengthy period. Oleg's suppression of Askold and Dir, to the extent that this story might be taken as historical example, provides us with an early instance of this kind of action. The *RPC* also records some later instances of exactly the same thing. In the civil war between Sviatoslav's two sons Yaropolk and Vladimir, new merchant settlements continued to be founded. Two Scandinavian leaders by the names of Rogvolod (Ragnvaldr) and Tury established their own trading centres at Polotsk and Turov. Their subsequent fate is not recorded,

first to appear, in the middle decades of the ninth century. Carolingian sources for c.800–20 mention in passing a whole series of regionally based small-scale political leaders at the head of their own groupings, as Avar dominion in central Europe unwound. One called Vojnomir supported the Franks against the Avars, a certain Manomir appears briefly, while a major revolt against Carolingian rule was led by Ljudevit. The sources are nothing like full enough for us to attempt a political narrative of how these different dynasties combined and eliminated one another to produce the much larger power that was Great Moravia. But that they did is clear enough and, again, we are given the odd snapshot. The first really prominent Moravian ruler, perhaps the real founder of dynastic pre-eminence, was called Mojmir, and Carolingian sources record what was clearly a highly significant moment in the 830s (the incident can be dated no more tightly than 833–6) when he drove a rival prince, Pribina, out of Nitra in Slovakia to bring a broader region under his direct control. Using this greater power base, the dynasty continued to extend its control, as and when it could. Carolingian power kept its westerly ambitions in check through most of the ninth century, but as that Empire waned in the early 890s the Moravians extracted the right to exercise hegemony over Bohemia. From then on, a still more exciting range of ambitions might have been open to this ruling line, had not its career been cut decisively short from 896 by the arrival of the nomadic Magyars as a major force in central Europe.[23]

If we had nothing but the available historical sources, the emergence of Piast Poland would be particularly mysterious. The Piast state suddenly jumps into Ottonian narrative sources in the 960s, already fully formed under the control of the Piast Miesco I. With a heartland west of the Vistula, beyond the immediate border region between the Elbe and the Oder, the new Polish state was simply too far away from imperial dominions for our chronicle sources to observe its growing pains. Not even the *Anonymous Bavarian Geographer* knew the political layout of lands so far to the east. Thanks to the wonders of dendrochronolgy, however, archaeological evidence – which is usually so much better at observing long-term development than the immediately political – has in this case brilliantly illuminated at least the final stages of the rise of the Piasts. Because the dynasty built its castles of wood – as pretty much everyone else in Europe was still doing in the first half of the tenth century – it has become possible, within just the last decade, to date their construction precisely. The results are revolutionary.

The emergence of the first Polish state used to be construed as a long, slow process of political consolidation, which gradually brought ever larger areas together under the control of a single dynasty. Long-term developments, as we shall examine in a moment, were certainly of critical importance to create the necessary conditions, but the archaeology has shown with striking clarity that the last stage of the Piast rise to power was sudden and violent. Piast castle construction clusters in the second quarter of the tenth century, demonstrating that the dynasty expanded its control over broader areas of Great Poland very quickly, from an originally narrow base (Map 20). More than that, in many of these localities Piast castles replaced a much larger kind of fortified centre, often dating back to the eighth century, many of which seem to have been destroyed exactly at the moment of Piast construction. The conclusion seems inescapable. The creation of Piast Poland, the entity that suddenly bursts into our histories in the mid-tenth century, involved the destruction of long-standing local societies and the imposition of Piast military garrisons upon them. How many of these local societies were 'tribes', for want of a better word – the kind of unit listed in the *Bavarian Geographer* for more westerly regions of the Slavic world – is unclear, as is how they were distributed across the landscape.[24] Just like Great Moravia and Bohemia, then, the new Polish state emerged by violent dynastic self-assertion, as the Piasts eliminated their rivals at the heads of those other, older political units.

Much is also obscure about the rise of the Riurikids. As we saw in the last chapter, the *Russian Primary Chronicle* (RPC) is both too Kiev-focused and too much composed from the hindsight of achieved Riurikid domination to provide a straightforward route into the complexities of early Rus history. Nor – at this stage at least – is the archaeological picture so arrestingly precise as that for Piast Poland. Nonetheless, the basic outlines of Scandinavian intrusion into Russia are clear enough, and hence too the key political developments that made possible the Rurikid state.

Amongst its other problems, as we saw, the *RPC* provides a thoroughly unconvincing account of both the date and circumstances by which political power came to be transferred to Kiev from Gorodishche in the north. The *Chronicle* both places the transfer a generation too early and seems to be hiding dynastic discontinuity or at least

disruption in its odd and seemingly sanitized account of the relationship between Oleg, the first major political figure associated with Kiev, and Igor, Riurik's son and heir. The story also presents Oleg as gathering an army in the north and taking control of the Middle Dnieper by force. Yet it closes by saying that, at the conclusion of these operations, an annual tribute of three hundred *grivny* was imposed by Oleg on Novgorod in return for peace. This, the *Chronicle* notes, was paid until the death of Prince Yaroslav in 1054, which is late enough to fall virtually within living memory of the compiler of the *Chronicle* in the early twelfth century. So the payment is presumably historical. But why would a ruler who came from the north to conquer in the south, as the story has it, end up imposing a tribute on the north?

There are two other big problems besides. First, the trade treaties with Byzantium confirm that, well into the tenth century, non-Riurikid Scandinavians ruled their own Russian settlements with a great deal of independent power, since they had to be represented individually in the negotiations. Second, before the end of the tenth century the *Chronicle* preserves only a very simplified version of Riurikid dynastic history. From that point on, the transmission of power from one generation to the next always involved many contenders and civil war, but before that date, even though we know that the early princes were multiply polygamous (as indeed were their successors), the *Chronicle* mentions only one son at each moment of succession and nothing but a smooth transition of power.

None of this is credible. The 944 trade treaty with Byzantium tell us that Igor had two nephews important enough to rate a separate mention. There is no record of them or their subsequent fate in the rest of the *Chronicle*, and it's hard to resist the conclusion that history has been edited to give an impression of secure and smooth Ruri dominance. Likewise, the Oleg story: was he a collateral relative Riurik who first conquered Kiev and then imposed his rule on north? Or was he a complete outsider who perhaps married into dynasty so as in some way to legitimize his rule after the fact? how, then, did power pass from him to Riurik's son Igor? It is hard to believe that Oleg didn't have heirs of his own, so happened to them? The politics of early tenth-century Russia clearly much messier than the *Chronicle* would have us believe independent Viking leaders and a self-assertive dynasty all join violently for position.

but both centres were among the twelve distributed in the next generation to the various sons of Vladimir, by which time their founders had clearly lost out. As part of the same civil war, another such locally dominant line, apparently a family established for a much longer period, that of Sveinald, also met its demise.[26] The full story of the suppression of the independent merchant lines is hidden from us, but it clearly happened, and it represented the final stage in the evolution of mercantile settlements into a fully fledged political union. Although the unique origins of the Rus state meant that the Rurikids began as one set of merchant princes among several – rather than as the leaders of one regional tribal group among several, as was the case with the Piasts and Premyslids – nonetheless violent dynastic self-assertion was central to the process of state formation.

The same was true of the last of these new states, Denmark, although here, too, the process differed substantially from that unfolding in the Avar successor states. In the small settlement of Jelling in central Jutland stands a not very substantial church and two huge mounds: the northern one 65 metres in diameter and 8 metres high, the southerly 77 by 11. Within the northern mound there is a wood-lined chamber dated by dendrochronology to 958, which was nearly the last resting place of King Gorm. Gorm's son and heir Harold Bluetooth originally buried him there, but transferred the body to the church when he himself converted to Christianity, probably around 965. Like the Mormons, Harold was taking no chances that his ancestors might be deprived of the joys of his new religion. Apart from shifting the corpse, he also erected a fabulous runestone whose inscription is still there to be read: 'Harold had these monuments erected in memory of Gorm his father and Thyre his mother, that Harold who won for himself all Denmark and Norway, and Christian-ized the Danes.'

The case of Denmark differs substantially from the other states, by providing a timely warning against the assumption that political developments always move in a straight line. As we have seen, a powerful centralizing political structure had existed in southern Jutland before the Viking Age, from at least the mid-eighth century when the Danevirke was first constructed. But this monarchy was destroyed by flows of new Viking wealth into Scandinavia. Wealth translated pretty much directly into warriors, and warriors into power, so that new wealth in sufficient quantities could not but generate political revolution. The old monarchy fell because so many 'kings' could now buy in so much military muscle that political stability evaporated.[27]

By the mid-tenth century, there are further signs of substantial change. For one thing, there seem to have been fewer kings. Viking-period sources demonstrate that a multiplicity of royals had existed in ninth-century Scandinavia. Apart from the one extended, or possibly two, separate dynastic lines found competing for power in southern Jutland (Godfrid, Haraldr and their descendants), there were more independent kings in the Vestfold west of the Oslo Fjord in Norway in the ninth century, and on the island of Bornholm. Birka and Sweden, further east, likewise, also had kings. A large number of other kings also appeared in western waters in the Great Army period, from the 860s onwards, and these must all have had their origins in some particular corner of Scandinavia. By my reckoning about a dozen of them are named, at different points: not enough to suggest that 'king' was a status that just anyone might claim, especially as we also meet men of slightly lesser status – jarls – who held back from claiming to be royalty. From the time of Harold Bluetooth, by contrast, the historical narrative throws up other 'kings' consistently in Sweden only, and occasionally in Norway. It would appear, therefore, that the word had undergone a change of meaning (as it did in other cultural contexts, too) from something like 'person from an extremely important family' to 'ruler of a substantial territory', the normal meaning of the word today.[28]

That said, the Jelling dynasty did seemingly build up its power by bringing under its control disparate territories that had had their own leaderships in the chaos of the later ninth century. It may have been the dynasty's success, of course, that brought about the substantive change of meaning in the word 'king'. Gorm's wife Thyre is called in another inscription 'the pride of Denmark'. It has been convincingly argued, on the basis of contemporary usage, that in c.900 the 'mark' element in 'Denmark' meant 'regions bordering the Danish kingdom'; in other words, somewhere other than the main centres of the Danish monarchy – perhaps northern Jutland or the southern Baltic islands. Like our other dynasties, therefore, despite the substantial differences in historical context, the political activities of the Jelling dynasty were fundamentally accumulative – putting together regions that had previously been independent. This process was begun by Gorm and carried on by subsequent members of the dynasty. Harold Bluetooth added control of southern Norway to the dynasty's portfolio of assets after the battle of Limfjord, but ruled it indirectly through the Jarls of Lade. Svein and Cnut maintained this hegemony through most of their reigns, and at times dominated the west coast of what is now Sweden as well. Even so, the heritage of old independence did not disappear overnight. From the narratives of Danish history in the eleventh century, it emerges very clearly that Jutland and the islands of Fyn and Sjaelland were still functioning on occasion as separatist power centres.[29]

The political processes behind all these new states, therefore, were similar. In each case, one dynastic line was able to demote or eliminate a peer group of geographically proximate rivals to bring a larger region under its control. The vagaries of this process further explain the propensity of the states it created to swap intervening areas amongst themselves. Given that all these areas were originally independent, it is easy to see why some of them might maintain a capacity for autonomous political activity long after they first accepted a new dynasty's domination, especially in a context where itineration and personal charisma rather than developed bureaucratic structures were being used to govern them. But while full of arresting stories and individuals of striking charisma, political narratives of achieved dynastic ambition do not remotely begin to tell the full story of state formation in the north and east at the end of the first millennium. History is littered with ambitious individuals trying to build their power and thereby eclipse every rival. In most cases, however, such ambition does not lead to new and impressively powerful state structures. Apart from looking at narratives of personal ambition, then, we also need to think about the broader structural transformations that made it possible for entirely ordinary ambitions to achieve such unusual outcomes.

State-building

Many of these changes were similar to those that had generated the larger political structures on the fringes of the Roman Empire in the first half of the millennium. Taking the long view, social and economic transformations of the most profound kind were structurally critical to the process of state formation in northern and eastern Europe. This is most obviously true of the Slavic-speaking world, but to a considerable extent applies to Scandinavia as well.

Up to the mid-first millennium, Slav or Slavic-dominated societies were characterized by little in the way of obvious social inequality. Whatever their exact geographical origins, the Slavic-speaking groups who burst on to the fringes of the Mediterranean in the sixth century had clearly emerged from the undeveloped, heavily wooded regions of eastern Europe, where settlements were small – no more than hamlets – and whose Iron Age farmers were operating at little above subsistence level and with few material markers of differing social status. This state of affairs had already begun to change radically in the sixth century, as a direct result of the migratory processes that brought some Slavic-speakers into a direct relationship with the more developed Mediterranean. From this, an unprecedented flow of wealth – the profits, more or less equally, of raiding, military service and diplomatic subsidy – quickly generated inequalities around which new social structures began to form. These showed themselves initially after c.575 in the rise of a new class of military leader, controlling quite substantial areas and groups several thousand strong – even if there is also some reason to think that other elements within Slavic society, represented by Korchak remains, retained older, more egalitarian social forms and were even using alternative kinds of migration, and in different directions away from the east Roman frontier, as a means to preserve them.[30]

The new Slavic states of the ninth and tenth centuries were constructed on a marked accentuation of these initial inequalities. This shows up most obviously in the existence of military retinues: that classic vehicle of social and political power, which had played such an important role in the transformation of the Germanic world. Presumably the new Slavic leaders of the sixth century had their henchmen, but large permanent retinues do not figure in any of the historical sources as a major force, military or social. The contrast with the ninth and tenth centuries is striking. The Arab geographers report that Miesco of Poland maintained a personal force of three thousand warriors – and this is just one account among many, stressing the importance of retinues at this time. In Bohemia, the fourteen dukes presenting themselves for baptism in 845 did so 'with their men', and the early Bohemian texts associated with Wenceslas refer both to his retinue and to that of his brother, Boleslav I. Frankish texts, similarly, mention the 'men' of both Mojmir and his nephew Zwentibald among the Moravians, and retinues were just as important in Russia. Again, Arab geographers pick out the four hundred men of the dominant Rus prince in the north in c.900, and retinues appear as important political pressure groups for several of the early kings in the narratives of the *RPC*. It was the need to satisfy the demands of his 'men', for instance, that led Igor to increase the tribute he customarily imposed upon the Derevlians. He may have regretted giving in, since, as we have seen, it led to his death at the hands of the aggrieved taxpayers. And as we saw among the Germani around the Roman Empire, the rise of permanent military retinues greatly increased the capacity of rulers both to bring rival dynasts into line and to enforce a range of obligations (such as army and labour services) upon the broader population. As such, it obviously played a critical role in the process of state formation, not least – again as among the Germani – in creating a much stronger dynastic component to power at the top. There is no sign among even the late sixth-century Slavs that power was in any sense hereditary, even if particular individuals could build up striking power bases.[31] But by the ninth and tenth centuries, dynasties dominated politics, and hereditary power was the order of the day.

But retinues were only one aspect of a broader process of social change. Part of the problem in understanding this bigger picture as fully as one would like to stems from uncertainties about its starting point. The idea of a highly egalitarian Slavic world in c.500 AD is strongly entrenched in both the scholarly literature and more popular mythology. It underlies the 'happy hippy' vision of Slavicization, and finds real support particularly from east Roman sources which note that sixth-century Slavic society was marked by a lack of structured social differentiation, as well as being unusually willing to take on prisoners as free and equal members. But such visions of Slavic equality need to be tempered with some caution. To echo a point made earlier in the case of the Germani, there are entirely non-material ways in which higher status can be all too real – if those enjoying it had to work less hard, enjoyed more food, and if their word counted for more when it came to settling disputes within the group.[32]

But even if we factor in a less egalitarian starting point for the evolution of Slavic society from c.500 (and as we have seen, any pre-existing egalitarianism was being undermined rapidly at this point by the twin processes of migration and development), much had clearly changed by the tenth century. Not only was political leadership now hereditary and its clout more wide-reaching thanks to its permanent military retinues, but Slavic society as a whole was marked by the evolution of clearly differentiated, and therefore presumably also hereditary, hierarchically arranged social categories.

At the bottom of the social scale, unfree population groups now played a prominent role in all our late first-millennium Slavic and Scandinavian societies. The slave trade was a major phenomenon of central and eastern Europe from the eighth century onwards. Likewise, as the new state structures evolved, a major component of their economic makeup (as we shall explore in more detail in a moment) became the unfree 'service village'. Given the available sources, it is not entirely clear whether the populations of these villages enjoyed a higher status than the slaves who were often exported – perhaps a status akin to those of the permanent 'freed' (better, 'semi-free') populations we encountered in the Germanic world. Either way, a large part of Slavic humanity had clearly been permanently relegated to hereditary lesser status (or statuses, if slaves and service villagers need to be distinguished) by the tenth century. However you model Slavic society in c.500 AD, the extent of change here can hardly be overemphasized.

Equally permanent at the other end of the social scale was a class of high-status individuals, often styled *optimates* in our sources. These men are recorded, for instance, attending assemblies in Bohemia to give their approval to their choice of Adalbert the Slavnik as Bishop of Prague in 982, and feature as the rulers of their own settlements under overall Rurikid rule in Russia (some of them independent enough to send their own ambassadors to Constantinople when the trade treaties were negotiated). Certain individuals with the same kind of high status also appear in the train of the King of Poland in the early eleventh century, and were presumably the kind of men Polish and other kings customarily offered hospitality to as they feasted their way round their kingdoms. Their existence a century earlier in Moravia is possibly also reflected archaeologically in the five so-called princely dwellings found in the hundred-hectare outer area at Mikulčiče, although these could have belonged to junior members of the ruling dynasty. The evidence suggests that this group coalesced out of originally three component elements. First, there were the close supporters of the new dynasty from within their home group. These were reinforced, second, by the elites of originally independent units (whether of Slavo-Scandinavian trading enterprises in Russia or 'tribal' units in Bohemia, Moravia and Poland), who accepted the new dynasty's domination; and, third, junior members of the ruling line. Before and even after they accepted Christianity, polygamy was usual – which made such junior royals a correspondingly numerous group, especially with a ruler like Vladimir who, as we know, numbered his concubines in the hundreds. Over time, the three became indistinguishable, between them eventually providing the nobility of the fully fledged kingdoms.

As in the earlier Germanic world, there was also an extensive free class operating in between the nobility and the unfree. They appear in some of the written legal sources from all the major kingdoms, except Moravia, which didn't last long enough to have any. Based on parallels with the rest of late first-millennium Europe, this group probably provided the bulk of military forces deployed by these kingdoms, beyond the specialist military retinues of the rulers. Elsewhere, it was customary for unfree populations to perform lower-status functions, such as providing the labour with which many of the striking monuments of these kingdoms were presumably constructed. Military service, by contrast, was higher-status, despite its obvious dangers.[33]

Even if you don't believe in an entirely egalitarian sixth century, Slavic society underwent a total restructuring between the years 500 and 1000. The sixth-century Slavic world evolved leaders who rose and fell in their own lifetimes, with no markedly hereditary element to their power. There is also no sign of a hereditary nobility, or of permanently unfree population groups. This might still have been true of at least some Slavic groups in the seventh century. The fact that a Frankish merchant, a complete outsider, like Samo, could still at this date be elected overall leader among a multiplicity of other Slavic *duces* would seem to indicate that such men were not sitting on top of a strongly hierarchical or hereditary social pyramid. But this had changed by the tenth century, and, equally important, the new states of the later period could never have appeared without these intervening transformations. Heritable dynastic power, the social and military clout of retainers and nobles, and a differentiated population who could be forced and/or persuaded to undertake necessary functions such as providing food and labour or military service: all of these were key elements of the new state structures, and none had existed in the sixth century.

The question of when the different elements of the restructuring had happened is very difficult to answer. The likeliest answer, as is so often the case, would appear to be mixed. Some of the change looks on balance to have had long roots. When it emerged from Avar control after Charlemagne's campaigns of the 790s, central European Slavic society already had the capacity to throw up powerful princes. Within a couple of decades, the chronicle sources give us a cluster of leaders – Voinomir, Manomir and Liudevit – capable of mobilizing significant military power for a variety of ends. This degree of control seems unlikely to have sprung up overnight and had probably been generated during the Avar period. This is also suggested by the fact that in both Moravia and Bohemia we find leaders – *duces* – with well-entrenched hereditary power over particular localities as early as the mid-ninth century. On the other hand, this observation needs to be balanced by the fact that most of the hillforts built in the Slavic world up to the ninth century appear to have been communally generated places of refuge. On excavation they characteristically lack any sign of an elite dwelling (often any permanent dwellings at all) or any other sign that the guiding hand of some great man was behind the project.[34] If a class of hereditary leader had emerged by c.800, then, it is important not to overestimate the extent of its social power.

Equally important, state-building had powerfully transformative social effects. Most obviously, the increasing wealth of particular dynasties led to the generation of retinues of increasing size and power. At the same time, much of the nascent nobility of the new states was a by-product of dynastic self-assertion, whether from the promotion of supporters and junior relatives, or the demotion of previously independent regional leaders. Castle-building in Bohemia, Moravia and Poland also involved the destruction of the old communal refuge-type hillforts and their replacement with new dynastic castles. And while the slave trade had certainly begun in the eighth century, it gathered pace dramatically in the ninth and tenth. Both of these latter developments probably played a major role in increasing the number of the unfree in the population (if not, perhaps, in initially generating them). My own best guess, therefore, would be that a longer, slower process of evolution had generated a body of hereditary group leaders by c.800 AD, but that the process of state formation after the collapse of the Avar Empire further revolutionized the situation.

How much of this broad model of social transformation is applicable to Denmark is a different question. State formation in Denmark differed substantially that in other cases, because there it was a question of state re-formation. A state-like structure comparable to the Slavic and Scando-Slavic ones already existed in southern Jutland and the islands from the eighth century, before being destroyed by the new wealth introduced into Scandinavia in the Viking period. As this would suggest and the sources confirm, Scandinavian society entered the last two centuries of the millennium with more entrenched social inequalities than was true of the Slavic world. Viking-period sources show us kings, jarls, freemen and unfree (thralls) already fully in existence. This makes good sense, not only because state-like structures already existed there but also because Scandinavia, or Denmark at the very least, had been part of the Germanic world (if belonging to its outer rather than inner periphery), interacting with the Roman Empire in the first half of the millennium, and had participated – as flows of Roman goods and bog deposits of weapons indicate – in some of the earlier processes of sociopolitical transformation. State formation in Denmark in the later ninth and the tenth century was probably much more a story of the promotion and demotion of existing power blocks and the dynasties at their heads, than of the fundamental social change that was central to the process among neighbouring Slavs.[35]

Social revolution, part cause and part effect, was absolutely necessary, therefore, to state formation at this time, at least in Slavic lands. But social change on this scale is never possible without parallel economic restructuring, and, again, there is plenty of evidence of this from contemporary central and eastern Europe. As with the social transformations, some of this preceded the formation of states and was a necessary precondition for their existence. Further change was then instituted by states themselves.

The hardest to document in a precise way is the development of the agricultural economy: food production. Not least, of course, is the fact that Slavic-speakers had come to dominate such a huge territory, with such a vast range of environments, that farming did not and could not have developed everywhere along a single trajectory. Nonetheless, the evidence indicates strongly, if at this point still rather generally, that farming outputs increased dramatically. At different speeds in different contexts, a revolution was under way that was bringing more land into production and instituting more productive farming practices, both in terms of the technologies employed and the management of land fertility. Most obviously, there was a substantial amount of forest clearance in central and eastern Europe in the second half of the millennium. In those parts of Poland with the right kind of lakes for taking pollen cores, the ratio of grass and tree to cereal pollen declined dramatically in these centuries from 3:1 to much more like 1:1, suggesting a doubling in the amount of cultivated land. This result can't be simply applied to the whole of Slavic Europe. I would expect the degree of change to have been less, for instance, as you moved north and east. Nonetheless, even within Mother Russia, the spread of Slavic-type cultures in northerly and easterly directions was closely associated with the spread of full-scale agriculture in the wooded steppe and reasonably temperate forest zones of the East European Plain. The phenomenon of general agricultural expansion is clear enough, then, even if it is impossible to put figures and dates on its impact in particular localities.[36]

The spread of more efficient farming techniques, too, is easy enough to document in outline. Initial contacts between Slavic-speakers and the Mediterranean world led some Slavs to adopt more efficient ploughs, types that turned the soil over so that rotting weeds and crop residues released their nutrients back into it. This increased both the yields that could be expected and the length of time particular fields might be cultivated. Further improvements had not yet worked themselves out fully by the time these states came into existence in the ninth and tenth centuries. The height of medieval sophistication in arable production, for instance, was the manor. Its advantages lay in the fact that it was a self-contained, integrated production unit with a large labour force, where farming strategies could be centrally directed towards greatest efficiency, particularly when it came to crop rotation for maintaining fertility, and costs (particularly of ploughing equipment) could be pooled and minimized. It was also an instrument for brutal social control, but that's another story. For present purposes, the point is that arable production in central and eastern Europe became fully manorialized only from the eleventh century onwards – after the new states had come into being. This finding came as a nasty shock for doctrinaire Marxists, since these were all supposed to be 'feudal' states, whose development was only made possible by manorializing agriculture, but the chronology is secure enough. Even if manorialization was still only nascent in the ninth and tenth centuries, however, we do have evidence that some key preparatory changes were under way. In particular, the amount of rye found in pollen cores in the second half of our period increases steadily. The use of rye, sown in the autumn rather than the spring, is associated with moves towards three-crop rather than two-crop rotation schemes. Three-crop schemes both increased the amount of land under cultivation at any one time (two-thirds rather than just a half) and preserved better fertility. This development perhaps also underlies the observation of the Arab geographers that Slavic populations gathered not one but two harvests each year.[37] There were important further developments yet to come, but much more food was being produced in central and eastern Europe by the year 1000 than had been the case five hundred years earlier.

This was critical to state formation in a number of ways. Until food surpluses were generated in substantial quantities, it was quite impossible (as was also the case with the Germanic world in the first half of the millennium) for leaders to maintain large specialist military retinues and other functionaries not engaged in primary agricultural production. Without economic surpluses, likewise, it was impossible for social differentiation to become entrenched. More food also meant more people,[38] and state formation could probably never have happened as it did without this increased population. It provided all the extra manpower required for ambitious construction projects. More important, but harder to measure in concrete terms, increasing the population density in central and eastern Europe increased substantially the competition for available resources. The need to belong to a group in order to flourish has always been a powerful reason why individuals are willing to accept the costs that usually accompany group membership. Put simply, one reason why peasants – or some of them – were willing to pay food renders and do labour service for rising dynasts was that they offered sufficient military organization to guarantee safe land retention in return.[39]

But much more changed in barbarian Europe between 500 and 1000 than the appearance of more people producing more food. Other economic developments were just as important to state-building, or nearly so. Military retinues, for instance, needed arms and armour, and, as we have seen, the new rulers of the tenth century controlled substantial reserves of precious metals (witness the gold cross erected by Boleslaw Chrobry over Adalbert's tomb) as well as all the other resources, apart from the mere physical labour, required to erect, decorate and furnish prestige projects such as the cathedrals and palace complexes that are such a marked feature of the period. In part, the rulers' capacity to do so arose from some general processes of economic development affecting the whole of central and eastern Europe at this time, processes beyond their control. In part, it was fostered by particular policies adopted by the dynasts themselves.

The biggest non-agricultural economic phenomenon in these years was the rise of an international trade network in furs and slaves. Some of its western axes began to operate in the seventh century, but it was in the eighth that it stretched into the Baltic, and the early ninth before it exploded more generally across eastern Europe. We've already encountered the role of waterborne Scandinavians in making this network's longer-distance connections work, and its central role in the whole Viking phenomenon. Consequently, the period saw a huge outflow of raw materials – largely people and furs – from northern and eastern Europe, and a flood, in return, of due payment. Large quantities of Byzantine silks were presumably one part of this, but few traces of these materials survive in our written sources, and none archaeologically. Beautifully finished silks were the main item Byzantium had to offer in exchange for imported items. What the archaeological record has produced in vast quantities, however, is silver coinage, above all from the Muslim world but from western Europe as well. These coins survive in quite astonishing quantities: as we saw in Chapter 9, over 220,000 Muslim coins in hoards of five or more according to recent estimates. This is all the more impressive given that silver has never been without value. The coins that survive are probably no more than the tip of a silver iceberg that has been reworked many times in the intervening centuries.

Kings, of course, were hugely interested in such massive flows of wealth. Not least, they could milk it for tolls, offering merchants safe places for making their exchanges and charging fees in return. Right at the beginning of the ninth century, the then King of pre-collapse Denmark, Godfrid, forced merchants who'd previously operated on Slavic territory at the trading centre called Reric to move to his newly built trading station at Hedeby in southern Jutland. The move is recorded in a contemporary source, and has found archaeological confirmation. The dendrochronology dates of early timbers recovered from Hedeby show that, as the text reports, it was built in c. 810 when Godfrid was at the height of his powers. The king's interests here can only have been financial. Prague, likewise, one of the key centres of the Premyslid dynasty of Bohemia, was also, as the Muslim geographers report, a major entrepôt in the slave trade. The tolls must greatly have swelled the coffers of the dynasty, and the trade's importance was such that one of the reasons given for Wenceslas' killing is that he was attempting to outlaw it.

Kiev, too, new home of the Rurikids in the tenth century, was a trading entrepôt of huge importance. Byzantine and Russian sources both confirm that it was the starting point for the Rus trade fleets that came to Constantinople every spring from the early tenth century onwards. And by the early eleventh, Thietmar of Merseburg tells us, the city boasted no fewer than eight markets. Only in the case of Poland do we lack explicit textual evidence of its participation in the new international trade networks, but this can only be an accident of (non-)survival. The territory of the Piasts has thrown up such a density of Muslim silver coin finds that there is no doubt that its population was in some way involved in the new trading networks.[40] So all of our new dynasties effectively tapped in to the new wealth being generated.

Nor was their role confined merely to taking tolls. They also sometimes interfered actively to reshape, as it were, the networks and maximize their own profits. This is easiest to demonstrate in the case of the Rurikids. In the tenth century the dynasty mounted collective military action on two occasions, in 911 and 944, to force the Byzantine authorities in Constantinople to grant Rus traders increasingly favourable trade terms, including the stipulation that they would get free bed and board inside the city for a month while conducting business. Not surprisingly, given the origins of Scandinavian interest in Russia, the treaties show us that members of the Rurikid dynasty (and not just its current head) were themselves active traders, so they had every reason to want to increase activity and market share. But I would not jump to the conclusion that this was an enterprise limited to the Rurikids. Byzantine–Rus trade connections just happen to be comparatively well documented, and it seems to me entirely likely, although undocumented, that other dynasts took an equally active interest in developing international trade links along the particular lines that best suited them.[41]

Even though we don't know as much about this as we would like to, a good case can be made that all these dynasties took a proactive role in organizing the economies of at least their dynastic heartlands.

The picture has emerged from a mix of historical sources and archaeological investigation. Looked at archaeologically, the tenth-century dynastic heartlands of Poland and Bohemia are striking for the relative density of their populations. Once again, this observation is based upon detailed knowledge of pottery chronologies, whose spread gives you a reasonable guide to the existence of settlements. The evidence suggests that this population density was not an accident, but the outcome of deliberate interference. As the archaeology has shown, a key moment in the rise of both Premyslid and Piast dynasties came when they destroyed the defended centres associated with the old sociopolitical structures, or 'tribes', and replaced them with their own chain of castles, in the later ninth and the earlier tenth century respectively. This process, it seems, was accompanied by the deliberate transfer of at least some of the subdued population groups to the dynasty's heartland. Some transfers are mentioned in our sources. The *RPC*, for instance, associates Prince Vladimir in the late tenth century with a mass transfer of various population groups – Slavs, Krivichi, Chud and Viatichi – to different places along the River Desna. Here, persuasion rather than force may have provided sufficient incentive for the move. In other cases – Poland and Bohemia – all we have is the archaeological reflection of the effects of such a process in the sudden creation of an unusually dense population cluster, but early texts (all gifts to monasteries) from both these kingdoms (and, indeed, from Russia too) make clear the purpose of these resettlements. The classic economic form to emerge from these early texts is the 'service village', already mentioned. These unfree villagers were required to fulfil particular economic functions for the king, such as bee-keeping or horse-breeding, in addition to providing standard food renders. The fact that they were unfree strongly suggests their origins lay in forced resettlements.[42]

Kings – or kings and their advisers – were economically alert enough, therefore, to maximize the exploitation of their dynastic heartlands. The way this was done suggests that they were operating in a world where there was little in the way of a functioning market economy in agricultural goods, since instead of simply being able to purchase the required items, specialist tasks had to be assigned to particular settlements. This is not surprising. The same was true of the ninth-century Carolingians who were still using service villages in certain areas, and is in line with both the coinage evidence and the fact that peasants owed the king food renders rather than cash taxes. North and eastern Europe was still at the time an economy lacking small change. Muslim silver coins were plentiful enough, but these were of much too high value to do your everyday shopping with. Use one of them at the baker's and you'd come home with a few sackfuls of bread, which would have gone stale long before you ate it. Likewise, although kings generally preferred cash taxes because they were infinitely more flexible, they could demand them only when the possibility existed for peasants to sell on any surplus production to merchants.

All of this adds another dimension, of course, to the tendency of these late first-millennium states to operate with a distinct centre–periphery dichotomy. Not only was this an accidental offshoot of the logistic limitations of itineration, but it reflected something more fundamental about the states' construction. Thanks to the policies of the triumphant dynasties, core and periphery were also distinct in population density and economic organization. In the case of the Kievan Rus state, because of the peculiarities of its origins, the process of core creation had an additional importance. Up to the mid-tenth century, the Riurikid dynasty displayed a distinct capacity to shift its centre of operations about. It first came to prominence in the north with an initial seat, it seems, at Gorodishche (old Novgorod) before transferring, as we have seen, south to Kiev on the River Dnieper in the late ninth century, when Oleg came south with his army.

The reasoning behind this transfer requires more than a little puzzling out. At first sight, it seems an odd move, since Gorodishche, as noted earlier, was better placed for controlling trade flows along the Volga to the Muslim world, which was a far richer trade route than its counterpart along the Dnieper to Constantinople. Kiev, however, had other advantages. Situated on the wooded steppe, Kiev was excellently placed to dominate the surrounding regions of what is now Ukraine, which had become home in the seventh and eighth centuries to a large Slavic agricultural population. This was organized into units with their own substantial political structures before the arrival of any Scandinavians. The Polian dominated the area immediately around Kiev, the other groups in the vicinity being the Derevlians, Severians, Radimichi and Dregovichi (Map 19). While less well placed in purely trade terms, Kiev offered Scandinavian dynastic wannabes far more in the way of exploitable human and economic resources. Already in the time of Oleg, the *RPC* tells us, the Grand Prince's army consisted not just

of Scandinavians, but of Slavic- and Finnic-speakers. For the Grand
Prince, who wanted to be much more than a merchant prince, Ukraine
had much more to offer than the north. Even so, the Riurikids were
not quite yet ready to give up the gypsy life. Oleg's son and heir
Sviatoslav engaged in wide-ranging campaigns, as far east as the Volga
and south all the way to the Caucasus. The *RPC* reports that, just
before his death, he was contemplating relocating the capital of the
dynasty a third time – to the Danube. The work of Sviatoslav's son
and eventual heir Vladimir, in generating a much stronger economic
core in and around Kiev along the Desna, had the particular effect of
rooting the Riurikid state once and for all in its Ukrainian heartlands.[43]

There is, of course, much else we'd like to know about how state-
building intersected with the economic development unfolding in the
later first millennium. One huge gap is the lack of detailed information
about iron-mining and steelworking. The kinds of armies deployed by
the new dynasts imposed a huge demand for these items, but there is
no good information on how this was satisfied. Nonetheless, the big
picture is clear enough in outline. State-building would have been
impossible without a number of pre-existing social and economic
transformations of fundamental importance: the generation of a much
more productive agriculture, the substantial population increase that
followed on from this, a huge increase in the amount of movable
wealth, and the more developed social hierarchies that formed around
its unequal distribution.

But if these huge changes provided the necessary backdrop, the
dynasts were themselves responsible for further social and economic
developments of profound importance. On the trading front, they not
only milked the new international networks for all the tolls they could
get, but actively extended them, as and when they could. At home,
likewise, agricultural production in the core areas of their kingdoms
was maximized. None of these processes was complete by the year
1000. Not only was agricultural production reorganized on a manorial
basis from the eleventh century, but the grandsons and great-grandsons
of the dynasts would also famously employ recruiting agents to bolster
output still further by making hundreds of thousands of German
peasants offers they could not refuse to move eastwards.[44] Even if all
this was still work in progress by 1000, we have arrived at a partial
answer to our question. Normal dynastic ambition produced entirely
abnormal results during this period because deeply rooted social and

economic transformation meant that the dynasts were pushing at a door that was already opening.

But this, at best, is still only half an answer. Dynastic ambition provided the activating mechanism for state formation; being both cause and effect of the political revolution we have been observing, massive social and economic transformation was its necessary backdrop. But what underlay those initial social and economic transformations that gave such free range to dynastic wannabes to remake the map of central and eastern Europe?

THE RISE OF THE STATE

In Soviet days, everything was so simple. From the fifth century onwards, often living alongside an existing population, egalitarian Slavs took possession of the landscape of what would become Slavic Europe. Then followed a long, slow process of internal social and economic evolution over the next four to five hundred years, until classes formed around manorial estate-based agriculture and the first states appeared, based on the unequal distribution of control of the agricultural means of production. This always looked more like a Marxist fairytale than anything to do with historical reality, and all the more recent work has only emphasized what a dramatic story state formation actually was. In many places, even initial Slavicization occurred maybe a hundred and fifty years later than used to be thought, manorialized agriculture followed state formation rather than preceding it, and archaeology has brought to light a sudden and violent final stage, where rising dynasts used military muscle rather than long-term socioeconomic evolution to destroy one sociopolitical order and replace it with their own. What we have to explain, all this emphasizes, is why in the ninth and tenth centuries dynasties were suddenly able to grasp the reins of power with so much vigour. Just as with the appearance of larger political structures in the Germanic world in the first half of the millennium, a key role was played by an increasingly dense network of contacts that grew up between Slavic Europe and its more developed imperial neighbours.

Empire Games

In central Europe, the Slavic world was in direct contact with two successive imperial states: the Carolingians in the eighth and ninth centuries, and the Ottonians in the tenth. Neither was as robust as their Roman predecessor, but each was more than powerful enough in their heyday for their military and diplomatic priorities to have major consequences for neighbouring Slavic and Scandinavian societies. Just as in the Roman period, the most immediate type of contact between Carolingians and Ottonians and their neighbours was imperial aggression. Both these late first-millennium empires built their internal political coherence around the distribution of gifts to local elites, who not only ran their localities but also provided emperors with military muscle.

But Carolingian and Ottonian emperors lacked not only the huge financial reservoir that control of the developed Mediterranean world had brought their Roman counterpart, but also any thoroughgoing powers of taxation even over such lands as they did control. As a result, their gifts to local elites often took the form of non-renewable assets such as land from the royal fisc, which generated a tendency for these empires to fragment from within, as power built up in local hands. The only thing that could square this circle was expansion, providing rulers with an alternative form of renewable wealth to large-scale taxation, and allowing them to reward local elites and maintain their own power at the same time. If anything, therefore, the maintenance of central power was actually predicated more upon expansion in both of these later empires than was the case even with Rome, and both preyed upon their neighbours as and when they could. In the ninety years separating the accession to power of Charles Martel in 715 and the later years of his grandson the Emperor Charlemagne, Carolingian armies were predatorily active in the field for all but five of them. In the first half of the tenth century, likewise, a steady *Drang nach Osten* on the part of what was originally the ruling ducal dynasty of Saxony was a key reason why Henry I and his son Otto I were able to beat off all comers and turn themselves into the Carolingians' imperial heirs.[45]

Predatory expansion always produced pillage, but its real benefit was more structured, long-term exploitation. Even territories not fully subdued were part of the story. From the time of Henry I, Bohemia had to pay an annual tribute, and after 950 owed military support to the dynasty. For about a decade from the mid-960s, likewise, Miesco I of Poland was cast in the role of tributary. For territories more thoroughly under the imperial thumb, the weight of exploitation was correspondingly heavy. Successful campaigns against the so-called Elbe Slavs (small-scale groups who lived broadly between the Rivers Elbe and Oder (Map 14) in the first half of the tenth century) allowed the Ottonians to establish nine major collection centres (called 'towns', *urbes* – or burgwards in our sources) east of the Elbe. These received what the charters euphemistically call 'annual gifts' from the Slavs, some of the proceeds of which were divided between the Ottonians' two favourite ecclesiastics, the Archbishops of Magdeburg and Meissen. It is the charters granted these sees by Otto that document the arrangement. Nor was it just ecclesiastical institutions that benefited from the flow of the new wealth. Frontier commands in what can only be called this colonial situation – called 'marches' – were in huge demand among Otto's magnates because of the opportunities for enrichment they offered. Their distribution was even the source of ferocious feuds within magnate families, when one branch received a nice juicy plum but another did not. Most famously, this was the origin of the bad blood between two brothers, Herrman and Wichman Billung. Herrman got the key appointment and was for ever loyal to Otto; jealousy led Wichmann throughout his long and bitter life into the camp of any opposition to Otto, whatever the issue.[46]

It is not the effects of expansionary policies upon the empires that we're primarily interested in, however, but how the exploited populations responded to this asset-stripping. They reacted exactly as you might expect, attempting to resist imperial expansion outright, and/or to minimize its effects when total resistance was impossible. In particular – and this is why imperial exploitation is so relevant to political consolidation, the subject matter of this chapter – uniting several originally independent, small-scale political units into a smaller number of large ones was one of the most effective strategies available to those seeking to fend off unwanted imperial attention.

The best example is provided by what proved in the long term to be a failed state formation: the Elbe Slavs again. As we've just seen, they felt the full weight of Ottonian imperialism. In 983, however, taking advantage of the dynasty's difficulties elsewhere, they rose in

massive and – in the short term – successful rebellion. Their resentment against Ottonian rule, and especially against the Church institutions that had been profiting so substantially from their exploitation,
manifested itself in a series of atrocities against churches and churchmen that are lovingly chronicled in our Christian sources. What's
particularly striking about this revolt, though, is the element of political
restructuring that was central to its success. When the Elbe Slavs fell
under Ottonian domination, they comprised a group of small-scale
political societies. The success of the revolt, however, was predicated
on the generation of a new alliance among them, manifest in the new
label 'Liutizi' which our sources start to give them in its immediate
aftermath. The Liutizi were not a new people, but old ones reorganized. As their counterparts among the Germani on the fringes of
the Roman Empire had come to appreciate so long ago, so had the
Elbe Slavs learned from bitter experience that hanging around in
larger numbers made it possible more effectively to resist imperial
aggression.[47]

Nor, when any initial phase of expansion and conquest was over,
was it necessarily any easier to have such imperial neighbours. The
reign of Otto III was marked by a spell of excellent relations between
himself and both the Bohemians and the Poles, culminating in the
Emperor's progress to the tomb of Adalbert. After Otto's death,
however, imperial policy changed dramatically. He left no son, and
from the accession of his cousin Henry I in 1003 there followed some
twenty years of pretty continuous warfare between the Empire and
the Piast state, in which the new Emperor was happy to use the pagan
Elbe Slavs as allies against his erstwhile Christian brothers. Henry,
of course, had his reasons, but this kind of inconsistency in policy
reflected the fact that populations beyond the Elbe were viewed as
substantially second-class citizens, which meant that the desire to
exploit them was always likely to be perceived as legitimate and hence
to reassert itself. This underlying attitude was not informed by such a
thoroughgoing, denigrating ideology as that of the Romans, whose
entirely coherent, and equally unpleasant, vision of 'barbarians' allowed
them to be treated like beasts, if that was convenient. The Poles and
Bohemians were partly protected by being Christian. It was no accident
that it was the pagans of the Elbe and Baltic regions, in later centuries,
who would eventually feel the full brutal weight of an ideologically
self-righteous form of imperialism: the so-called Northern Crusades

which saw Christian Teutonic knights, amongst others, burn and kill their way north and east. Nonetheless, even Christian Slavic states suffered from second-class status, and could never be sure that the instinctive imperial desire to profit from the exploitation of outsiders might not reassert itself at their expense.[48] And, in fact, the diplomacy of the ninth and tenth centuries throws up many examples of the same kind of change in imperial policy that the Poles suffered from in the early eleventh.

In the later eighth century, for instance, when Charlemagne was engaged in his long and tortuous conquest of Saxony, one particular group of Elbe Slavs, the Abodrites, were key allies. Established on the Saxons' eastern border, where the Carolingians were attacking from the south and west, the Abodrites provided an extremely useful second front, and Charlemagne was duly grateful for their support. In return he gave them extra territory, direct military and diplomatic support and trading privileges. Once Saxony had been absorbed into the Empire, and particularly when it became the seat of Empire, the Abrodites found themselves surplus to strategic requirements. Instead of useful allies, they started to look like potential subjects, and imperial policy swung 180 degrees. Even when they were not being conquered outright, aggressive diplomacy became the order of the day, reaching its culmination in a murderous banquet organized by one frontier commander, the Margrave Gero, at which thirty of their leading men were assassinated. This dwarfs the dinner-party assassinations even of the Roman era, which were regular events but usually took out only one leader at a time. The uncertainty of life on the edge of the Empire was central to the experiences of successive rulers of the Moravians in the next century, too. Take the early years of King Zwentibald. He first came to power with the help of the east Carolingians in 870; then, in just three years, as imperial policies shifted, found himself imprisoned for several months before we finally see him raiding Bavaria in retaliation for his treatment.[49] It was living close to a powerful Empire but holding – in the view of a large section of its citizenry – second-class status that laid you particularly open to these kinds of dangerous changes of policy. It was always possible for some faction within the Empire's ruling circle to make political capital for itself by championing a harder – and more profitable – line towards you.

Examples could be multiplied, but there's no point. What's interesting here is the overall effect of imperial predation on the societies at the sharp end. The Elbe Slavs' revolt provides a particularly arresting example of well-founded resentment in action, but the effects were similar elsewhere. The natural suspicion of the Moravian dynasty, for instance, shows up in the religious sphere. Along with many of our new dynasties, the new ruling line quickly decided to opt for Christian conversion. Rather than simply accepting it from the Carolingians, however, they explored every other possible avenue, famously importing the Byzantine missionaries Cyril and Methodius, with papal blessing, in 863. This was done in the teeth of sustained Carolingian resentment, however, and shows very clearly the suspicion in which the Empire and all its doings were held. Eventually, after Methodius' death, the Moravians were forced into religious line and his remaining disciples expelled in favour of Frankish clergy in 885, but the expectation of imperial exploitation remained, manifest not least in the incident of 882 with which this study began, when Zwentibald the Duke of the Moravians and his men captured the Frankish Counts Werinhar and Wezzilo and cut off their tongues, right hands, and genitals.

They were out for revenge because of the way Engelschalk had treated them when he had been in charge of the same frontier region, and were trying to prevent Engelschalk's sons from seizing their father's old job. The Moravians had an entirely coherent agenda here, and their revenge was very symbolic. I am obviously not privy to the mindset of your average ninth-century Moravian, but this was clearly a case, in the best Mafia tradition, of mutilation with a message. My best guess would be that cutting off the right hand and tongue emphasized that neither deed nor word could be trusted, while removing the genitals expressed the hope that this line would have no further progeny. Taken to these lengths, natural resentment against imperial military and diplomatic aggression could become a building block, both practical and ideological, by which new dynasties could extract consent for their rule. Becoming part of a larger entity always involved taking on obligations of service, but these might be acceptable if, as a result, the worst effects of predation were fended off. And although the Elbe Slavs and the Moravians provide the most explicit instances, there is every reason to suppose that imperial predation had the same effect in all of the border states: Poland, Bohemia and Denmark as well.

If this, what you might call 'negative benefit', was the main effect of military and diplomatic contact with Empire on the capacity of our dynasties to build their state structures, there were also more positive ones. On occasion, when imperial policy was in your favour, there were great photo opportunities. Otto III's great progress to the tomb of Adalbert was a stupendous international occasion, and Boleslaw Chrobry, like many a modern leader at a summit meeting, must have been extremely happy to have his subjects see how highly he was regarded by the reigning Emperor. On the other hand, of course, it may just have been the sight of that massive cross of solid gold hanging over the tomb that set calculators whirring in the brains of some of the Emperor's entourage as they worked out exactly how much wealth might be won from a successful war against the Piasts (leading eventually to twenty years of warfare, but that's another story).

Not, of course, that the violence ran only in one direction. Just like their imperial contemporaries, these new dynasts had to win political support from their magnates to rule effectively, with gift-giving just as much the order of the day east of the Elbe as west of it. Their marriage policies, if anything, made the problem worse. Under the influence of Christianity, they did begin the move from outright polygamy to serial monogamy, but multiple wives and plentiful offspring were the rule – if not quite on the scale of the seventh-century Samo, who ended up with twelve wives and thirty-seven children. This marital profligacy meant that succession disputes and dynastic infighting were extremely common. Yaroslav the Wise, son of Vladimir, for instance, secured his power in 1018 only after a lengthy civil war against his half-brother Sviatopolk that saw many ups and downs and the deaths of at least three other brothers and half-brothers. And Vladimir himself had had to fight a similar war in the 970s against his half-brother Yaropolk, with similar numbers of dynastic casualties. These kinds of wars could be won only by mobilizing a wide range of support among magnates and retinues, which required wealth distribution on a considerable scale. And, just as was the case in the Roman period, leading successful raids on to the richer and more developed soil of an imperial neighbour was an extremely effective mechanism when it came to securing that perfect gift without bankrupting yourself. Accounts of the counter-raids of the Elbe Slavs, not surprisingly, focus on their propensity for smash-and-grab, but the same was true, in only a slightly more structured way, of all our other frontier dynasties. Each outbreak of trouble with the Moravians in the ninth century, or the Bohemians and Poles in the tenth, was accompanied by its due measure of wealth liberation.[50]

Aside from wealth and prestige, close contact with an imperial neighbour also helped secure the power of new dynasties in some more precise ways. Two leap out of the source material. First, the tenth-century Slavic states were entirely up to date in their modes of warfare, possessing armoured knights aplenty. This had not been the case before 800 AD. In the ninth century, even Saxon contingents within east Frankish, Carolingian armies at first took the form only of infantry and light cavalry. The heavily armed, mailed Saxon cavalry of the Ottonian period emerged only in the late ninth and the early tenth century, as the Saxons finally caught up with the times. Against this backdrop, it is very striking that tenth-century Bohemian and Polish armies also boasted at least some heavy armoured cavalry. We know little, if anything about Slavic warfare before 800, but it's a pretty fair bet that if even the Saxons didn't have the latest military hardware at that point, then neither did the Slavs. So where did knowledge of it, and access to it, come from over the next hundred years? The likeliest answer is that it actually came from the Empire, slipping eastwards over the Elbe. Already in 805, in a capitulary issued at Thionville, the Emperor Charlemagne attempted to limit trade with the Slavic world to a number of designated points along the Elbe frontier, including Bardowick, Magdeburg, Erfurt, Hallstadt, Forchheim, Regensburg and Lorch, not least because he professed himself worried about arms shipments. This immediately makes you think that arms were flowing pretty freely across the frontier, since even imperial states of the first millennium lacked the bureaucratic machinery to maintain effective border controls. Obviously, the idea of state-of-the-art Carolingian hardware was highly attractive for Slavic groups who might have to fight off Frankish armies, but such imports also had important internal political effects. Not for nothing did early modern European populations associate standing armies with royal autocracy. Acquiring the kind of military equipment that made his forces militarily dominant also put a nascent dynast in the perfect position to face down internal rivals and suppress dissent. Importing imperial military technology, therefore, directly advanced the process of state formation in the periphery.[51]

With this in mind, the economic organization of the core areas of these new states is also interesting. As we have seen, all quickly evolved a loose pattern of great estates, where particular service villages fulfilled specialist functions in addition to providing basic food supplies. This mode of organization was also prevalent in the ninth-century Carolingian Empire, particularly in its less economically developed reaches east of the Rhine. This was, perhaps, just a sensible way to ensure vital products in pre-market-economic conditions, and arrived at entirely independently east of the Elbe. There must be at least some chance, however, that we are looking here at further, slightly more unexpected fruits of close contact between Empire and periphery.

Compared with the Roman period, what's missing from this cocktail of contacts is the kind of diplomatic manipulation in which Roman emperors excelled, systematically promoting particular dynasts by rearranging prevailing political geographies in their favour because they seemed to promise the best hope of medium-term frontier security. Carolingian and Ottonian emperors did at times promote their particular favourites, such as the Abodrites, but there is no sign in the sources that they attempted consistently to interfere with the political structures of their neighbours. There is a good chance, however, that the diplomatic agendas of a different Empire may have played an important role in the earlier stages of these processes of transformation. Moravia, Bohemia and, to an extent, Poland, can all be seen as successor states to the Avar Empire destroyed by Charlemagne just before the year 800. We don't know a huge amount about the internal running of the Avar Empire, but what there is would suggest that it functioned very much along the lines of that of the Huns. Certainly, like the Huns, the Avars operated an unequal confederation where the military power of their originally nomadic core was mobilized to hold a range of initially unwilling subjects to an Avar political allegiance. There was a range of more and less favoured statuses that subjects might occupy within this overall pattern. The most interesting snapshot of its operations that we have describes how one group, descendants many of them of Roman prisoners, achieved free (as opposed to slave) status, and were granted thereby their own group leader. This does sound like the Hun Empire too, and would suggest that, lacking any complex government machinery, the Avars tended to rule their subjects through trusted allied princes. If so, it is very likely that Avar rule will have cemented further the power of the kind of leaders who were appearing anyway in some Slavic groups by c.600, as they began to control the flows of new wealth coming across the east Roman frontier in particular. This combination – of sixth-century development reinforced by subsequent Avar diplomatic manipulation – is the likeliest explanation, in my view, of why the collapse of Avar rule was marked by the swift appearance of a series of Slavic leaders of seemingly substantial and established authority.[52]

Overall, military and diplomatic contact between these new states and the adjacent Empire thus took many forms. Imperial attentions were in general predatory, resulting in a huge groundswell of aggression flowing across from the imperial side of the frontier. This was matched, when conditions were right, by a countervailing tendency on the part of the new states, or factions within them, to raid the rich assets available west of the Elbe. So much is only what you might expect, but both phenomena had a strong tendency to advance state formation, giving nascent dynasts ideological cement or just plain cash to employ in advancing the process. Alongside these major themes of contact went some subthemes that also pushed the process forward: exports of military and other technologies, and occasional moments when benevolent imperial attention advanced the capital of particular dynasts.

Looking at the broader patterns of development from the ninth century onwards, two other points are worth making. First, as regards the two effects of proximity to an empire, the 'negative benefit' – using its aggression as ideological cement for your own state formation – and the 'positive benefit' – being able to raid it as a source of ready funds – a comparison of the fate of the Elbe Slavs with the Piast and Premyslid states suggests that the former was the more important. While the Elbe Slavs were in the better position for raiding, being situated right on the imperial frontier, and indulged in it aplenty, this also meant that it was too easy for the Empire to get at them in return. And, of course, the whole point about an imperial power is that, when it put its mind to it, and other factors were not interfering, it was that much more powerful than any surrounding states. There could only be one victor, therefore, in a head-to-head collision between the Empire and the Elbe Slavs. Poland was significantly further away, insulated geographically from immediate imperial aggression, while the upland basin of Bohemia enjoyed the stratigraphic protection of the Bohemian Forest, the Ore and the Krkonoše Mountains. By itself, then, ready access to raiding was not a sufficient basis for state formation. It was a useful additional resource, but only if you could survive imperial counterattack and use to your benefit all the resentment that this would generate.

Second, these different types of contact, both positive and negative, pushed the target societies in the same direction in the longer term – providing, that is, you were in a position to survive attempted imperial conquest. The unifying force of the struggle to survive, the effects of occasional imperial approbation, the flows of raided funds, exports of military hardware and administrative acumen: all strongly facilitated the ability of nascent dynasts to advance towards regional domination. Nor was this pro-dynastic effect limited just to military and diplomatic interaction.

Globalization

It is clear that for their state-building operations to be successful, the dynasts needed the consent of some of the population groups caught up in the process. In the cases of Moravia, Bohemia and Poland, at least, the new dynasts initially rose to prominence within their own local grouping, or 'tribe', and were then able to win consent for their wider regional ambitions, depending, presumably, on the degree of success they had already achieved. Even when the larger state structures had come into being, rulers still needed that consent, certainly from the *optimates* of the core heartland, and probably also from a wider free class, if we are right in seeing such a social grouping as playing a major role in Slavic society at the turn of the millennium. At the same time, other dimensions of state-building were based on the exercise of brute force. Not least, extending power beyond your original group involved destroying the hillfort refuges of nearby populations and resettling many of them in your own core areas. Large and well-equipped military retinues were a key component in the new state structures, and it is hard to conceive of these shock forces not playing a major role in the destruction of the old political order and accompanying population displacements.

What all this highlights is the overwhelming importance of the dynasts' ability to accumulate wealth in unprecedented concentrations. Military retinues used up huge amounts of it. Obviously, they required feeding, lots of feeding. All the comparative evidence on warrior retinues, and some specifically relating to the new states, suggests that being fed on a heroic scale was a basic expectation. This was not just a matter of greed. They tended to spend mornings hitting large bits of wood with double-weight swords (to build skill and muscle strength) and engaging in mind-expanding activities, all of which used up a lot of calories. Feeding the brutes, however, was not the half of it. As we have seen, a striking feature of the retinues of these new kingdoms is their state-of-the-art arms and armour, especially the defensive armour, which was massively expensive, whether bought – as one suspects it was in the first instance, given the Thionville capitulary – from Frankish gun-runners or produced at home, as it eventually must have been. It was the military potential of the retinues that allowed the rapid, violent expansion that is so characteristic of Premyslid, Piast and even Moravian state formation. But creating them required huge amounts of cash. The obvious questions, therefore, are where did it all come from, and how did the dynasts manage to get their hands on it?

Looking at central and eastern Europe between 800 and 1000, far and away the most likely answer is that they were drawing funds from the new international trade networks in furs and slaves. Again, there are some similarities between this phenomenon and the processes that earlier transformed the largely Germanic societies on the fringes of the Roman Empire. There, the Roman standing armies were a constant source of demand for agricultural products and for labour of all kinds, whether in the form of extra soldiers or just as slaves. As we have noted, the steady flow of payments back across the frontier then helped create the new social structures that underpinned the larger Germanic confederations in the later Roman period. There are, however, some key differences between the two contexts. First and foremost, there is the size of the operation. The fur and slave trades of the later period operated on a much greater geographical and monetary scale than any Roman counterpart. Slaves, of course, were always expensive items, but the fur trade, unmentioned in sources from the first half of the millennium, was much more valuable than any Roman trade. Also, there is no sign that slaves were coming from as far north and east during Roman times. I am not inclined to think it accidental, therefore, that the operations of the later trade networks should have left more trace in our sources, both historical and archaeological, than any of the earlier commerce.

Second, and one of the reasons why the scale was so much larger, the late first-millennium network operated with multiple sources of imperial demand for its high-value goods. Demand seems to have originated in western Europe, with goods even from northern Russia being shipped there from the mid-eighth century. The trading station at Staraia Ladoga came into being a couple of generations before any Muslim connection had been established. This makes perfect sense, since increasing demand in western Europe at this point coincides with the rise of the Carolingian dynasty. But an Islamic dimension soon came into play. Not long after 800, Muslim silver coins started to flow north in vast numbers, part of the trade having now been diverted to a second set of customers, the elites of the Abbasid Empire. This was the greatest state of its age, and so demand from there soon dwarfed the west's, to judge at least by the amount of Muslim silver that ended up in the Baltic region. The Muslim connection was not broken even when the Abbasid Caliphate collapsed in the early tenth century, since a great successor state quickly arose under the control of the Samanid dynasty of eastern Iran whose silver mines made them fabulously wealthy. Sometime in the mid- to late ninth century, finally, Constantinople came into the picture. Much less wealthy than the Muslim world, it was nonetheless a distinct third centre of elite demand.[53]

The relative proliferation of sources also allows us to explore the operations of this trading network in more detail than was possible for its Roman-era counterparts. We have already come across some of the major waterborne routes that Scandinavian adventurers opened up in the ninth century: particularly, down the Volga and its tributaries to the Muslim world, and down the Dnieper and across the Black Sea to Constantinople. There were also land routes running through central Europe into the west, on which Prague was a major staging post. We can also, importantly, say something about where the slaves were generally being captured. The Arab geographers report that the Rus raided westwards for their victims, while the 'western Slavs' raided eastwards. Confirmation of this picture is provided by the distribution of the Muslim silver coins that came back north in return for all the slaves and furs. Striking concentrations emerge. Two are where you might expect: along the Volga and its tributaries, and in Scandinavia. A third, however, lay between the Oder and the Vistula, right in the heartland of the Piast state. Even more arresting is the complete absence of coins in the immense tracts of territory east of the Vistula and north and west of the Dnieper. Pretty straightforwardly, then, the coin distributions confirm the reports of the Arab geographers. The areas without coins are precisely those from which the slaves were being extracted, caught between the rock of the Rus and hard place of the west Slavs.[54]

This suggests some further thoughts about how, precisely, the new dynasts were making money out of these international networks. All were busy extracting tolls, but the Rurikids, as we have seen, were doing much more than that. Active participants in the networks, they were also to be found developing markets, not just taxing them. And given that much of what was being traded was actually slaves, there might well have been an intimate link between the evolution of the new networks and those eminently important military retinues. Violence and terror are the order of the day with slave trading, not just because individuals resist capture, but also because the cowed and terrified are that much easier to transport. I remember as an undergraduate picking up the standard textbook on medieval slavery and glancing through it in an idle way because it was written in French and the subject was not absolutely central to that week's work. But my attention was attracted by a map that appeared to have a series of battle sites marked by the usual crossed-swords symbol. This seemed odd. On closer inspection, the symbols were not crossed swords, but scissors, and the legend read 'points de castration'. This does not need translation. Nor did women fare much better. The Arab geographers certainly enjoyed the barbarous nature of the northern societies they were describing and deliberately underlined the total ghastliness of the Rus slave traders. Ibn Fadlan describes them as the filthiest of God's creatures, emphasizing the unpleasantness of their personal hygiene habits. He also refers only to females and children among the slaves being sold down the Volga, taking a voyeur's delight, too, in how much sex went on between the slavers and their victims.

It's hard to know quite what to make of this. The literary accounts could make you think that the trade with the Islamic world was entirely in women, but I don't know whether to believe this or not. Perhaps the huge distances involved meant that shipping males was just too dangerous, since, although moving on water, the potential refuge of riverbanks was never that far away, unlike in the later Atlantic slave trade. I don't have any doubt, however, that sexual exploitation was a major feature of the action. It always is, in the case of women and slavery, and you have to wonder where Vladimir obtained the three hundred concubines he kept at Vyshgorod, the three hundred at Belgorod and the two hundred at Berestovoe.[55]

The real point, though, is that highly trained, well-equipped military retinues were an excellent tool not only for state-building, but for capturing slaves as well. Some of the raiding was done by intermediaries, but the Rus did a fair amount of their own dirty work, and there is every reason to suppose that this was also true of the west Slavs, probably the retainers of both Piasts and Premyslids. As we have seen, many Muslim coins have turned up on Piast territory, and their lands were conveniently near to the areas from which both texts and the absence of coins tell us that slaves were being taken. To my mind, it is not too much of a stretch to suppose that, like their Rurikid peers, the Piasts built up the military capacity of their retinues not only from toll revenues but also by actively participating in the international slave trade.

The point about the new trading connections is not just that they generated new wealth. At least as revolutionary as the wealth itself was the multiplier effect stemming from the fact that new power structures evolved to maximize and control the direction of the flow. Just as in modern globalization, new connections generated big-time winners but also decided losers. The biggest winners were the new dynasties and their chief supporters: the leading men behind them and their military retinues. The chief losers were of course the slave-producing populations, but also the ascendant dynasts' near-neighbours, who lost their independence and became the occupants of the unfree service villages. And, again like today's globalization, the new interconnections between the more and the less developed were not just economic. Ideas, too, crossed the frontier, and here also the transformative effect of the new contacts was extremely powerful.

The most important set of ideas to bridge the gap in these centuries was undoubtedly, as has long been recognized, the Christian religion. Christianity had been formally adopted by rulers across most of Scandinavia and central and eastern Europe by the year 1000. The Piast dynasty converted in the 970s, the Danes under Harold Blue-tooth at more or less the same time, the Premyslids a generation or so earlier, and the Rurikids half a generation later under Vladimir. The Moravians, of course, had picked up Christianity in the mid-ninth century. For all its triumphant progress among them, however, the new dynasts of non-imperial Europe found one dimension of their new religion potentially problematic. From the person of Charlemagne onwards, although it was not then a new idea, the imperial title carried the connotation that its possessor wielded the highest authority, having been personally chosen by God to rule in His stead on earth. To accept Christianity, therefore, was implicitly to recognize the legitimacy of imperial overlordship, and this naturally made the dynasts hesitate. There was also the practical consideration, if you didn't have an entirely independent ecclesiastical province, that part of any revenues generated (by tithes, for instance) for religious purposes in your domain would in practice pass outside of your control, since they were owed to the archiepiscopal see. Archbishops also, at least notionally, had a strong say in the appointment of bishops, so an 'imperial' archbishop could interfere with the choice of bishops within your territory.

These potential problems certainly got in the way of the Moravian dynasty's acceptance of Christianity. They tried to resolve them by getting their Christianity via a combination of the papacy and Byzantium, rather than the all too adjacent Franks. It was perhaps for similar reasons that Anglo-Saxon missionaries, not the nearby imperial churchmen, played a key role in the early stages of Christianization in Scandinavia. In the long run, however, nearby imperial patronage usually proved too hard to resist, and the best option was to accept your Christianity from that quarter, but – like Poland – extract the right to your own archbishop, thus insulating yourself from the worst hazards.[56]

But why accept Christianity at all? We have already met one obvious benefit. Accepting the religious orientation of rich, imperial, developed Europe was an important move if you wanted to escape the category of 'barbarian' and win admittance to the club of Christian nations. Even if you then faced possible problems of imperial hegemony – or the claims of it, at least – this was probably still a better option than remaining in the barbarian category, where no holds would be barred whenever it seemed a good idea to some influential faction within the Empire's structures to look to profit at your expense. This, of course, was the problem that led to the longer-term demise of the Elbe Slavs, even if, at the start of the eleventh century, they briefly benefited from Henry I's desire to curb Piast power. It has also long been canonical to identify a series of specific advantages for ambitious dynasts in adopting Christianity when it came to the internal administration of their realms.

These fall into three broad categories. First, conversion to Christianity brought kings and rulers a degree of ideological promotion. It was a commonly accepted Christian idea in the first millennium that no ruler could win power without God's will. Converting to Christianity thus allowed rulers to claim to be God-chosen, putting ideological blue water between themselves and their nearest rivals. This was potentially useful in the political context, most career-minded dynasts having risen only recently above a pack of peers, and mostly by brute force. Second, Christianity was a religion of the Book: all its basic texts, the commentaries on them, and the practical rules that had evolved over the centuries to organize its operations, came in written form. Christian churchmen as a whole, therefore, operated at a higher degree of literacy than the average even elite population of early medieval Europe. Clerics could thus make useful royal servants, and in all the cases we know about came to be employed as such by their converted rulers. In the longer term, it would in fact be the literacy of churchmen that would make it possible to sustain more bureaucratic forms of administration – particularly useful when it came to assessing and raising tax in the form of cash. Third, and this flows on naturally from the second, Christianity was a high-maintenance religion. Buildings, books, full-time clergymen: all this was very expensive. So the institution of Christianity always involved establishing new taxes – by the late first millennium, often in the form of tithes – by which the religion's activities could be funded. Everything suggests that kings kept some of these revenues for their own purposes, sometimes directly by appropriating part of the tithe, sometimes indirectly. The indirect method worked well because kings often kept the right to appoint leading churchmen such as bishops and abbots, could then appoint their personal supporters to these positions, and thus be sure of their financial and other compliance.[57]

I've always been suspicious of this list. Claims, for instance, always have to be tested. Just because converted rulers claimed extra respect by styling themselves as God-chosen, this doesn't mean that anyone actually gave it to them, and in most documented cases conversion made precious little difference to prevailing political cultures. Post-conversion kings were just as likely to be opposed, deposed and murdered as their pre-conversion predecessors. It was particularly rich, for instance, that Boleslav II chose to take out the Slavniks on St Wenceslas day. It was surely deliberate, and you might be tempted to think that doing it on the name day of the Premyslids' royal saint gave the act a kind of legitimacy, notwithstanding its brutality. But Boleslav II was the son and heir of Boleslav I, Wenceslas' brother, murderer and replacement, so maybe the line of Boleslav I just liked to kill its rivals in late September, with the choice of day a reminder to all potential rivals of how they had always dealt with them. The second area of proposed advantage, likewise, was far too long-term to have been in the forefront of any converted dynast's calculations. Given that the time lag between Christian conversion and the appearance of a convincingly developed literate administration in the well-documented Anglo-Saxon case, for instance, was several centuries, it does seem unlikely that initial converts were much seduced by visions of a potential revolution in government.[58]

Of the advantages generally seen in conversion, therefore, only the third seems to carry much weight, and this, like escaping barbarian status, was a real factor in the minds of dynastic converts of the ninth and tenth centuries. By then, the forms of Christian taxation were so well established in imperial Europe that extending them to a new area was an entirely straightforward move, and one that held considerable potential advantages for kings.[59] I strongly suspect, however, that both of these advantages paled into relative insignificance next to another dimension of conversion that is not so often discussed. Its importance emerges, slightly paradoxically, from contexts in which the new religion was actively resisted.

As the literate religion of the developed imperial world, backed by all the ideological cachet that perceived success imparts by association, Christianity usually 'won' in the culture clashes of the early Middle Ages: in much the same way, I suspect, that Levi's and McDonald's have been adopted the world over because of their association with the winning world brand that is America. Just occasionally, though, exposure to Christianity generated a violent and opposite backlash (as sometimes does American success in the modern world). We came across one example earlier, when the leadership of the fourth-century Gothic Tervingi started to persecute Christians because they associated the religion with Roman hegemony. Another couple of examples of the same phenomenon can be observed six hundred years later. An overt anti-Christian ideology was central, for instance, to the revolt of the Elbe Slavs against Ottonian rule after 983, when churches and monasteries were robbed and burned and even dead bishops exhumed, despoiled and insulted. Given that the Church was an instrument of colonial exploitation in these marcher lands, the degree of anger is perhaps not surprising. A slightly different, ruler-directed anti-Christianity had also surfaced in Russia at more or less the same time. Although Igor's widow Olga had converted to Christianity under Byzantine influence and was perhaps baptized on a visit to Constantinople in 957, two of her sons, Sviatoslav and Vladimir, ruling successively after her death, positively championed the claims of non-Christian religion against their mother's choice. Here the issue would appear to have been more cultural than practical, since no colonial Byzantine Church structure had yet reared its ugly head in Kiev.[60]

What's fascinating about these examples of aggressive anti-Christianity, however, is that even to begin to compete with the Christian challenge, the nature of prevailing non-Christian religion had itself to change. To unite his many and varied peoples against Christian influence, Vladimir did not outlaw all their different gods, but he did elevate Perun, an old Baltic and Slavic god of thunder and lightning, into a supreme god, and force his subjects to pay homage. Vladimir was pulling together Scandinavians, Slavic-speakers, Finnic-speakers and goodness knows who else, so any impulse towards an anti-Christian religious unity was bound to involve picking one from the no doubt wide range of cults being followed within his highly disparate following. And even among the culturally much more homogeneous Elbe Slavs, anti-Christianity involved major religious change. Again, it was not that all other cults were outlawed, but the new confederation of the Liutizi was held together now by common adherence to one overarching cult, that of Rethra. Dues were owed by everyone to the god's priests and temples, and Rethra was consulted before any and every act of war and presented with a tithe of any spoils. We don't know a huge amount about Slavic paganism before conversion, but as this need to generate a new overarching cult to fight off Christianity confirms, everything suggests that there was a wide variety of cults, each sociopolitical grouping – 'tribe' – having its own.[61]

Against this backdrop, Christianity offered another powerful attraction to dynasts seeking to unite the unprecedentedly large territories under their control. The huge variety of non-Christian cults with which they were faced was part of a cultural structure belonging to the previous and long-established political order. An attractive feature of Christianity in this context was its licensed intolerance: the refusal to accept the validity of any other religious cult. Adopting Christianity permitted a ruler to stamp out pre-existing cultic practices, whether or not he yet had enough Christian priests around to substitute for them a fully functioning Christian Church. As such, it allowed him to break down one of the main cultural barriers that might otherwise have restricted his attempts to create a new political order. Alongside the other more 'positive' attractions, Christianity brought with it an entitlement to destroy existing religious structures, which made it the perfect ideological accompaniment to a process of political unification.

PEERS AND PERIPHERIES

The new states that appeared in northern and eastern Europe towards the end of the first millennium were the products of long and complex processes of transformation, some of whose roots went back a very long way. The migratory expansions of the late fifth and sixth centuries kick-started the appearance of substantive social differentiation among Europe's Slavic-speakers. The Avar Empire then seems to have established a new kind of hereditary leadership amongst those Slavic groups subject to its dominion, and the new states of the ninth and tenth centuries were able greatly to increase food production and population levels in parts of what had been barbarian Europe's least developed region. At least some of these transformations may have resulted in larger sociopolitical units, built largely on consent, individuals accepting the burdens inherent in being part of a larger mass of humanity for the economic and political security that such allegiances offered. This much, at least, is indicated by the kinds of hillfort being erected up to the year 800, which look essentially communal in inspiration, refuges born of common need, not fortifications built at the orders of some grandee.

Thus far, the process of state formation is reasonably compatible with the models for social change that are sometimes given the jargon heading 'peer polity interaction'. What this means, translated from the original Martian, is that you're looking at a world where change is brought about through a gradual process of competition between social units that are more or less of the same size and power.[62] This evolutionary process was rapidly overtaken, however, in the last two centuries of our period by a series of dramatic developments for which the catalyst was increasingly complex contacts with the outside world. First, Charlemagne destroyed the Avar Empire, letting loose a power struggle among its dependent subjects. And while this struggle was being fought out, new trade networks, combined with military and diplomatic ties, brought a vast amount of new wealth into eastern and northern Europe, not least in the form of precious metals. Cornering the market in this wealth then allowed the most successful dynasties to arm themselves to a degree far beyond any yet seen in the region, and to spread their domination suddenly and by force.

Within this two-stage process, your key move, as a participating dynast, was to establish yourself in a position – geographic and/or economic – from which to maximize profits from the new wealth flowing up and down the international trade networks. Of the four dynasties that flourished so dramatically in the last two centuries of the millennium (not counting the shorter-lived Moravians), three, certainly, emerged in perfect positions to benefit. Prague, home of the Premyslids, was a major entrepôt in the overland slave routes established across central Europe. The Rurikids were intimately involved in the slave and fur trades, while hints in the Arab sources and a suspiciously high density of Arab coin finds indicate that the Piasts, too, had their fingers firmly in the pie. The same may well have been true of the Moravian ruling line, since the routes intersecting at Prague also ran through their heartlands, but for this there is no explicit evidence. Within mainland largely Slavic Europe, there is a strong correlation between a type-A position in the trade networks and successful state formation.

The biggest mystery in this respect is perhaps the Jelling dynasty of Denmark, for whose involvement in the new commercial set-up we have no specific evidence. State formation in Jutland and its attendant islands had much deeper roots than its counterparts in northern and eastern Europe. Given that there was some kind of state there in the pre-Viking period, state formation in Denmark may have been about reactivating something that had never quite died, and was hence less dependent on cornering new wealth so as to build military capacity. It can be argued, however, that the fate of the Jelling dynasty too was intimately bound up with the international trade networks. About the time that Sviatoslav, Grand Prince of the Rus, launched his aggressive campaigns east towards the Volga in the 960s, silver stopped flowing northwards into Scandinavia, although it continued into Russia. It is hard to escape the conclusion that, as with their assaults on Constantinople, the Riurikids' wars here were linked at least in part to market share, designed, amongst other things, to cut Scandinavian traders out of the Volga route. After the brief hiatus, Scandinavian merchants clearly found a new route to the south, and the silver began flowing again for a decade or so. Then, in the 980s, the flows of Muslim silver into the Baltic came to a definitive halt.

It is precisely at this point that Scandinavian raiders started to trouble western European waters again, particularly the prosperous Anglo-Saxon kingdom of Ethelred the Unready, from whom the Scandinavians consistently demanded silver coins and bullion. We owe our detailed knowledge of Ethelred's coinage to these tenth-century Vikings, in fact, since tens of thousands of them survive in Scandinavian contexts. This pattern suggests that the drying-up of Muslim silver, arguably the result of Rurikid intervention on the Volga, led northern Scandinavians to look for alternative sources, and the Jelling dynasty put itself at the head of this enterprise. In doing so, it avoided the fate of the Godfrid dynasty in the first half of the ninth century, whose established power base was undermined by the first flows of Viking-era wealth back to the Baltic. More pointedly, the fact that the Jelling dynasty led the new western attacks also suggests that its power was dependent in some way on the Muslim silver flows that had just been cut off. This came probably from the income from tolls, but also from direct trading of tribute goods for its own benefit, just like the Rurikids.[63] If so, the Danish dynasty was not so different from our other tenth-century success stories.

Other forms of contact with imperial Europe were also important to dynastic success. Exploiting imperial aggression, unless – like the Elbe Slavs – you were too close to it, was an excellent mechanism for generating internal consent if your dynasty was able to offer effective leadership. New military technologies, economic advances, imperial Christianity, not to mention wealth derived from raiding richer, neighbouring imperial lands – all of these were forms of interaction with imperial Europe that drove the engine of state formation, and provided a crucial catalyst for the transformation of largely Slavic-dominated eastern and northern Europe in the ninth and tenth centuries. And, of course, even the beginnings of social stratification in the sixth century, and the importing of better ploughing techniques, can be traced to an earlier round of such contacts. In analytical jargon, a centre–periphery model – where you're dealing with partners to an exchange who are substantially unequal in power – better fits the data from the late first millennium than 'peer polity interaction', and two particular character-istics of this second model need to be stressed. First, the interactions encompassed a wide range of different contacts. This is not a model, as some earlier varieties tended to be, predicated on economic exchange – trade – alone. Political, ideological, even technological contacts all played a role, and all pushed sociopolitical change in broadly the same direction. Second, as with the Germani earlier in the millennium, non-imperial Europe should be cast as anything but passive receptors of imperial gifts. Quite the opposite: northern and eastern European populations, or elements among them, were active agents in all these exchanges, seeking to maximize their beneficial impacts and minimize drawbacks.[64]

It remains, finally, to think a bit further about the role of migration in this unfolding drama. Compared with the generation, say, of the western successor states to the Roman Empire in the fifth century, migration played only a small part in the final stages of state formation in northern and eastern Europe. Of the five state structures explored, only one – that of the Rurikids – took on its distinctive shape because of the intervention of immigrants, and even there, as we have seen, Scandinavian immigrants arrived only in relatively small numbers. It is hard to see Novgorod and Kiev becoming interlinked, even loosely, without the decisive intervention of Scandinavian traders and their determination to take cuts from each other's operations. But there were not enough of them to dominate even the military forces of the new state, which drew on Slavs, Finns and everyone else besides. This was migration operating on a much smaller scale, even, than in those Roman successor states created by partial elite replacement. And the Danish, Polish, Bohemian and Moravian states were all created from entirely indigenous population groups. Indeed, the migration element would seem to be smaller even than that involved in the appearance of larger Germanic powers on the fringes of the Roman Empire in the second and third centuries AD.

In that first revolution among Europe's barbarians in late antiquity, migration sometimes played a larger role, and sometimes a smaller one, and always sat alongside processes of socioeconomic and political transformation. But there was usually some kind of population transfer, characteristically in the direction of the Roman frontier, where wealth-producing contacts with the developed Mediterranean world could be maximized. The Slavic era began with an analogous pattern of migration in the late fifth and sixth centuries, as Slavic-speakers moved into contact with the east Roman Empire and found ways of prospering from that proximity, which set off a profound transformation of their own societies. But when state formation accelerated so dramatically in the ninth and tenth centuries, this kind of migration pattern is conspicuous only by its absence. The new Slavic and Scandinavian states formed where they stood, with no drift towards the magnetic pole of more developed imperial Europe.

This leaves us with one final problem to explore as this study of barbarian Europe comes to its conclusion. Why was it that long-established patterns of migration, so common in the first two-thirds of the millennium, ceased to operate in its last centuries?

11

THE END OF MIGRATION AND THE BIRTH OF EUROPE

In the mid-890s, the latest nomad menace burst into the heart of Europe. Following in the footsteps of Huns and Avars, the Magyars shifted their centre of operations from the northern shores of the Black Sea to the Great Hungarian Plain. For the most part, the results were everything that past experience of nomad powers would lead you to expect:

> [The Magyars] laid waste the whole of Italy, so that after they had killed many bishops the Italians tried to fight against them and twenty thousand men fell in one battle on one day. They came back by the same way by which they had come, and returned home after destroying a great part of Pannonia. They sent ambassadors treacherously to the Bavarians offering peace so that they could spy out the land. Which, alas!, first brought evil and loss not seen in all previous times to the Bavarian kingdom. For the Magyars came unexpectedly in force with a great army across the River Enns and invaded the kingdom of Bavaria with war, so that in a single day they laid waste by killing and destroying everything with fire and sword an area fifty miles long and fifty miles broad.[1]

The populations of the Great Hungarian Plain and surrounding regions, not least Great Moravia centred in Slovakia, were quickly subdued, and an orgy of equine-powered aggression saw Magyar raiding parties sweep through northern Italy and southern France with a ferocity not seen since the time of Attila, while full-scale Magyar armies defeated their East Frankish counterparts three times within the first decade of the tenth century.

But one element in the usual mix of nomad pastimes is missing from the Magyars' European tour. Five hundred years before, the two pulses of Hunnic movement westwards – first on to the northern shores of the Black Sea in the 370s, and then on to the Great Hungarian Plain a generation later – had thrown semi-subdued, largely Germanic-speaking clients of the Roman Empire across its frontiers in extremely large numbers. Two hundred years later, the arrival of the Avars west of the Carpathians would prompt the departure of the Lombards for Italy and a widespread dispersal of Slavic-speakers in every direction: south into the Balkans, west as far as the Elbe, north towards the Baltic, and even eastwards, it seems, into the Russian heartland. Destructive as it was in so many ways, the arrival of the Magyars generated no documented population movements whatsoever (apart, of course, from those of the Magyars themselves). Why not? The answer lies in the dynamic interaction between migration and development which had played itself out across the European landscape over the previous thousand years.

MIGRATION

The absence of any secondary migration associated with the Magyars is all the more surprising because it is one of the central findings of this study that, contrary to some recent trends in scholarship on the period, migration must be taken seriously as a major theme of the first millennium. This trend has not eliminated migration entirely from accounts of first-millennium history, but certainly incorporates a powerful tendency to downplay its importance. In some quarters even the word itself is avoided wherever possible, because 'migration' is associated with the simplistic *deus ex machina* of the 'invasion hypothesis' model of explanation, which was so prevalent up until the early 1960s. In this view, migration meant the arrival of a large mixed group of humanity – a 'complete' population: men and women, old and young – who expelled the sitting tenants of a landscape and took it over, changing its material cultural profile more or less overnight. This model was massively overused, and trapped the developing discipline of archaeology into migrato-centric models that crippled creativity. Besides, as many archaeologists have since pointed out, the model didn't really explain anything anyway, because it never properly addressed the issue of *why* large groups of human beings might have behaved in such a fashion. This being so, it has been both reasonable and natural for subsequent archaeologists to concentrate upon other possible reasons for material cultural change. And these are legion. Everything from religious conversion to agricultural innovation and social development can have profound effects upon material cultural profiles. A highly suspicious attitude towards migration has also crossed the boundaries between disciplines. Some early medieval historians are now also so convinced that nothing like the old invasion hypothesis could ever have happened, that they are happy to suppose that historical sources must be misleading whenever they seem to be reporting possibly analogous phenomena.

A central aim of this study, however, has been to re-examine the evidence for first-millennium migration with a more open mind, and above all to reconsider it in the light of everything that can be learned about how migration works in the modern world. And from this point of view, one of its key conclusions is that the evidence for migration in the first millennium is both much more substantial and much more comprehensible than has sometimes been recognized in recent years. A deep-seated desire to avoid mentioning migration (a more successful version of Basil Fawlty and the war) has thus been wrenching discussion of some pivotal moments of first-millennium history away from the most likely reconstruction of events, and, in so doing, hampering analysis of the broader patterns of development that were under way.

It is an inescapable conclusion from all the comparative literature that a basic behavioural trait of *Homo sapiens sapiens* is consistently to use movement – migration (mentioned it again, but I think I got away with it . . .) – as a strategy for maximizing quality of life, not least for gaining access to richer food supplies and all other forms of wealth. The size of migration unit, balance of motivation, type of destination, and other detailed mechanisms will all vary according to circumstance, but the basic phenomenon is itself highly prevalent. In practice, two particular migration models have been retained in even the most minimizing of recent discussions: 'elite replacement' for larger-group movement, and 'wave of advance' for smaller migration units. Part of the attraction of both has been that they are safely different from the old invasion hypothesis. Elite replacement suggests both that not very many people in total were involved in the action, and that their migratory activity didn't really have that much effect. If you just replace one elite with another, what's the big deal? The wave-of-advance model employs mixed migratory units – essentially families – but their colonization of landscapes is piecemeal, slow, by and large peaceful, and decidedly not deliberate – intention being one of the elements of the old invasion model which revisionists find most problematic. How much of first-millennium European migration can be successfully described by employing these models?

Migration Modelling

Some of it, certainly. Cheating only slightly in chronological terms, the classic, superbly documented example of elite replacement is the Norman Conquest of England in 1066. In the following twenty years or so, as *Doomsday Book* shows, an immigrant, basically Norman elite took over the agricultural assets of the English countryside, evicting or demoting the existing landholders. But the overwhelming majority of the indigenous population remained exactly where they had been before the Normans arrived. Likewise, at least some elements of Wielbark expansion in the first and second centuries AD and of its later Slavic counterpart, particularly the spread of Korchak-type farmers through the largely unoccupied central European uplands, probably had a wave-of-advance quality about them. Looking at the millennium as a whole, however, these models are both too simple and too narrow to describe the totality of recorded migratory action.

First, the models themselves need a substantial overhaul. They either collapse different situations into undifferentiated confusion, or are of such limited applicability as to be more or less useless – at least for first-millennium Europe. As currently construed, elite replacement fails to distinguish the particularity of a case such as the Norman Conquest, where the invading elite could fit easily into existing socio-economic structures, leaving them intact, and any broader effects on the total population remain correspondingly small, if not so minimal as those wanting to undermine the importance of migration might think.[2] But this kind of elite replacement applies only when the incoming elite was of broadly the same size as its indigenous counterpart, and I strongly suspect, even if I could never prove it, that, over the broad aeons of human history, this will have been true only in a minority of instances.

Certainly the first millennium AD throws up more examples of a different kind of case, where the intruding elite, if still a minority – and even quite a small one – compared to the totality of the indigenous population, was still too numerous to be accommodated by redistributing the available landed assets as currently organized. In these cases, existing estate structures had to be at least partially broken up and the labour force redistributed. As a result of this process, the entire balance between elite and non-elite elements of the population was restructured, and the overall cultural and other effects of the migration process were likely to be correspondingly large. This kind of elite migration could not but have huge socioeconomic consequences, and potentially also much greater cultural ones as the indigenous population came into intense contact with an intrusive elite, which was more numerous than its old indigenous counterpart. It was this intense contact, seen in Anglo-Saxon England and Frankish Gaul north of Paris from the fifth century, and perhaps to a lesser extent the Danelaw after 870, that generated substantial cultural, including linguistic, change, as the indigenous population was forced into modes of behaviour dictated by a new and relatively numerous foreign elite living cheek by jowl among them.[3]

Different again were cases of only partial elite replacement, particularly common in more Mediterranean regions of the old Roman west in the fifth and sixth centuries. Here there was some economic restructuring to accommodate the intruders – Goths, Vandals, Burgundians and others – but considerable elements of the old Roman landowning elites survived. In the longer term, it was the immigrants in these cases who struggled to hold on to their existing culture, and long-term linguistic change moved in the other direction. That is not to say, however, that this – the most limited form of migration on display in the first millennium – had only negligible consequences for the areas affected. In the first instance, high politics were dominated by the intrusive elites at the expense of their indigenous counterparts, at least when it came to matters like royal succession, and the overall political effect was sufficient to initiate major structural change. The disappearance in the medium to longer term of large-scale, centrally organized taxation of agricultural production, and the consequent weakening of state structures in the post-Roman west, are best explained in terms of the militarization of elite life that followed the creation of those structures at the hands of intrusive new elites.

The wave-of-advance model requires an equally substantial theoretical overhaul. The basic problem with it, even with ostensibly relevant cases such as Slavic Korchak expansion in the fifth and sixth centuries, or Wielbark expansion in the first and second, is that the Europe of the first millennium AD retained few if any uncontested landscapes of the kind that may have existed when the first farmers had been operating four thousand years before. By the year 1000, there were still plenty of forests, and we take our leave of European history at a moment when a further wave of agricultural expansion was in the process of hacking great swathes through them. But farmers had been clearing the landscape for millennia by this date and many of the best spots had long since been claimed. In this kind of context, random, uncontested expansion, even by small groups, was rarely an option. Korchak-type family or extended family groups probably did spread in largely uncontested fashion, but they did so by moving in a thoroughly non-random fashion through less sought-after, more marginal habitats of upland central Europe. And even here, the total subsequent subjugation of landscapes to the Slavic cultural model, combined with the documented aggression of Slavic groups in other contexts, strongly suggests that a degree of coercion might still have been involved. The same was probably also true of earlier Wielbark groups. Early Wielbark expansion seems to have been carried forward by small social units, but adjacent northern Przeworsk communities certainly came into Wielbark cultural line as a result of their activities. This could have been voluntary, but I suspect that examples of small-scale migration from the Viking period give us a more likely model for what was going on.

Small-scale Scandinavian migration units began carving out territories for themselves in northern Scotland and the northern and western isles of Britain from pretty close to the start of the ninth century. In this case, the logistic problem of getting access to shipping imposed constraints that did not apply in the Korchak or Wielbark cases. Hence, as is documented in subsequent Scandinavian expansion into Iceland and Greenland, the migration units, even if small, did have to be organized by jarls or lesser landowners (*holds*) who had sufficient wealth to gain access to shipping. But whereas Iceland and Greenland were more or less unpopulated landscapes, northern Scotland and the isles were not, and, even if the migrating units were individually small, Scandinavian expansion into these territories was certainly aggressive. Older suggestions that the result was ethnic cleansing are outdated, but the indigenous population was forcibly demoted to lesser status, and, over time, absorbed into the invaders' cultural patterns. Small-scale migration need not, therefore, necessarily mean peaceful migration. As long as they confronted an indigenous population who did not have larger-scale, regionally based political structures, small migration units could still insert themselves successfully by aggressive means. Alongside a wave-of-advance model for small-scale migration that was random in direction and peaceful in nature, therefore, we need to add small-scale migration flows that were non-random or aggressive, or both. This kind of model is potentially highly applicable to the generally already-occupied landscape of first-millennium Europe, relevant not only to Wielbark, Korchak, and some Viking expansions, but also perhaps to the early stages of eastern Germanic expansion towards the Black Sea in the third century, of Elbe Germani into the *Agri Decumates*; or of Slavic groups north and east into Russia in the seventh to the ninth centuries.

What also emerges from the evidence is that too clear a line cannot be drawn between wave-of-advance and larger-scale migration. Just because an expansion began with small-scale migration units, does not mean that it stayed that way. The best-documented case here is provided by the Vikings. Initial Scandinavian raiding and settling, in the late eighth and early ninth centuries, were all carried forward by small groups. The earliest recorded violence involved the crews of three ships – perhaps a hundred men – and there is no reason to think that the settlements around Scotland and the isles need have been carried forward by groups much larger than this. But, as resistance and profits both built up, and the desire eventually formed to settle more fertile areas of the British Isles, where larger political structures in the form of Anglo-Saxon kingdoms barred the way, more important Scandinavian leaders became involved in the action, and larger coalitions formed among the migrants. This reached its climax during the Great Army period from 865, when coalitions formed with the idea of carving out settlement areas first in Anglo-Saxon England and then in northern Francia. If the early raids were undertaken by groups of no more than a hundred strong, the series of Great Armies each comprised much more like five to ten thousand men. The water-borne nature of the action in the Viking era always needs to be kept in mind because it imposed logistic problems that did not apply in other cases, but the evolution from raiding parties to great armies nonetheless provides a well-documented model of how – on the back of evident and growing military and financial success – originally small-scale expansion might eventually suck in much larger numbers of participants. The evidence is not so good for some of the earlier expansions, and these were not affected by the problems of water transport. Nonetheless, the expanding momentum of Viking-era migration provides a helpful model for understanding a series of other first-millennium migratory phenomena, not least the second- and third-century Gothic, and fourth- and fifth-century Lombard expansions, which again, it seems, started small, but grew in scale until forces large enough to fight major battles against Roman armies and regional competitors (such as the Carpi), came to be involved. Anglo-Saxon expansion into former Roman Britain can also be partly understood with such a model in mind, and it is potentially applicable to the third-century Alamanni.

Even without venturing into really contentious areas, therefore, the full range of first-millennium evidence suggests some major revisions to now-standard migration models. But in addition to small-scale migration, elite replacements, and migration flows of increasing momentum, first-millennium sources do periodically report large, mixed groups of human beings on the move: 10,000 warriors and more, accompanied by dependent women and children. Not only do such reports arouse suspicion because they seem uncomfortably close to the old invasion hypothesis, but this particular type of migration unit does not figure in modern migratory patterns, where large, mixed groups of migrants are seen only when the motivation is political and negative – when populations are fleeing oppression, pogrom and massacre, as in Rwanda in the early 1990s. This is not what is reported in first-millennium sources, which describe both a more positive motivation and a greater degree of organization among groups intruding themselves in predatory fashion into other people's territory. Can we believe what the sources seem to be telling us? Should we retain large, mixed and organized groups of humanity as part of the overall picture of first-millennium migration?

Invasion

Even when employing the most up-to-date methods – DNA or steady-state isotope analysis – the kind of evidence that archaeological investigation brings to this debate is at best only a blunt tool. It remains hotly disputed whether much usable DNA will ever be recovered from human remains of first-millennium vintage laid down in the damp and cold of northern Europe. And too much has happened in demographic terms since the first millennium for the percentage distributions of modern DNA patterns to give much clear insight into the relative proportions of their progenitors 1,500 years ago, except perhaps in the highly exceptional case of Iceland, where there was no human population before the Viking era.[4] Steady-state isotopes, likewise, only reveal where someone came to dental maturity. The children of two immigrants will have fully indigenous teeth, and this kind of analysis will always carry an in-built tendency to underestimate the importance of migration. Arguments based on more traditional types of archaeological investigation – the transfer to new regions of items or customs originally characteristic of another – are also unlikely to be any more conclusive.

The reasons are straightforward. By the birth of Christ, most of Europe had been settled and farmed, after a fashion, for millennia. And since even the most aggressive and dominant of immigrants usually still had a use for indigenous populations as agricultural labour, migration did not tend to empty entire landscapes. Furthermore, as all comparative study has emphasized (and modern experience shows), when migrants move into an occupied landscape, the result – in material and non-material cultural terms – is always an interaction. There are only a relatively few items in any particular group's material cultural profile that are so loaded with meaning that they will be held on to, for good or ill, in the longer term. Everything else is open to change under the stimulus of new circumstances, so you can hardly expect migration to involve the complete transfer of an entire material culture from point A to point B in normal first-millennium European conditions. There will always be some elements of continuity in the material cultural profile of any region subject to migration, and this makes it entirely possible, if you are so inclined, to explain any observable change in terms of internal evolution. Goods and ideas can move without being attached to people, and if what you observe archaeologically is no more than a limited transfer of either, it will always be possible to explain it in terms of something other than population displacement. But the fact that it will always be *possible* to do this does not mean that it will necessarily be *correct* to do so, and

the inherent ambiguity of archaeological evidence is sometimes misin-
terpreted. Ambiguity means exactly that. If the archaeological evidence
for any possible case of migration is ambiguous – which it usually will
be – then it certainly does not prove that migration played a major
role in any observable material cultural change – but neither does it
disprove it. What all this actually amounts to is that archaeological
evidence alone cannot decide the issue. It is important to insist on this
point because there has been a tendency in some recent work to argue
that ambiguous archaeological evidence essentially disproves migration, when it absolutely does not. Overall, of course, this forces us
back on to the historical evidence. How good a case can be made from
historical sources for the importance of large, organized and diverse
groups of invaders on the move in the first millennium?

The answer has to be complex. There are some clear instances
where a migration topos, a misleading invasion narrative, has been
imposed on more complex events. Jordanes' account of Gothic expansion into the northern Black Sea region in the late second and third
centuries is a classic case in point, as is the picture of the fourth- and
fifth-century Lombard past to be found in Carolingian-era sources and
beyond. But in other cases, the historical evidence in favour of distinct
pulses of large-scale migration involving 10,000-plus warriors and a
substantial number of dependants is infinitely stronger: the Tervingi
and Greuthungi who asked for asylum inside the Roman Empire in
376, for instance, or the movement of Theoderic the Amal's Ostrogoths
to Italy in 488/9. In both these cases, attempts have been made to
undermine the credibility of our main informants, respectively Ammianus and Procopius, but they lack conviction. Ammianus described
many different barbarian groups on the move on Roman soil in the
course of his historical narrative and only on this one occasion does he
refer to very large mixed groups of men, women and children. The
idea that he was infected by some kind of migration topos in this
instance, but not elsewhere, takes a lot of believing. Likewise Procopius: he is not in fact the only source to describe Theoderic's
Ostrogoths on the march to Italy as a 'people' in a quasi-invasion-
hypothesis sense of the term (a large, mixed grouping of men, women
and children). One contemporary commentator even described them
as such in person to Theoderic and other actual participants gathered
at his court. You wouldn't want to hang anyone in a court of law on
this kind of evidence, but its credibility is pretty much as good as

anything else we get from the first millennium. To reject it on the basis of a supposed migration topos is arbitrary.[5]

Not quite in the same category of solidity, but still well within the usual limits of first-millennium plausibility, likewise, is a range of evidence indicating that moves of a similar nature were made by large, organized Vandal and Alanic groups on to Roman soil from 406, and by Radagaisus' Goths in 405.[6] And while more argument is certainly again required, by far the likeliest reconstruction of Alaric the Visigoth's career indicates that it was founded on mobilizing the Tervingi and Greuthungi of 376, settled in the Balkans by treaty in 382, into a series of further moves from 395 onwards. These are all instances of large, mixed-group movement that pass muster on all the normal rules of first-millennium evidence. There are also enough of them to require us not to dismiss too quickly a series of other cases, where the evidence is a notch or two weaker: in particular the population movements associated with the rise and fall of the Hunnic Empire, which saw the gathering-in of armed, largely Germanic groups, and their subsequent departures from the Great Hungarian Plain as competition built up among them in the era of Hunnic collapse. Here the evidence for large group migration is either partial (the cases of the Rugi or Heruli), or implicit rather than explicit (those of the Sciri, Sueves, and Alans). Although we can find some convincing cases where the action has been mistakenly cast in the form of an invasion-hypothesis-type population movement, therefore, there are many others where there is no good reason to think that this has happened. And, in fact, even the Goths and Lombards are worth a closer look.

In both instances, we are dealing with highly retrospective miscastings of the action. Jordanes was writing about events that happened three hundred years before his own lifetime, and the Lombard authors in the ninth century and beyond about migratory activity that was then four to five hundred years in the past. On one level, it is easy to see why mistakes might have crept in, but there is more to say here than just that. For in neither case was it complete fantasy to be thinking in terms of migration of some kind. The totality of the evidence for both the second- and third-century Goths and the fourth- and fifth-century Lombards does indicate that substantial population displacement played a major role in these eras of their respective pasts.

The evidence is better for the Goths. Here we have contemporary accounts locating Goths in northern Poland in the first and second centuries, but north of the Black Sea from the mid-third. There was also a major material cultural revolution north of the Black Sea in the third century, in the course of which a whole series of customs and items became prominent in the region, which had not previously been part of its characteristic profiles. Some of the more distinctive among the new features, moreover, had been well-established aspects of life and death in first- and second-century Poland. These archaeological indications cannot prove that Gothic migration took place from the Baltic to the Black Sea regions, but, taken in conjunction with the contemporary historical evidence, they amount to a very serious argument to that effect. And while that historical evidence clearly indicates, as we have seen, that, even if there were many separate groups involved in the action rather than one 'people', and that some of them were perhaps originally numerically challenged, this did not remain the case throughout the migratory process. The third-century Goths provide, in fact, an excellent case of a migration flow of increasing momentum, which didn't really stop until the Gothic Tervingi had displaced the Carpi as the dominant grouping between the Danube and the Carpathians in the decades either side of the year 300. Though much less detailed, the Lombard evidence is similar.

Lombards are well attested in the Lower Elbe region, just south of modern Denmark, in the first and second centuries AD. In this case, there are no contemporary historical indications at all of any major population displacements from this region in the Roman period, and what archaeological evidence there is might suggest only a series of relatively small ones, like the first Gothic flows towards the Black Sea. Yet again, however, Lombards were present on the Upper Elbe in sufficient numbers by the 490s to move in and destroy the hegemony of the Heruli in the western half of the Great Hungarian Plain through main force. Whatever its earlier forms, therefore, Lombard expansion towards the Danube, like that of the Goths towards the Black Sea, eventually took the form of much larger pulses of population. Neither Jordanes nor our Lombard sources invented the concept of large-scale migration from nothing, therefore, even if they miscast its form. And, just to push their rehabilitation one stage further, subsequent Gothic and Lombard migrations, occurring between these initial flows and our sources' composition, had taken the form of large, composite group moves, both in the direction of Italy: the Ostrogoths in 488/9 and the Lombards some eighty years later.[7]

When examined more closely, therefore, neither Jordanes nor the Lombard sources give us reason to deny the reality of the large-group migrations recorded in other sources. That said, it is very important to recognize that even our *echt* examples of large-group first-millennium migration do not conform exactly to the old invasion-hypothesis model. Not even the largest groups were whole 'peoples' moving from one locality to another untouched by the process. They could both shed population and gain it. This was presumably even more true of drawn-out migration flows, such as the second- and third-century Goths and fourth- and fifth-century Lombards, but the pattern is only explicitly documented for some of the large-group moves. Decisions to move on such a scale were never lightly, and often caused splits. The Tervingi who crossed into the Empire in 376 left behind them north of the Danube a significant minority of the old group's membership who adhered to the old leadership. Theoderic the Amal's father caused another split when he moved the then Pannonian Goths into the Roman Balkans in 473, and Theoderic himself left behind at least some elite Goths who were absorbed into the military-political hierarchies of the east Roman state. When it comes to gathering recruits, the Lombards were joined by a mixed group of 20,000 Saxons for their move to Italy, together with descendants of much of the flotsam and jetsam left over by the post-Attilan struggles for power in the Middle Danube. Theoderic the Amal, likewise, added a body of Rugi to the Gothic following built up over two generations by his uncle and himself. Similarly, the relationship between the two Vandal groups and the Alans who crossed the Rhine together became much tighter in the face of Roman counterattack in Spain, so that, by the time they invaded North Africa in 429, the survivors, united behind the Hasding monarchy, were much more of a cohesive political unit than the loose alliance they had been twenty-three years before. As much recent work has emphasized, there was as much of the snowball to these migratory movements as the billiard ball.

Another significant departure from the old invasion model is the fact that, when looked at closely, these large groups were mixed not only in age and gender, but also in status. Visions of the Germanic *Völkerwanderung* produced in the great era of nationalism in the nineteenth and early twentieth centuries had in mind large invasion groups of free and equal warriors with their families in tow. But within the larger groups, two separate status-categories of warrior are documented, and there is reason to suppose that non-militarized slaves also participated in at least some of the moves. It is only the higher warrior class that fell into the 'free' category, and the fact that they were by definition an elite class suggests that this group was some kind of minority. The key decisions about migration in the period were being taken, therefore, only by a minority of the participants, with lesser warriors and slaves having little if any influence. Recognizing the reality and significance of these status distinctions also imposes clear limits on the extent to which currently fashionable ideas about the freedom with which group identities could be chosen and discarded can really have applied in practice. What kind of idiot would have chosen to be of lesser-warrior or slave status if group identity was entirely a matter of individual choice? By extension, this also indicates that we need to be careful as to how far we suppose snowball-type phenomena to have operated. Since much of the population of barbarian Europe was not in control of its own destiny, the right to join or not join in large-group migration must have been exercised only by certain, more elite elements among participating populations.[8]

The final modification that must be made to the old invasion-hypothesis model of large-group migration concerns its supposition that large-scale intrusions drove out existing populations. There are several good examples of large-scale invasion in the first millennium, but none where the evidence suggests mass ethnic cleansing. Indigenous populations were often faced with a choice between accepting subjugation or moving on, a choice which would have felt particularly brutal to indigenous elites who had most to lose from the arrival of a new set of masters. But there is no convincingly documented case where the response to this choice led to the complete evacuation of an extensive landscape. At the very least, indigenous populations supplied good agricultural labour, and many of our immigrant groups anyway had lower social-status categories into which newly subjugated indigenous populations could easily be slotted.

These alterations are important, but they remain modifications rather than denials of the basic proposition that the evidence for large, mixed, and organized migrant groups from the first millennium is, ostensibly, periodically convincing. Nationalist visions of whole ancestral 'peoples' clearing out new landscapes for themselves to enjoy can be consigned to the recycle bin of history. The groups documented in our sources were political entities, which could grow or fragment, which contained individuals occupying lesser- and higher-status categories, and which inserted themselves in correspondingly complex ways into new but already thoroughly inhabited environments. Though fair enough on the basis of the available historical evidence (and not denied by the archaeological), can this proposition still be maintained in the face of the non-appearance of such phenomena in modern migratory patterns? The answer to this question is bound up, in my view, with that to a far larger one: Why did European migration take the forms it did in the first millennium? Answering this question requires us to set the observable patterns of demographic displacement between the birth of Christ and the year 1000 against all that comparative study can teach us about migration as a general human phenomenon.

Migration Mechanics

There are a myriad detailed ways in which the mechanics of first-millennium migration correspond to what has been observed in better-documented case studies of early modern and more recent migration. Not least of these, the crucial importance of active fields of information in dictating precise destinations is just as prominent in the first millennium as in later eras. Germanic expansion towards the Black Sea in the third century was clearly exploiting information about the region which had built up through the operations of the Amber Route. Slavic groups first came to know the Roman Balkans as raiders before exploiting that knowledge to turn themselves into settlers as and when political conditions permitted. Scandinavian expansion to the west in the Viking era likewise operated on the back of intelligence acquired by participation in the emporia trading networks of the eighth century, while those working to the east took a generation or so to find their way down the river routes of western Russia to the great centres of Islamic demand for northern goods, having originally opened up the eastern hinterland of the Baltic to feed western markets. To these entirely uncontroversial examples, I would also add some others. A major contributory factor to the apparently odd stop/start migratory patterns of some of the groups entering Roman territory either side of the year 400 was the need to acquire information about further possible destinations before hitting the road again. The Goths, especially the

Tervingi who entered the Empire in 376, already knew about the Balkans, for instance, but not about Italy and Gaul, to where they moved on in the next generation. It took twenty years (and their participation in two Roman civil wars that took some of them lengthy distances in that direction) before they were ready to take the next step. Likewise the Vandals and Alans: Spain marked the end of their original migratory ambitions, and it again took twenty years and some exploratory sea raids before they were prepared to venture across the Straits of Gibraltar to North Africa. More generally, the whole broader phenomenon of migration flows of increasing momentum is clearly a product of growing knowledge. It was precisely the fact that exploratory expansionary ventures into a new region produced profitable outcomes for the pioneers that encouraged others to participate. In some modern cases, such as the spread of the Boers northwards from the Cape, the pioneers were deliberately recruited scouts, sent to check the viability of larger-scale expansion. The same effect could also be achieved, however, by a less formal grapevine.

The study of modern migration also devotes much effort to the key issue of why some people from any particular community choose to move, whereas others in more or less identical circumstances stay put. Tackling this complex issue fully requires the kind of detailed information which is simply unavailable for the first millennium, but it is worth pointing up the relevance of the issue. In the cases of large-group migration reported in any detail in our sources, there is no instance where the decision to move did not generate some kind of split among the affected population group. The same is true, only more so, of the more extended migration flows. For all the Germani of Polish origin who ended up by the Black Sea in the third century, there were many others who stayed behind, shown by the fact that the Wielbark and Przeworsk cultural systems continued to operate. Likewise, many Angles and Saxons did not relocate to England in the fifth and sixth centuries, and Scandinavia was not emptied in the Viking period. Such divergences of response were only natural, of course, given the magnitude of the decisions involved, and first-millennium populations clearly felt the same stress of migration as modern ones, even if we can't explore their reactions in detail.

Stress also manifests itself in the modern world in the phenomenon of return migration. Looked at closely, all modern migration flows see substantial numbers of immigrants returning to their original homelands. Again, the level of information is not sufficient to allow us to discuss this topic properly for the first millennium, but aspects of the Viking period emphasize that it, too, needs to be recognized as a real phenomenon. The initial phases of Scandinavian expansion were all about gathering wealth, whether by raiding or trading, or both. Having gathered their wealth, different individuals then made different choices about how to invest it. Some chose, even early on, to stay put at their points of destination in the east and west (as shown by the early settlements in northern Scotland and the isles), whereas others chose to take their new wealth back home to Scandinavia, eventually prompting a massive shake-up in Baltic politics. With this example in mind, I (as others) would be happy to believe reports that some incoming Anglo-Saxons also eventually chose to return to the continent.[9]

Closely related, too, to the stress of migration, but this time something we can explore in greater detail, is the highly significant influence on patterns of movement of an ingrained migration habit. In modern migration flows, an existing tradition of mobility often plays a vital role in dictating which individuals within a particular group of people will decide to move. Individuals who have moved once are more prone to move again within their own lifetimes, but, less intuitively obvious, the habit is also passed on between generations. The children and grandchildren of migrants are much more likely than the average to move again themselves. A tradition of personal or familial mobility clearly generates a greater propensity to attempt to solve life's problems, or look for greater opportunity, by moving to new localities. Anyone might move if the stimulus to do so is large enough, but the required stimulus is smaller for those with established migration habits.

The effects of this factor can be seen at work on at least two different levels in the first millennium. First, at least two of the broader population flows, those of the Wielbark and Przeworsk Germani in the second and third centuries, and of the early Slavs three hundred years later, involved populations whose farming techniques were then insufficient to maintain the fertility of any individual piece of arable land for more than a generation or two. A general, periodic local mobility was simply a fact of life for these populations, and there is every reason to suppose that this facilitated the eventual transformation of a more random wave-of-advance-type expansion into a channelled migration flow when information began to filter back about the opportunities available at an entirely new set of longer-distance destinations. Second, a more specific tradition of distinct, longer-distance relocation clearly built up among some particular first-millennium populations. The fourth-century Gothic Tervingi are probably most famous for the fact that a majority of them decided to seek asylum inside the Roman Empire in 376. That decision was greatly facilitated, however, by active memories of recent migrations. This same Gothic group had taken possession of their existing lands in Wallachia and Moldavia between the Lower Danube and the River Dniester only in the decades either side of the year 300, and a generation or so later, in the 330s, had attempted to move bodily to new locations on the fringes of the Middle Danube region. It was the children of those who had moved into Moldavia and Wallachia who were on the move again in the 330s, and their children and older grandchildren who decided to seek a new life inside the Roman Empire in 376. Similar observations apply to many of the other groups caught up in the rise and fall of the Hunnic Empire, both those who fled inside Roman borders in the crises of 376–80 and 405–8, and those who moved first to the Middle Danube under Hunnic influence and/or duress, and then out of it after Attila's death. The willingness of some Norse to move on to Iceland and Greenland in the later ninth century was likewise facilitated by the fact that they were the immediate descendants of Viking immigrants to Scotland and the isles. In fact, examples like the Goths or Slavs demonstrate how moves that were initially generated by general traditions of local mobility could then spawn the more specific traditions of larger-scale mobility that underlay the move of many Tervingi on to Roman soil in 376, in the same way that internal migrants within the European landscape provided many of the recruits for the trek to North America in the nineteenth century.

Aside from the emotional costs of migration, financial ones were also a major factor in any migrant's calculations. Most first-millennium migration that we know anything about was a question, more or less, of walking and wagons. It involved no major transportation costs, apart from wear and tear to animals, peoples and wheels, and participation was consequently open to many. It nonetheless involved many indirect costs, above all the potential food shortages that were bound to result when movement disrupted normal agricultural activity. As a result, food stocks had to be maximized before moving, unless circumstances were completely overwhelming, and this meant that autumn was the classic moment to make a move – just after the current year's harvest had been gathered and while there was still a chance of some grass growing to feed the oxen pulling the wagons and other animals. Alaric's Goths moved into Italy in both 401 and 408 in the autumn, Radagaisus' Goths in autumn 405. The Vandals, Alans, and Sueves who crossed the Rhine at the very end of 406 likewise presumably began their trek from the Middle Danube in the autumn of that year.[10]

As usual, there is little information about the impact of migration costs beyond this very basic point, but logistic problems do show themselves from time to time in the available data. Above all, extended periods of movement left groups particularly vulnerable in economic terms. Flavius Constantius was able to bring Alaric's Goths – now led by Athaulf and Vallia – to heel by starving them out in 414/15. By that date, they had been living off the land without planting crops for six or seven years. Later in the fifth century, similarly, after the collapse of the Hunnic Empire, the surviving sources give us just a little insight into the logistic strategies adopted by Theoderic the Amal. His grouping journeyed around the Balkans with wagonloads of seedcorn in the 470s, and one dimension of its diplomatic negotiations with the Roman state involved providing it with agricultural land. Even on the march, noticeably, this group always sought to establish more regular economic relationships with Balkan communities, rather than merely robbing them. This meant that the communities could keep on farming and producing surpluses, from which the Goths could siphon off a regular percentage, whereas destroying them by pillage would only have fed Theoderic's followers once.

Logistic factors had a still bigger impact upon population flows requiring more than land transport. Mass access to sea transport did not even become a possibility until the advent of steerage class in the enormous transatlantic liners of the later nineteenth century. Before that point, travel costs necessarily limited participation in any kind of maritime-based expansion. Again, the Viking period provides the best-documented first-millennium example. Ships were highly expensive, and even specialist cargo ships could carry only limited numbers of people and their goods. Thus Viking raiding required the less well-off to come to some kind of joint arrangement for funding the purchase or hire of a ship (though how many shipowners, I wonder, would be willing to hire out shipping for raiding ventures?), or to attach themselves to a leader of higher status.[11] Logistic limitations figured even more strongly when it came to the settlement phases, when so many more types of people and a wider range of bulky farming equipment were required. The relevant shipping costs alone make Stenton's suggested phase of large-scale Norse peasant settlement in the Danelaw pretty much inconceivable. Who would have bothered to pay for this when there was a subdued, low-status Anglo-Saxon labour force already in situ? We also see the effects of logistics on the Icelandic settlements, where each immigrant unit was headed by a higher-status individual who could presumably cover the large initial investment costs of transportation. Logistics may also have limited the number of Scandinavian women who participated in the Norse migration flow compared to the other land-based movements of our period. Modern DNA patterns suggest that only one-third of immigrant women to Iceland came all the way – directly or indirectly – from Scandinavia, with the rest moving a shorter distance from the British Isles. This may reflect the fact that it was too expensive for more than a minority of warriors to bring their Scandinavian sweethearts with them. On the other hand, given that the men involved were relatively wealthy, and that they were pagan and polygamous, it may be that women outnumbered men early on, with each Scandinavian male bringing with him not only his Scandinavian sweetheart but a couple of British or Irish babes besides.

As frustratingly limited as all the Viking-period evidence is, our other major first-millennium examples of maritime movement – Anglo-Saxon moves across the North Sea and the expedition of Vandals and Alans across the Straits of Gibraltar – are not even illuminated in this much detail. The impact of logistic demands must, though, have been similar. Possibly the Vandals and Alans were in a position to use intimidation to requisition some transport, but I doubt that they were able to avoid having to meet most of their costs themselves. And there is reason to think that the fact that they could only move a relatively few people at a time dictated their initial choice of landfall: Morocco, far away from the better-defended heartland of Roman North Africa, which was their ultimate destination. The logistic impossibility of moving many people at once was probably also a major factor in the drawn-out nature of Anglo-Saxon migration into southern Britain.

So far, so good: there are a whole series of ways in which the comparative literature sheds light on first-millennium migration, even given the limited data set available. But this still leaves some big questions to answer. Above all, can we in fact accept what the sources appear to be telling us: that the first millennium occasionally saw large, mixed, and organized population groups take to the road? And, if so, how are we to explain this phenomenon both in its own context, and the fact that it has not been observed in more modern and better-documented eras? Again, modern migration studies, in my view, can help answer these questions, but a fully satisfactory explanation also requires us to explore first-millennium migration patterns against the backdrop of a much wider set of transformations that were unfolding simultaneously in barbarian Europe.

MIGRATION AND DEVELOPMENT

Comparative studies provide two basic points of orientation when thinking about the likely causes of any observable migration flow. First, it is overwhelmingly likely that a substantial difference in levels of economic development between adjacent areas will generate a flow between the two, from the less-developed towards its richer neighbour. What 'adjacent' means will vary enormously in different eras according to what transport is available, and a situation that would otherwise generate a 'natural' flow of people may be interfered with by political structures at either end, or by the availability of information. All things being equal, however, a flow of population will be one result of different levels of development, the result of *Homo sapiens sapiens'* inherent tendency to use movement as a strategy for maximization. The second point is equally basic. In the vast majority of cases, the precise motivation of any individual migrant will be a complex mixture of free-will and constraint, of economic and political motives. There are exceptions, not least when political refugees are driven forward by fear of imminent death, but most migrants are motivated by some combination of all four factors. Taken together, what both of these observations stress above all is that migration will almost always need to be understood against prevailing patterns of economic and political development. Taking this approach, in my view, provides satisfactory explanations for both the general geographical 'shape' of first-millennium migration, and the seemingly odd nature of its characteristic migration units.

Migration in Roman Europe

At the start of the first millennium, the most highly developed region of Europe – in both economic and political terms – was the circle of the Mediterranean united under Roman domination, to which the largely La Tène landscapes of the south and west had recently been added. La Tène Europe featured developed agricultural regimes whose surpluses could support relatively dense populations, together with considerable production and exchange in other sectors of the economy. The Romans did not only conquer La Tène populations as they moved north, but it is a fact – and not an accidental one – that their conquests ran out of steam more or less at the outer fringes of La Tène Europe. The reason was simple: beyond that zone, the profits of conquest ceased to be worth the costs. Beyond La Tène Europe lay the territories of the largely Germanic-speaking post-Jastorf world. There was a large degree of variation in economic patterns across this zone, not least because some of its populations had been in substantial engagement with their La Tène neighbours for a considerable period. In general terms, however, political units were smaller in scale here than those of La Tène Europe, even before the latter was incorporated into the Roman Empire, and agricultural productivity was lower. General population density was therefore less, and there are fewer signs either of non-agricultural production and exchange, or of marked differences of wealth (at least as expressed in material cultural terms). Beyond Jastorf Europe, the northern and eastern reaches of the European landscape were still host to Iron Age farming populations (as far as ecological conditions permitted), but they operated still less productive agricultural regimes, their settlements were smaller and even more short-lived, and they possessed little in the way of material cultural goods.

Faced with this distinctly three-speed Europe (Map 1), comparative migration studies would lead you to expect flows of population from its less developed regions to the more developed (i.e. in broadly southerly and westerly directions). And in the Roman period – the first three centuries AD – this is essentially what occurred. The economic and sociopolitical structures of more developed Roman Europe sucked in population from its less developed neighbours in a variety of forms, particularly from adjacent, largely Germanic-dominated post-Jastorf Europe. Many individuals entered the Empire as voluntary recruits for Roman armies on the one hand, or involuntary slaves for a variety of economic purposes on the other. These population flows are well known and require no further comment here. But the larger and more contentious Germanic population flows of the second and third centuries also fall into line with this pattern, in the general sense that they too moved broadly south and west towards more developed Europe. A full understanding of their particular history, however, also requires an understanding of how a broader set of interactions with the Roman Empire had in the meantime been transforming the three-speed pattern which prevailed at the birth of Christ.

For one thing, the military and political structures of the Roman Empire fundamentally explain the geographically asymmetrical outcome of Germanic expansion in these years. The forces behind the expansion seem to have been operating very generally in Germanic-dominated central Europe, but the resulting population flows had much more dramatic effects in the south-east, and particularly north of the Black Sea, than in the south-west. Where Germanic immigrants took over no more than the *Agri Decumates* in the south-west, further east Dacia was abandoned and political structures north of the Black Sea were entirely remade. There may have been some difference in the scale of the migratory flows in operation in each direction, but this, too, was reflective of the more fundamental cause of the different scale of outcome. Flows south and east were operating against the clients of Rome's inner frontier zone, rather than directly against the military power of the Empire itself. As a result, the likelihood of success was that much greater than in the south-west, where the Empire's military power had to be tackled directly.

Why the successful leaders of these expansion flows were generally willing to settle for outcomes that left them largely on the fringes of Empire, instead of pressing on permanently across the frontier, also comes down in part to the longer-term effects of interaction with the Roman world in the first two centuries AD. The operations of trading mechanisms – both in longer-distance luxury goods and shorter-distance largely agricultural products – opportunities for raiding richer

Roman territories, and the Empire's own propensity for bolstering the power of its clients with diplomatic subsidies all meant that in the first two centuries of the Empire's existence new wealth built up at the courts of Germanic kings in the immediate vicinity of the frontier. Three-speed Europe thus developed a fourth gear in the form of an inner zone of clients whose wealth outstripped those of their former peers in what now became the outer periphery of post-Jastorf Europe. Not only was it militarily much less dangerous for the leaders of Germanic expansion to restrict their operations to areas beyond the imperial frontier, but two centuries of interaction with the Empire, and the subsequent accumulations of wealth, had made the frontier zone an attractive target for predatory expansion in its own right. Before these processes had unfolded, there would have been little point for ambitious Germanic warlords in moving, say, from northern central Europe to southern central Europe, or from north of the Carpathians to the south-east, since the potential material gains for such efforts would have been minimal.

Understanding the action of the later second and third centuries in this way also explains the apparently odd form taken by at least some of the units participating in these migration flows. The first recorded attempt from the outer periphery to tap into the new wealth building up closer to the imperial frontier took the form of a raid. As the power of King Vannius of the Bohemian-based Marcomanni weakened in his dotage, an ousted political rival was able to organize a warband from central Poland (and possibly northern Poland too) to ransack the movable wealth around his court, much of it the proceeds of diplomatic subsidies and his cut from the activities of Roman merchants. Although I cannot prove it, I would be willing to bet that this was but one example of a far from uncommon phenomenon. Hit-and-run raids were not, however, the most effective way to tap into all the new wealth accumulating in the immediate hinterland of the Empire. For entirely structural reasons to do with trade, Roman diplomatic methods and even ease of raiding Roman territory, the best opportunities to benefit from the new wealth-generating interactions with the Roman Empire were all limited geographically to the immediate frontier zone, and any greater ambitions towards wealth acquisition among groups and leaders in the outer periphery required their permanent relocation towards the frontier. It is therefore hardly surprising that raiding gave way to migration in the second and third centuries as more ambitious leaders and followings from the outer periphery looked to win control of the new Rome-centred wealth flows operating in barbarian Europe.

But by the end of the first century AD, there was no potentially lucrative spot along the frontier that was not already occupied by a warlord of some kind, and no sitting tenants were likely to surrender their highly advantageous position without a fight. Any permanent relocation towards Rome's frontier therefore necessarily required the destruction of existing political structures, and this explains why the second- and third-century migration flows eventually encompassed substantial military forces numbered in the thousands, rather than warbands of just one or two hundred men. Warbands might raid effectively enough, but their power was insufficient to remake an entire political structure, so that ambitious wannabes from the outer periphery had no choice but to recruit larger expeditionary forces to achieve their aims.

It is worth pausing to consider this pattern of migratory expansion in the light of more recent and better-documented examples. This kind of intentional, predatory intrusion on the part of thousands of armed individuals is not generally seen in the modern world, and this is sometimes put forward as an objection to supposing that it ever occurred in the past. Half of the answer to this objection is that, though not common, this kind of activity has indeed been seen in the relatively modern world: it is exactly the same basic kind of migratory pattern observable among the Boers of the Great Trek. In that case, the intrusive units could be smaller because the Boers enjoyed a massive advantage in firepower over their Zulu and Matabele opponents. In the second and third centuries, any technological advantage was probably more likely to have lain with the groups of the inner periphery being targeted, since they may well have been buying Roman weaponry, so that the intrusive forces from the outer periphery had to be more or less as large as those deployed by the sitting kings of the frontier region.

The other half of the answer comes from thinking about precisely why modern migratory flows, even if cumulatively large, tend to operate on the basis of small individual migration units of just a few people at a time. They do so because the migration-unit size is dictated by the way in which modern migrants seek to access wealth from the more-developed economies to which they have been attracted. In the modern context, wealth is accessed by individual immigrants finding employment in the industrial or service sectors of an economy, which is well-paid at least from the relative perspective of the immigrant him- or herself. The underlying principle here is not that migration-unit sizes are always likely to be small, but, rather, that they will be appropriate to the means by which the wealth of the more developed economy is going to be accessed. All the economies of first-millennium Europe were essentially agricultural, and extremely low-tech. As a result, even in the developing periphery of the Roman Empire, they did not offer many even relatively well-paid jobs for individual migrants, except for a few who could attach themselves to the military followings of frontier kings. For those with ambitions to unlock the wealth of this world on a much larger scale, coming as an individual immigrant, or merely within a small group, was a pointless exercise. In such a context, you had to arrive with enough force to defeat the sitting tenant, and prompt the Empire to identify you now as its preferred trading and diplomatic partner on your particular sector of the frontier. Although this kind of migrant group is not commonly seen in the modern world, therefore, it actually accords with the fundamental principles behind all observed migration flows. Large-scale predatory intrusion was as appropriate to wealth acquisition via migration in the first millennium, as individual movement is now.

Levels of development also explain the other fundamental oddity of these second- and third-century population flows: that many of the warriors were accompanied by women and children. Germanic-dominated Europe of the early centuries AD was a world of low-tech, small-scale farms producing only limited food surpluses. As a result, the economy could not support large warrior retinues; the kind of food renders available even to fourth-century kings could support only one or two hundred men. Again like the Boers, therefore, the kind of larger military expeditions that were required to take over a revenue-producing corner of the Roman frontier could never have been mounted using just the small numbers of military specialists that existed in the Germanic world. Recruits were required from a broader cross-section of society, many of whom already had dependants. These participants would obviously not have wanted to leave their dependants behind in the long term – aside, perhaps, from a few of the younger teenage ones – but even to have left them in the short term, while the expedition reached a hopefully successful conclusion, would have been to expose them to substantial risks. In context again, therefore, it was only natural for Germanic expeditionary forces of more than one or two hundred men to be accompanied by numerous familial dependants.[12] There were a few women even on the Boer scouting expeditions, but the larger trekking parties were always mixed, and the women, in fact, were far from bystanders when it came to fighting; they loaded the flintlock rifles and even shot them when necessary. Germanic women of the second century had no rifles to load, but they no doubt had their own key roles to play, even on substantially military expeditions. Although the recorded nature of these Germanic migration flows looks odd, both in size and composition, in the light of some of the comparative literature, it does accord with the fundamental principles behind observed migratory behaviour, once due allowance is made for differences between the first and third millennia.

Völkerwanderung and Beyond

The evolving patterns of development and migration unfolding in the Roman era came to a head in the so-called *Völkerwanderung*. In the later fourth and fifth centuries, documented European history is marked by the appearance of a whole series of migrant groups comprising 10,000 or more warriors and a large number of dependants, which were powerful enough to survive direct confrontation with the military and political structures of the Roman imperial state. Seen in the broadest of terms, these extraordinary pulses of large-group migration were produced by the intersection, at a critical moment, of a number of related lines of development. First, by the mid- to late fourth century, processes of economic and political development among the Germani had reached a point where political structures had sufficient strength to hold together such enormous groups of warriors and their dependants within a reasonably solid edifice. But, second, these structures had been generated by the expansionary processes of the second and third centuries, and were close enough in time to those events to retain a tradition of migration that could be mobilized when circumstances were appropriate or demanded it. And, third, perhaps the other side of the same coin, their economic structures were not yet so rooted in the arable cultivation of any particular landscape that it was impossible for them to conceive of shifting their centre of operations to another locality.

Viewed against the backdrop of long-term development in the Germanic world, and particularly against the more immediate events of the third-century crisis, the existence and activities of these very large migrant groups are certainly explicable, but that should not take away from the extraordinary nature of the action. For, though larger and more cohesive than their counterparts of the first century, none of the groups that initially emerged from the imperial periphery was in itself large enough to confront the Roman Empire with success, and yet the aggregate outcome of their collective activities, as we have seen, was the destruction of the west Roman state. This highly unpredictable outcome was itself the result of further intersections between contingent historical events and longer-term patterns of development.

First, it took the unintentional stimulus provided by the Huns to get sufficient numbers of these largely Germanic groups from beyond Rome's Rhine and Danube frontiers moving on to Roman soil at broadly the same time to make it impossible for the Roman state merely to destroy them. Had these groups – even given that they were larger and more cohesive – arrived separately on Roman territory, the result would eventually have been their destruction, and there were still far too many of them to organize any unified plan for the Empire's destruction. The key element missing from the Germanic world of the imperial periphery, as opposed to its Arab counterpart, was the lack of a Muhammad to provide an alternative and unifying ideology to that of the Roman state. But, second, once established on Roman soil, the processes of political amalgamation that had been unfolding over the long term beyond the frontier reached a relatively swift climax. This key point was missed in much of the traditional nationalist historiography. By insisting on treating the groups who eventually founded successor states to the western Roman Empire as ancient and unchanging 'peoples',[13] this historiography missed the fact that most of them were explicitly documented as new coalitions which formed on Roman territory out of several groups – usually three or four – and who had been independent of one another beyond the frontier. Visigoths and Ostrogoths, Merovingian Franks, the Vandal–Alan coalition – all represented a further step-change in the organization of barbarian political structures, and it was this further evolution which really produced groups that were large enough (deploying now 20,000 warriors and more) to destroy the western Empire.[14]

Contingent as much of this was – there is no sign that there would have been such an influx on to Roman soil without the intrusion of the Huns – one dimension of the action was far from accidental. The new and much larger political formations that became the basis of the successor states could not have come into being on the far side of the frontier. The level of economic development prevalent in the periphery of the Empire in the fourth century did not produce sufficient surplus to allow political leaderships enough patronage to integrate so many followers in that context. Only when the economy of the Empire could be tapped directly for extra wealth, and when the Roman state was providing extra political stimulation towards unification in the form of a real outside threat, was there a sufficient economic and political basis for these larger entities to come into existence. Political structures were the product of, and limited by, prevailing levels of development, and the new state-forming groups could not have emerged in a purely barbarian context.[15]

But if there is a real sense in which the *Völkerwanderung* can be seen as the culmination of Roman-era patterns of development in barbaricum, its outcomes nonetheless revolutionized broader patterns of development across Europe as a whole. To start with, the new states that emerged on former Roman territory made imperial Europe considerably less imperial. The epicentre of supraregional power in western Europe shifted decisively north around the year 500, the second half of the millennium being marked not by Mediterranean-based imperial power, but a series of broadly Frankish dynasties whose prominence was based on economic and demographic assets located north of the Alps between the Atlantic and the Elbe. Again, this can be seen as a culmination of trends of development set in place in the Roman period. The fact that the new imperial power of western Europe should be based on a combination of a chunk of former Roman territory with a substantial part of its ex-periphery is a clear sign of how profoundly that periphery had been transformed by its interaction with Roman power in the preceding centuries. At the birth of Christ, this landscape on either side of the Rhine could never have supported an imperial power, not being remotely wealthy or populous enough, but Roman-era development on both banks of the river radically transformed this situation. At the same time, the political structures of post-Roman Frankish-dominated western Europe, particularly the militarization of its landed elites, meant that this new imperial state was different in kind to its Roman predecessor. Lacking the power to tax agricultural production systematically, it was a less dominant and less self-sufficient kind of entity, which required the profits of expansion to provide its rulers with enough patronage to integrate its constituent landowners. And when broader circumstances did not allow for expansion, fragmentation followed, with power quickly seeping away from the centre to the peripheral localities. Periods of great central authority and external aggression – the hallmarks of empire – thus alternated with others of disunity in the second half of the millennium, where Roman imperialism had previously presented a more consistently cohesive face. There is a real sense in which the pre-existing inequalities of the first half of the millennium were in part eroded from the top, as it were, by the fact that imperial Europe became less consistently imperial.

More fundamentally, and also more interesting given that it has been so much less discussed, is the effect of the *Völkerwanderung* upon barbarian Europe. By the sixth century, Germanic-dominated Europe as it had stood in the Roman era had almost completely collapsed. Where, up to the fourth century, similar socioeconomic and political structures had prevailed over a huge territory from the Rhine to the Vistula in the north and to the River Don at their fullest extent in the south, by c.550 AD, their direct descendants were essentially restricted to lands west of the Elbe, with an outlying pocket on the Great Hungarian Plain, which was about to be terminated by the arrival of the Avars (Map 15). The *Völkerwanderung* had played a central role in this revolution, though not by actually emptying these landscapes of all their inhabitants. Settlement did completely disappear in some restricted localities, but, even making maximum assumptions, the exodus from Germanic Europe from the fourth to the sixth century was not on a large enough scale to denude central and eastern Europe of its entire population. What the *Völkerwanderung* clearly did do, however, was empty much of the old inner and outer peripheries of the Empire of the armed and organized, socially elite groupings which had previously run them. From the perspective of barbarian Europe, the period saw not just the collapse of the Roman Empire, but also the collapse of the larger state-like structures and organizations of its periphery, the vast majority of which relocated themselves, in the course of the migrations, on to parts of just the old inner periphery – between the Rhine and the Elbe, and the Great Hungarian Plain – and actual, largely western Roman territory.

This first extraordinary revolution in barbarian Europe marked a caesura in over half a millennium of broadly continuous development over large parts of central and eastern Europe. It also allowed a second and equally dramatic transformation. In the aftermath of Germanic collapse, population groups from the third zone of Europe as it stood at the start of the Roman era started to develop, for the first time as far as we can see, substantial political, economic and cultural inter-actions with the rest of Europe. The Romans had some kind of knowledge of the Veneyi who inhabited that part of Europe's low-speed zone closest to them. Tacitus in the first century knew that they were out there, beyond the Vistula and the Carpathians; Ptolemy a couple of generations later could add the names of a few of their broader social groupings. But, remarkably, there is no evidence at all that these populations were sucked into the political events of the first half of the millennium in any shape or form. Venedi mounted no known raids into Roman territory, find no mention in narratives of the Marcomannic War or the third-century crisis, and do not even seem to have participated in the structures of Attila's Empire, which incor-porated so many of the other population groups of central and eastern Europe. Nor do the distribution maps of Roman imports suggest that these European population groups from east of the Vistula and north of the Carpathians played a major role in any of the trade networks stretching out into barbaricum in the Roman era, though some of the routes surely passed through their territories.

More or less immediately after the collapse of Germanic Europe, however, Slavic-speakers started to emerge from the low-speed zone to take an increasingly important role in recorded narratives of broader European history. By about 500 AD, they had moved south and east of the Carpathians into direct contact with the east Roman frontier, and were beginning to raid across it. Their capacity to do so may have been the result of preceding interactions with Goths and others of the more organized groups of the Germanic periphery to the Roman Empire, which pass more or less unmentioned in our historical sources.[16] Be that as it may, the new contacts with the east Roman Empire massively accelerated any nascent processes of development already operating among those Slavic groups involved, as raiding and

diplomatic subsidies brought in unprecedented quantities of movable
wealth, and stimulated among them both militarization and the
formation of larger political structures, both of which allowed profits
from the new relationship with Constantinopolitan territories to be
maximized. All this ran parallel to some of the kinds of transformation
seen in the Germanic world in the early Roman period, and, following
the collapse of Germanic Europe, Slavic-speakers had already emerged
by 550 as the main barbarian 'other' confronting east Rome's civilization in south-eastern Europe.

At this point, a second nomadic 'accident' bent existing processes
of development substantially out of shape, and acted as a crucial
catalyst in the further transformation of barbarian Europe. Like the
Huns, the Avars swiftly built a powerful military coalition in central
Europe, one of whose main effects was to siphon off still larger
amounts of Mediterranean-generated wealth into now largely Slavic-
dominated central Europe. This, of course, further stimulated the
competition for control of that wealth, which had already been
producing a new kind of military kingship in the Slavic world even
before the Avars appeared. Equally important, and just like the Huns,
the Avars lacked the governmental capacity to rule their large number
of subject groups directly, operating instead through a series of
intermediate leaders drawn in part from those subject groups. We lack
much in the way of detailed information, but there is every reason to
suppose that this would have had the political effect of cementing the
social power of chosen subordinates, further pushing at least their
Slavic subjects in the direction of political consolidation.[17] The third
major effect of the Avars was both to prompt and to enable a wider
Slavic diaspora, as some Slavic groups moved further afield to escape
the burden of Avar domination. Large-scale Slavic settlement in the
former east Roman Balkans – as opposed to mere raiding – only
became possible when the Avar Empire (in combination with the
Persian and then Arab conquests) destroyed Constantinople's military
superiority in the region. But at least some of these Slavs were as
much negatively motivated by a desire to escape Avar domination
as they were by a positive desire to move on to Roman territory.
Elsewhere we lack historical narratives, but the same desire to escape
Avar domination surely played a substantial role in the widespread
further dispersals of Slavic groups from c.550 onwards: westwards
towards the Elbe, northwards to the Baltic, and even eastwards into

the heart of Russia and Ukraine. It remains unclear to what extent this eastern expansion represented the first intrusion of Slavic-speakers into western Russia, or whether we are really looking at the expansion of particular groups of Slavic-speakers who had been made more politically organized and militarily potent through their interactions with the East Romans and Avars, and were thus able to assert their dominance over fellow Slavic-speakers who had not participated in the same process.

Either way, the process of Slavicization – the establishment of the dominance of Slavic-speaking groups across vast areas of central and eastern Europe – again combined processes of migration and development in intimate embrace. Interaction with the Roman Empire's more developed economy generated new wealth flows which prompted political consolidation and militarization among at least some Slavs. But the groups who benefited from this new wealth were only able to do so because they had already physically moved into a tighter Roman orbit after the collapse of the Hunnic Empire, presumably in order to make precisely these kinds of gain. The sociopolitical revolution they experienced as a consequence then pre-prepared them, especially under the extra stimulus provided by the Avars, to spread their domination by further migration across broad swathes of central and eastern Europe. Some of this certainly involved the absorption of the clearly numerous indigenous populations that had survived the processes of Germanic collapse. Some of that absorption will have been peaceful, as some east Roman sources suggest, but at the same time many Slavic groups were becoming increasingly militarized, and the results of Slavicization were strikingly monolithic. If some Slavic groups, particularly of the Korchak type, remained peaceful small-scale farmers up to the year 600 and beyond, many others were undergoing rapid transformation as new wealth brought social differentiation and militarization. Much of the subsequent Slavicization of Europe was clearly brought about by the armed and dangerous Slavs, not the Korchak farmers – not least in those parts of Russia where Slavic domination was advanced by communities of a few hundred pushing one fortified settlement after another into clearly hostile territory.

The Birth of Europe

East Roman wealth and Avar interference marked only the beginning of a much broader development process, which unfolded right across the vast area of Slavic-dominated Europe in the second half of the millennium. By the tenth century, this had produced the first state-like dynastic structures that much of northern and eastern Europe had ever seen. These new entities still operated with major limitations by the year 1000, distinct patterns of centre and periphery being discernible across the vast territories notionally under their control. A governmental mechanism based on itineration was not capable of governing such large territories with even intensity, and this shows up in their regular propensity to swap control of very large intermediate territorial zones between them. Nonetheless, these states were capable of centrally organized activities that are straightforwardly impressive. Much bigger in geographical scale than the Germanic client states that emerged on the fringes of the Roman Empire in the fourth century, they were also capable of greater acts of power. They built more and bigger buildings, supported larger, better-equipped, and more professional armies, and quickly adopted some of the cultural norms of more developed, imperial Europe: above all the Christian religion.

Everything suggests that the transformative mechanisms that produced these new entities were similar in nature to those that had generated the larger Germanic client states of the fourth-century Roman periphery. In both cases, a whole range of new contacts – via trading, raiding, and diplomacy – led to unprecedented flows of wealth into the non-imperial societies. The internal struggle to control these flows of wealth then led to both militarization and the emergence of pre-eminent dynasts, who eventually used their domination of this wealth to generate permanent military machines that could institutionalize their authority by destroying and/or intimidating pre-existing, more local authority structures. As a result, potential rivals were steadily eliminated and power was increasingly centralized.

But if the basic processes were the same, the second half of the millennium saw the Slavic world develop further and faster than its largely Germanic counterpart had done in the first. The explanation for this disparity in part lies in the broader range of stimuli operating in barbarian Europe after 500 AD. Western parts of the Slavic world established a full range of economic, military and diplomatic contacts with a sequence of Frankish imperial powers in western Europe. At the same time, two hundred years of Avar imperial domination at the heart of central Europe had important effects on a broader Slavic clientele, as did interaction with a further, if lesser, European imperial power: the Byzantine Empire. Equally, if not more important, more distant parts of the largely Slavic-dominated barbaricum were interacting with a fourth and still greater imperial power in the form of the Islamic Caliphate. There is no sign of any large-scale trade networks in either slaves or furs operating out of central and eastern Europe to feed Near Eastern as well as Mediterranean sources of demand in the first half of the millennium, so these later networks represented flows of wealth with no precedent in the Roman era. And to judge both by the staggering numbers of Islamic silver coins that survive and their correlation with the core areas of the new Slavic states, there is every reason to suppose this extra-European imperial stimulus played a major role in the transformation of Slavic Europe.

The other obvious explanation for the faster development of Slavic Europe is the impact of the new military technologies of the last two centuries of the millennium – notably armoured knights and castles – which made it much easier for those dynasts who could establish control over the new wealth flows to intimidate potential opponents. For even if the new states all encompassed less intensively governed peripheries, the power that they could exercise in dynastic core territories is (horribly) impressive. The brutal power inherent in the destruction of old tribal strongholds and their replacement with new dynastic ones – in both Bohemia and Poland – emerges strikingly from the dramatic archaeological evidence that has become available in recent years. Dynastic power is equally apparent in the movement of subdued populations into core zones of the new states, and their general economic organization, illustrated this time by a combination of archaeological evidence and the earliest strata of documentary evidence preserved from the new states.

The nature and overall significance of these processes of development could hardly be clearer, and their consequences were myriad. In broadest terms, the most important of these might well be the first emergence of Europe as a functioning entity. By the tenth century, networks of economic, political and cultural contact were stretching right across the territory between the Atlantic and the Volga, and from the Baltic to the Mediterranean. This turned what had previously been a highly fragmented landscape, marked by massive disparities of development and widespread non-connection at the birth of Christ, into a zone united by significant levels of interaction. Europe is a unit not of physical but of human geography, and by the year 1000 interaction between human populations all the way from the Atlantic to the Volga was for the first time sufficiently intense to give the term some real meaning. Trade networks, religious culture, modes of government, even patterns of arable exploitation: all were generating noticeable commonalities right across the European landscape by the end of the millennium.

For the purposes of this study, however, the processes of development are more immediately important for the role they played in bringing to an end the kind of conditions that had generated the large-scale often predatory forms of migration – whether in the concentrated pulse form of the *Völkerwanderungen* or the more usual flows of increasing momentum – which had been a periodic feature of first-millennium Europe. Inequalities of development across the European landmass had not completely disappeared, but they had been greatly reduced. Essentially, the new trade networks, combined with more general agricultural expansion (the latter still very much a work in progress), meant that politically organized power structures in central and eastern Europe were now able to access wealth in large quantities in their existing locations. Agricultural and broader economic development also meant that they were busy entrenching themselves in some entirely new ways in some specific geographical zones of operation, at least in their core territories.

As a result, the kinds of positive stimulation that had periodically prompted large-group migration had been structurally removed, or at least massively eroded. Migration was never an easy or universally prevalent option in first-millennium Europe, but rather a strategy that was sometimes adopted when the gains were worth the stress of mounting expeditions into only partly known territory with no absolute guarantees of success. Once social elites could access wealth without the extra insecurity of relocation, they became much less likely to resort to that strategy. And, of course, the less they did so in practice, the less they were ever likely to, as previously ingrained migration habits unwound both among themselves, and among the broader population under their control as more intense patterns of arable farming were generating more permanent patterns of cultivation. Overall, both elites and broader populations within barbarian Europe were becoming much more firmly rooted in particular localities, and, as a result, were much less likely to respond by migration even when faced with powerful stimuli that might in other circumstances have led them to shift location.

This, to my mind is the underlying explanation of the particular problem with which this chapter began. Where many Goths and other Germani (though certainly not all) responded to the Hunnic menace, and the Slavs to its Avar counterpart, by seeking new homes elsewhere, the arrival of the nomadic Magyars on the Great Hungarian Plain engendered no known secondary migration. The actions, nature and eventual fate of the Moravian state encapsulate the difference. Rather than run away, the Moravians stood and fought the Magyars, just like the armies of Frankish imperial Europe. They lost (as, initially, did many of their Frankish counterparts), but the fact that the Moravians stayed put reflects the deeper roots they had sunk in their own particular locality, and the fundamentally different nature of political power in barbarian Europe as it had developed by the end of the first millennium. Earlier, the prevailing limitations of agricultural technique in barbarian Europe generated a broad local mobility, and large disparities in levels of wealth and development had encouraged the more adventurous periodically to attempt to take over some more attractive corner of the landscape, closer, usually, to imperial sources of wealth. The Moravians, by contrast, built castles and churches in stone, on the back of wealth generated by more intense agricultural regimes and wider exchange networks. With so much invested where they stood, it was not going to be easy to shift their centre of operations. The same was true of the other new dynasties of the late first millennium too. All were much more firmly fixed in particular localities than their earlier counterparts, both because of developing agricultural technique and because trade networks made other types of wealth available well beyond the imperial borderlands. In overall terms, processes of development had both eliminated the massive inequalities that had previously made long-distance, large-group migration a reasonably common option for Europe's barbarians, and rooted central and east European populations more deeply in particular landscapes.

Not, of course, that any of this really spelled the end of migration.

Some human beings are always on the move in search of greater prosperity or better conditions of life, and European history from the tenth century onwards is still marked by migration on a periodically massive scale. From late in the first millennium onwards, however, medieval migration generally took one of two characteristic forms. On the one hand, we see knight-based elite transfers. The Norman Conquest is a particularly large-scale and successful example of this phenomenon. Much more usual were bands of one or two hundred well-armed men looking to establish small principalities for themselves by ousting sitting elites and/or establishing their rights to draw economic support from a dependent labour force. The productive rootedness of peasantry and the empowering effect of new military technologies were key factors in dictating the characteristics of this particular migratory form. Castles and armour allowed them to establish a form of local domination based on quite small numbers of men that was extremely hard to shift. The other common form of migration was the deliberate recruitment of peasantry to work the land, with lords offering attractive tenurial terms to provide the incentive, and employing agents to run recruiting campaigns. Again, new patterns of development were of crucial importance here, since the extra agricultural productivity of the new arable farming technologies being put into practice in the late first millennium made it highly desirable for the masters of the landscape to secure sufficient labour to maximize agricultural outputs. Though they had come a long way, the new Slavic states still lagged behind western and southern Europe in levels of economic development. They therefore figured among the chief customers for the new peasant labour being mobilized from more developed parts of Europe where higher population levels reduced opportunities for ambitious peasants to get more land on better terms. As a result, hundreds of thousands of peasants from west-central Europe would be attracted eastwards by the offer of land on much better terms than could be secured at home, and the Slavicization of much of old Germanic Europe that had occurred in the early Middle Ages was partly reversed by an influx of Germanic-speaking peasants.[18]

NEWTON'S THIRD LAW OF EMPIRES?

Both of these later medieval forms of migration are very well evidenced, operating, as they did, in an era when literacy was intensifying across most of Europe, so their importance within developing European history cannot be contested in the way that that of their earlier counterparts of the first millennium has come to be. The prevalence of these different forms in a later era, however, is no objection to the broader argument of this book, that larger-scale, socially more broadly based predatory forms of migration than knight-based expansion had played a hugely important role in the making of Europe in the first millennium. The later migratory forms were entirely appropriate to the economic and political conditions prevailing across the Europe of the central Middle Ages. The kinds of large-scale predatory migration flow studied in this book – essentially combining peasants and elites within the same migrating groups, where the later Middle Ages saw them move separately – were equally appropriate to their own area. In the first millennium, highly disparate patterns of development then combined with a lack of agricultural rootedness and relatively low agricultural outputs. This meant that the economy of barbarian Europe could support only very few military specialists, so that it was necessary and possible for ambitious leaders to put together large and hence necessarily broad-based expeditions to secure wealth-generating positions on the fringes of more developed, imperial Europe. This in turn generated forms of migration that were different from those operating in the central Middle Ages, and different again from those we are used to in the modern world. Migration in the first millennium looks thus not because our sources were infected with a distorting cultural reflex, but because prevailing circumstances contrasted in some key ways from those operating subsequently. They entirely conform to the basic principles of modern migration, however, in that direction of movement and form of the migration unit were both largely dictated by prevailing patterns of development.

In short, there is every reason to respond to the limitations of the old invasion-hypothesis model not by rejecting migration as an important explanatory factor in first-millennium history, but by bringing a series of more complex migration models back into the picture. Deployed in more analytic fashion, migration ceases to be a catch-all, simplistic alternative to 'more complex' lines of explanation focusing on social, economic and political change. Understood properly, and this is the central message screaming out from the comparative literature, migration is not a separate and competing form of explanation to social and economic transformation, but the complementary other side of the same coin. Patterns of migration are dictated by prevailing economic and political conditions, and another dimension in fact of their evolution; they both reflects existing inequalities, and sometimes even help to equalize them, and it is only when viewed from this perspective that the real significance of migratory phenomena can begin to emerge. A further line of thought that follows from this is that prehistorians should perhaps not be too quick to reject predatory migration either as a periodic contributor to the shaping of Europe's deeper past. If the argument is correct that the predatory forms of migration observable periodically in the first millennium were generated by a reasonable degree of geographical proximity between zones of highly disparate levels of development, combined with the existence of societies where those who farmed also fought and were not deeply rooted in one particular patch of soil, then these are conditions which are likely to have existed in many other ancient contexts too, and periodic predatory migration could reasonably be expected as one natural consequence.

That is no more than a side issue for this study, however, and thinking about the transformation of barbarian Europe in the first millennium in overall terms, there is no doubt that development played a profoundly more important role in the process than migration. Old narratives had this the other way round, emphasizing the arrival of named peoples at their assigned places across the map of Europe at different points within the millennium, until all the modern nations were in place. In this view, movement and arrival were the events of key historical importance, and what happened subsequently was so much detail. This was deeply mistaken. Much more important than these occasional moments of arrival, many of which led precisely nowhere, were the dynamic interactions between the imperial powers of more developed Europe and the barbarians on their doorstep: Germanic, largely, in the first half of the millennium, then Slavic, largely, in the second. It was these interactions, not acts of migration, that were ultimately responsible for generating the new social, economic and political structures which brought former barbarian Europe much more to resemble its imperial counterpart by the end of the millennium. This is not to say that these transformations were inherently a good thing, or that there was something inherently better about imperial Europe, but the evidence leads directly to the conclusion that it was new connections with imperial Europe, and the responses to those new connections on the part of elements within barbarian societies, that ultimately demolished the staggering disparities in development that had existed at the birth of Christ. This in a nutshell is the second major argument I have been attempting to make. Not everywhere in Europe was Christian and full of states built around castles, knights and a productive peasantry by the year 1000, but this was true to an extent that would have astonished Tacitus in the first century AD. He thought that eastern Europe was home to creatures with 'human faces and features, but the bodies and limbs of beasts'; in his terms, barbarian Europe was barbarian no longer.[19]

Migration had played a role – sometimes a very major one – in this unfolding story. Especially if you take the definition of mass or significant migration offered in the comparative literature – and I have found this extremely helpful – migration can be understood as central to the action at various key points in the millennium. Perhaps above all, the Hunnic 'accident' threw enough more organized Germanic groupings on to Roman soil in a short enough space of time both to undermine the central Roman state and to generate a massive collapse in the old power structures of barbarian central Europe. This in turn allowed for an extraordinary Slavic diaspora whose cultural effects – the widespread Slavicization of central and eastern Europe – remain a central feature of the European landmass to this day. These are hardly minor phenomena. Even so, migration should generally be given only a secondary position behind social, economic and political transformation when explaining how it was that barbarian Europe evolved into non-existence in the course of the millennium. For one thing, aside from particular and unusual moments like the Hunnic or Avar accidents, patterns of migration were entirely dictated by and secondary to patterns of development. It was only when nomadic intruders added a much stronger shade of politically motivated migration into the picture that the relationship was reversed, and migration started to dictate patterns of development, undermining both the west Roman state and Germanic Europe in one fell swoop.

Even without the Huns, moreover, these processes of development would eventually have undermined the Roman Empire. Looked at in the round, what emerges from the first-millennium evidence is that living next to a militarily more powerful and economically more developed intrusive imperial neighbour promotes a series of changes in the societies of the periphery, whose cumulative effect is precisely to generate new structures better able to fend off the more unpleasant aspects of imperial aggression. In the first millennium, this happened on two separate occasions. We see it first in the emergence of Germanic client states of the Roman Empire in the fourth century, and again – this time to more impressive effect – in the rise of the new Slavic states of the ninth and tenth. This repeated pattern, I would argue, is not accidental, and provides one fundamental reason why empires, unlike diamonds, do not last forever. The way that empires tend to behave, the mixture of economic opportunity and intrusive power that is inherent in their nature, prompts responses from those affected which in the long run undermine their capacity to maintain the initial power advantage that originally made them imperial. Not all empires suffer the equivalent of Rome's Hunnic accident and fall so swiftly to destruction. In the course of human history, many more have surely been picked apart slowly from the edges as peripheral dynasts turned predator once their own power increased. One answer to the transitory nature of imperial rule, in short, is that there is a Newtonian third law of empires. The exercise of imperial power generates an opposite and equal reaction among those affected by it, until they so reorganize themselves as to blunt the imperial edge. Whether you find that comforting or frightening, I guess, will depend on whether you live in an imperial or peripheral society, and what stage of the dance has currently been reached. The existence of such a law, however, is one more general message that exploring the interactions of emperors and barbarians in the first millennium AD can offer us today.

MAPS

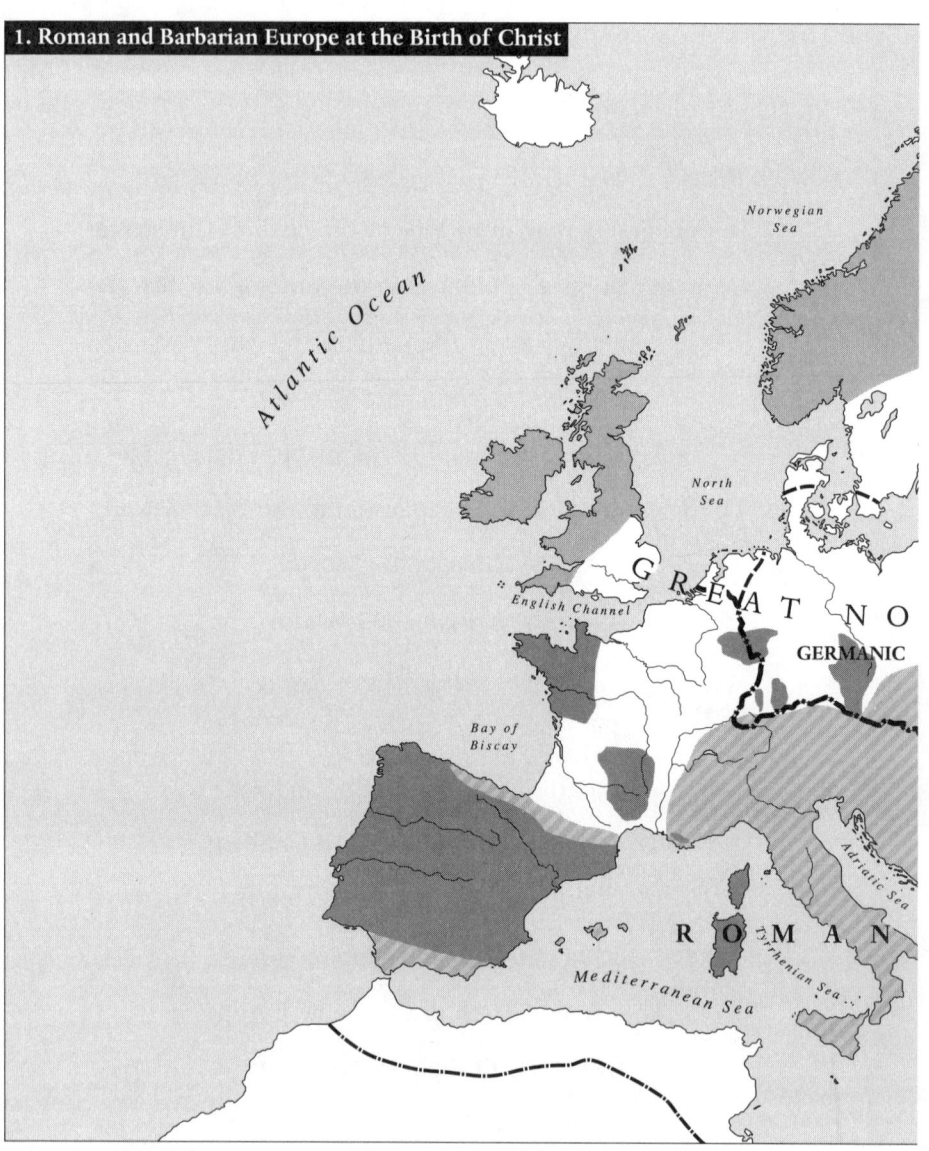

1. Roman and Barbarian Europe at the Birth of Christ

Roman Empire

Germanic-dominated Europe

Major geographical features

Northern boundary of conventional farming

North-west highlands

Old massifs

Alpine mountains

Baltic Sea

SCYTHIA

NORTH EUROPEAN PLAIN

EUROPE

Sea of Azov

Black Sea

Caspian Sea

EMPIRE

Aegean Sea

Ionian Sea

800 kilometres

500 miles

2. Germanic political groupings in the time of Tacitus

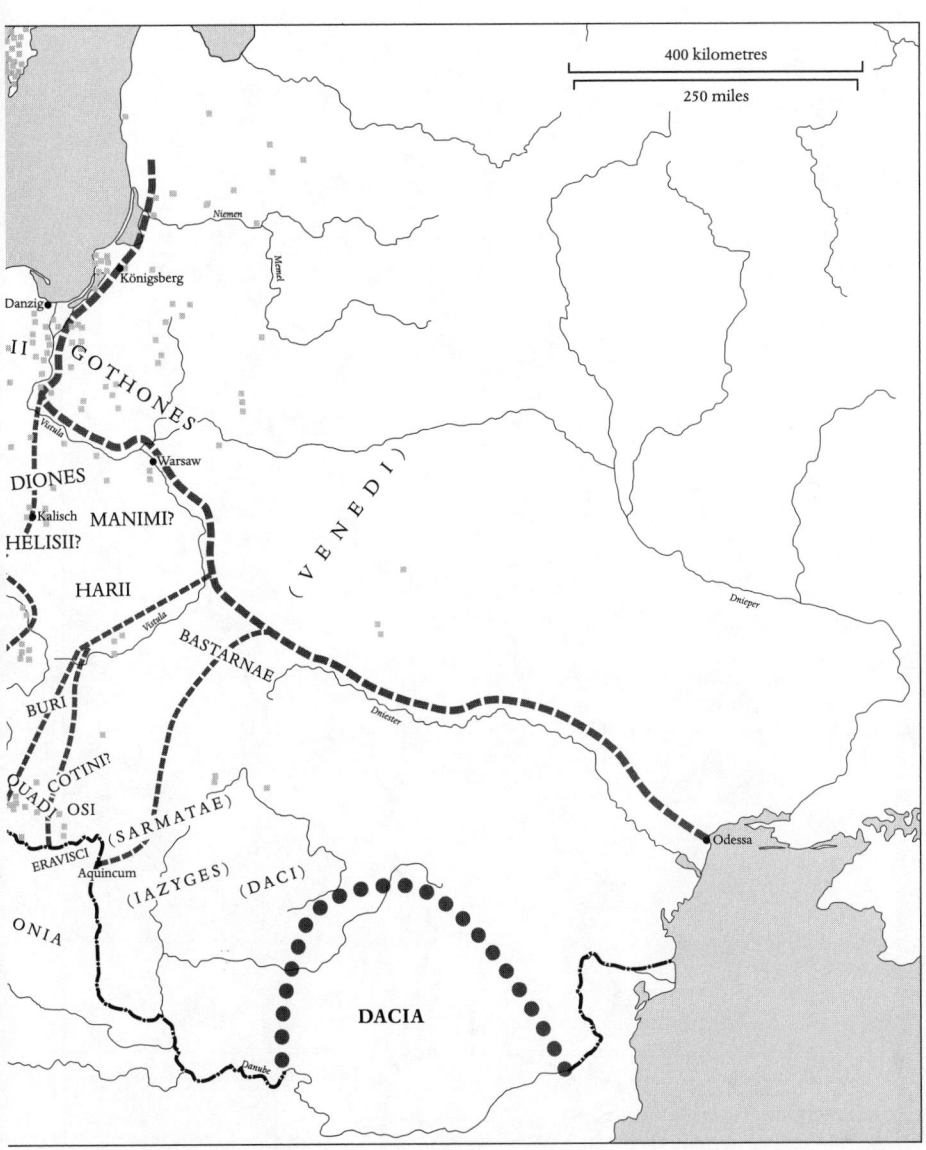

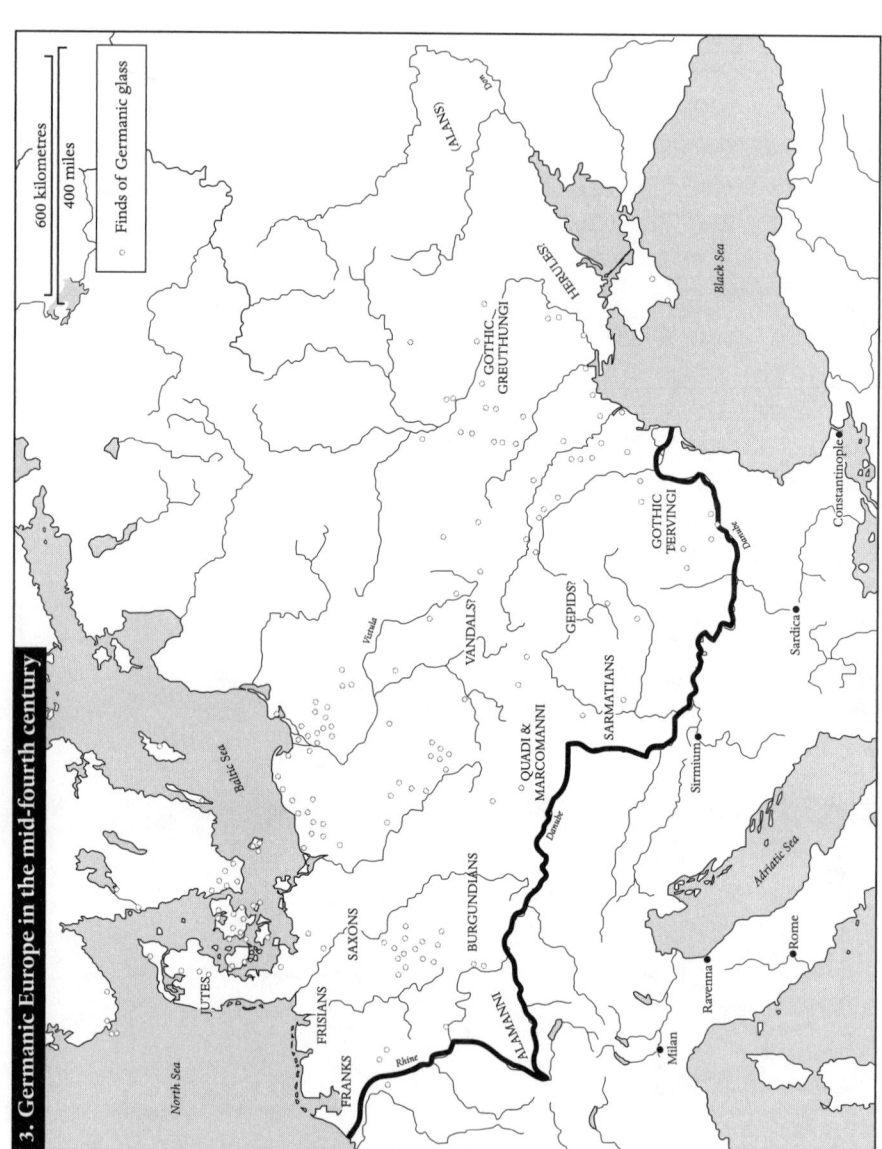

3. Germanic Europe in the mid-fourth century

4. The Marcomannic War

300 kilometres

200 miles

RAETIA   Roman province name

━·━·━ Roman Empire

╌╌╌╌ Roman province

QUADI   Tribal names

Wielbark period 1

Przeworsk period 1

Wielbark period 2

Przeworsk period 2

North Sea

Baltic Sea

WIELBARK

LOMBARDS & UBII

PRZEWORSK

Vistula

Bug

Therouanne

GERMANIA INFERIOR

Bavay

Amiens

BELGICA

GERMANIA SUPERIOR

MARCOMANNI

Elbe

Rhine

Vistula

NARISTI

VANDALS

166-7

QUADI

LUGUDUNENSIS

RAETIA

Danube

COSTOBOCI

NORICUM

ALPES PUENIAE

PANNONIA INFERIOR

IAZYGES

ALPES COTTIAE

PANNONIA SUPERIOR

NARBONENSIS

ITALIA

ALPES MARITIMAE

DACIA

ROXOLANI

DALMATIA

Danube

Ligurian Sea

Adriatic Sea

MOESIA SUPERIOR

MOESIA INFERIOR

Black Sea

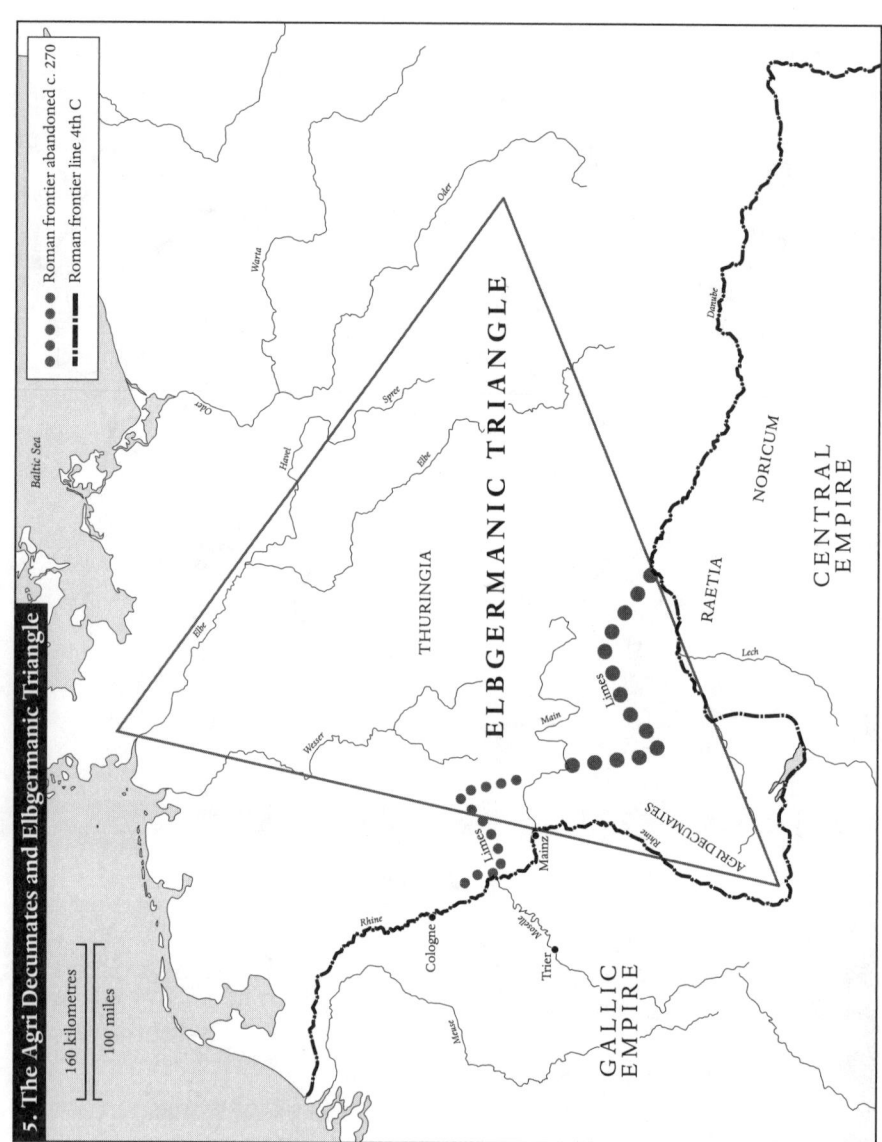

5. The Agri Decumates and Elbgermanic Triangle

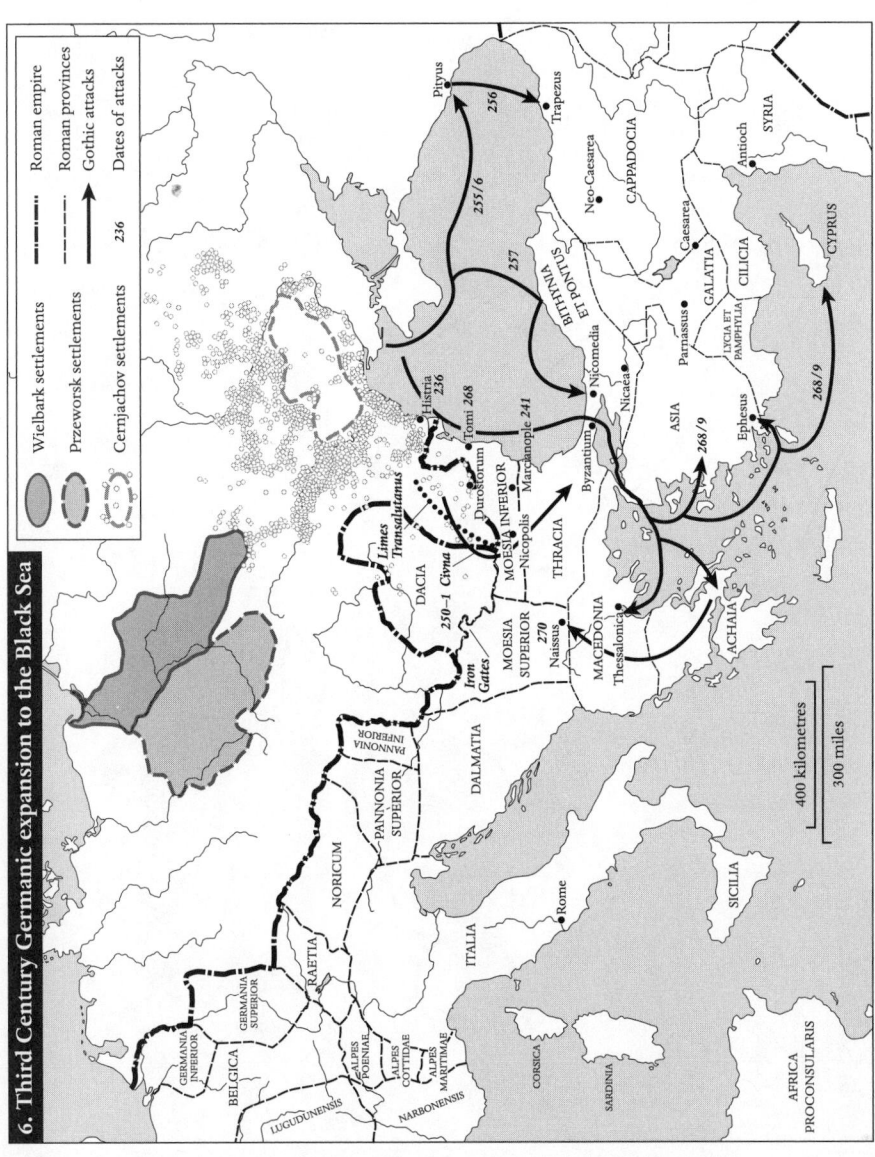

6. Third Century Germanic expansion to the Black Sea

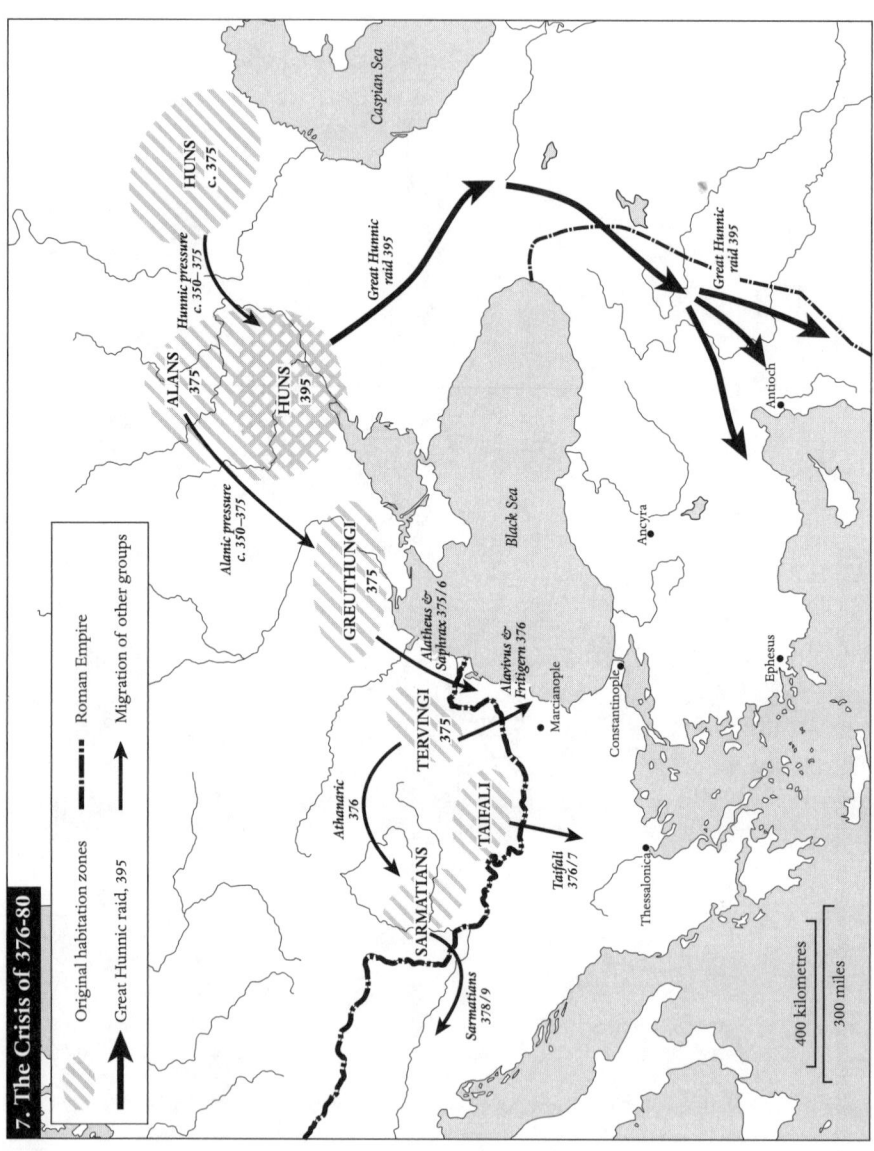

7. The Crisis of 376-80

Original habitation zones

Roman Empire

Great Hunnic raid, 395

Migration of other groups

HUNS c. 375

Hunnic pressure c. 350-375

ALANS 375

HUNS 395

Great Hunnic raid 395

Great Hunnic raid 395

Alanic pressure c. 350-375

Caspian Sea

Antioch

Ancyra

Black Sea

GREUTHUNGI 375

Alatheus & Saphrax 375/6

Alavivus & Fritigern 376

Marcianople

Constantinople

Ephesus

TERVINGI 375

Athanaric 376

TAIFALI

Taifali 376/7

Thessalonica

SARMATIANS

Sarmatians 378/9

400 kilometres

300 miles

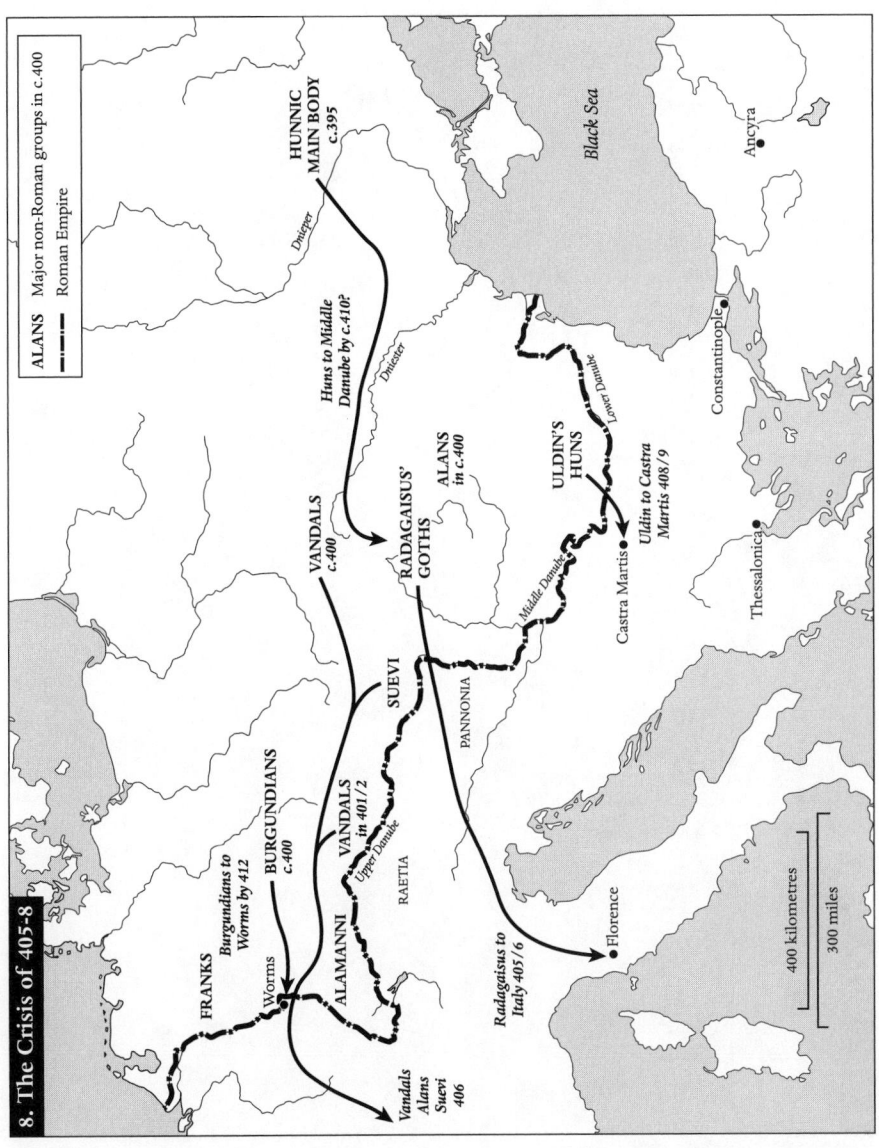

8. The Crisis of 405-8

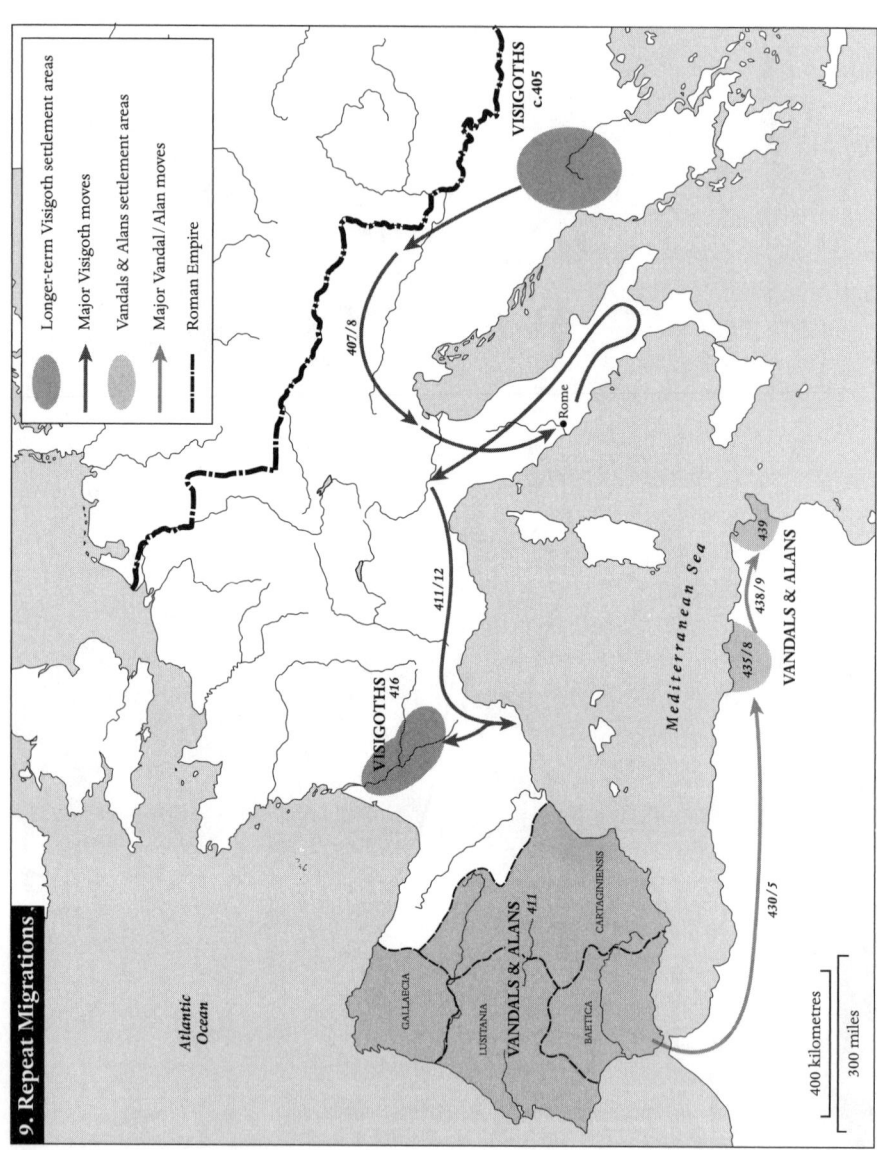

9. Repeat Migrations

Longer-term Visigoth settlement areas

Major Visigoth moves

Vandals & Alans settlement areas

Major Vandal/Alan moves

Roman Empire

VISIGOTHS
c.405

407/8

411/12

Rome

Mediterranean Sea

VISIGOTHS
416

VANDALS & ALANS
439
438/9
435/8

VANDALS & ALANS

430/5

Atlantic
Ocean

GALLAECIA

LUSITANIA

VANDALS & ALANS
411

BAETICA

CARTAGINIENSIS

400 kilometres

300 miles

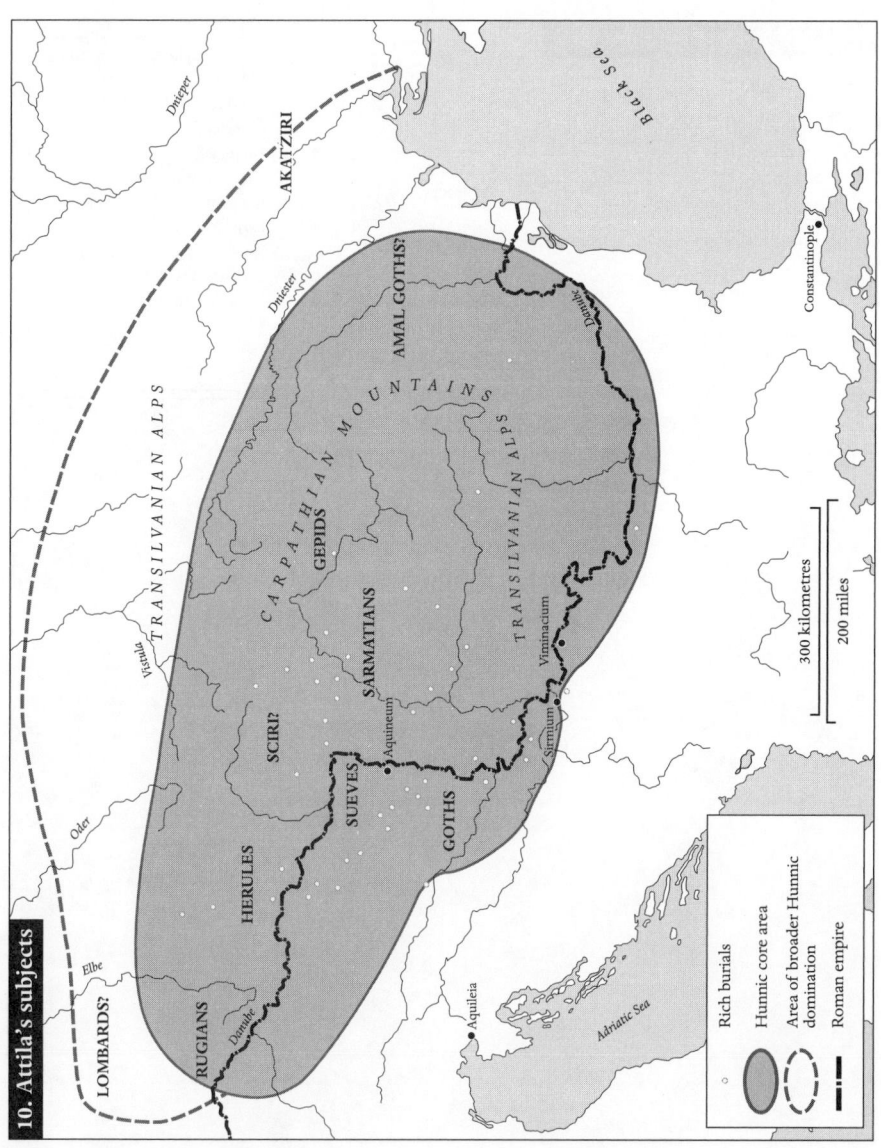

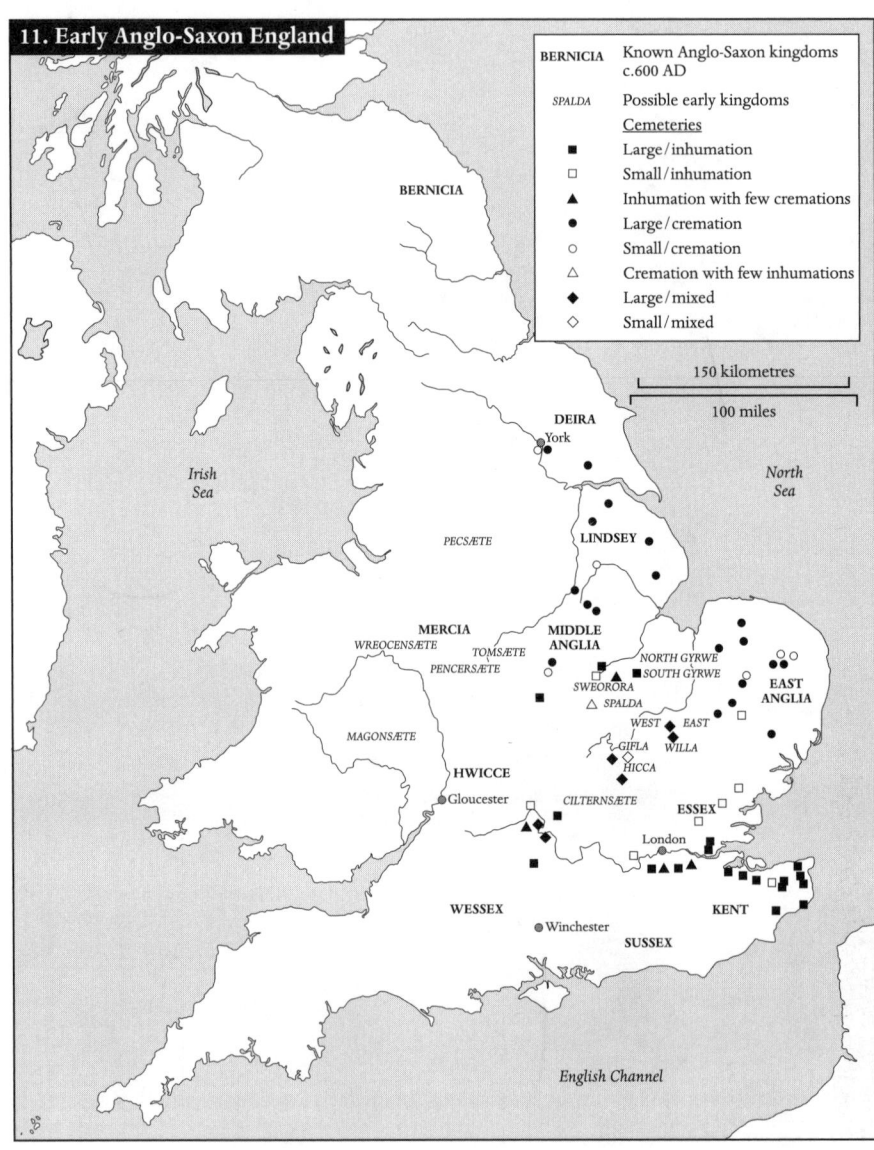

11. Early Anglo-Saxon England

BERNICIA   Known Anglo-Saxon kingdoms c.600 AD

SPALDA   Possible early kingdoms

Cemeteries

■   Large/inhumation
□   Small/inhumation
▲   Inhumation with few cremations
●   Large/cremation
○   Small/cremation
△   Cremation with few inhumations
◆   Large/mixed
◇   Small/mixed

150 kilometres

100 miles

BERNICIA

Irish Sea

North Sea

DEIRA
York

PECSÆTE   LINDSEY

MERCIA   MIDDLE ANGLIA
WREOCENSÆTE   TOMSÆTE
PENCERSÆTE   NORTH GYRWE
SWEORORA   SOUTH GYRWE
△ SPALDA
WEST   EAST
GIFLA   WILLA
HICCA

EAST ANGLIA

MAGONSÆTE

HWICCE
Gloucester   CILTERNSÆTE
ESSEX
London

WESSEX   KENT
Winchester
SUSSEX

English Channel

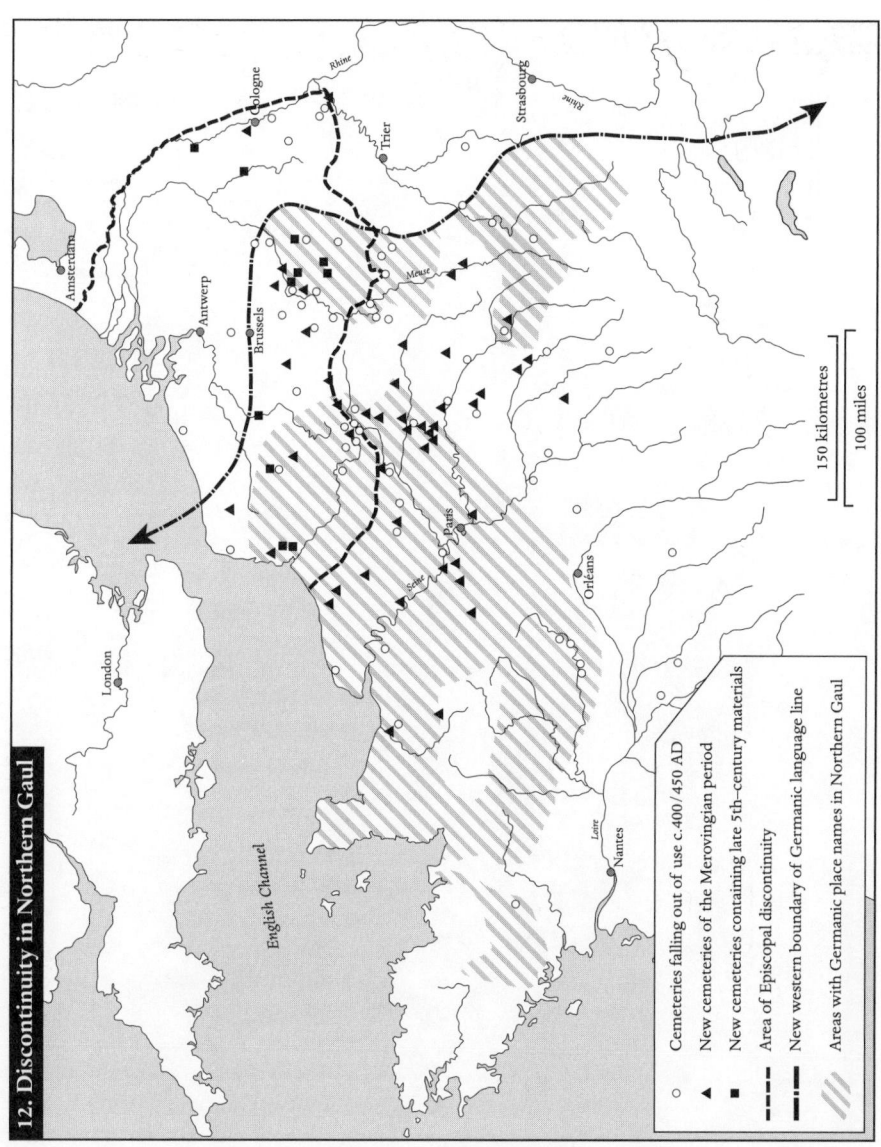

12. Discontinuity in Northern Gaul

○   Cemeteries falling out of use c.400/450 AD

◀   New cemeteries of the Merovingian period

■   New cemeteries containing late 5th–century materials

Area of Episcopal discontinuity

New western boundary of Germanic language line

Areas with Germanic place names in Northern Gaul

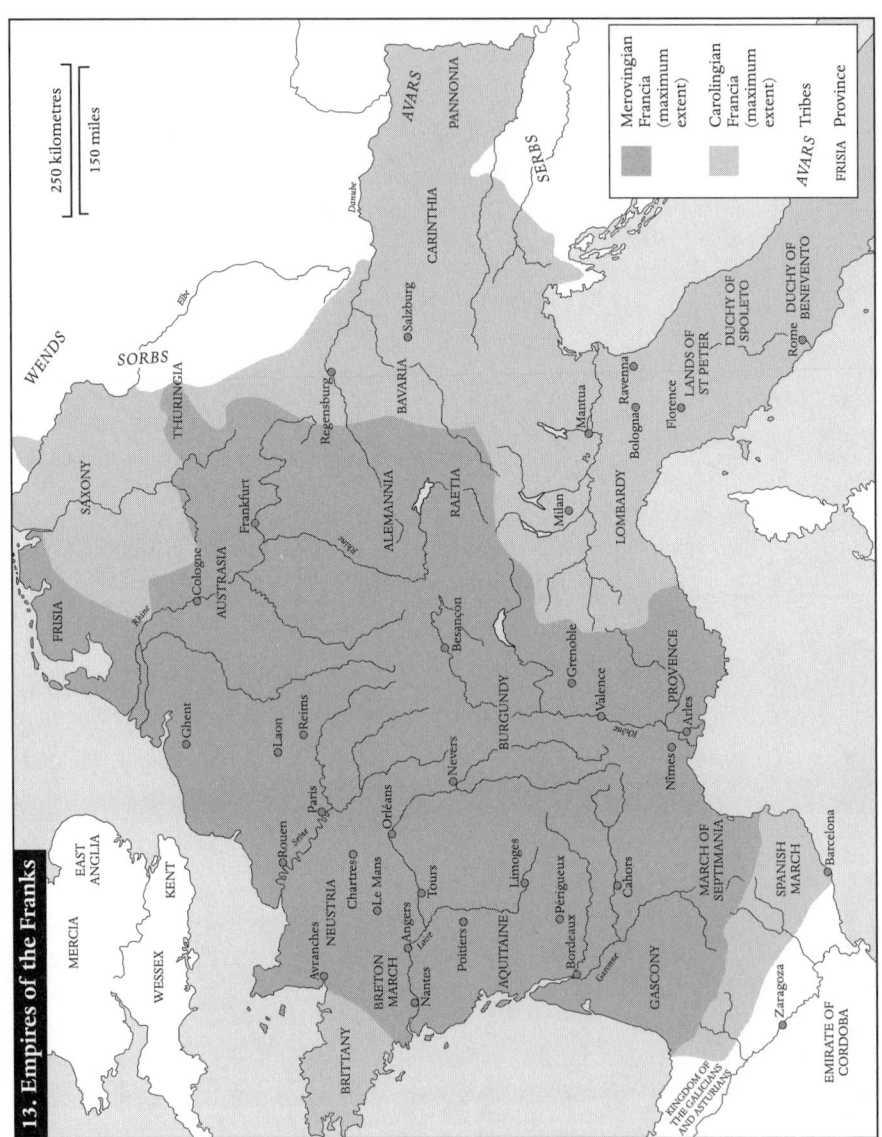

13. Empires of the Franks

14. The Ottonian Empire

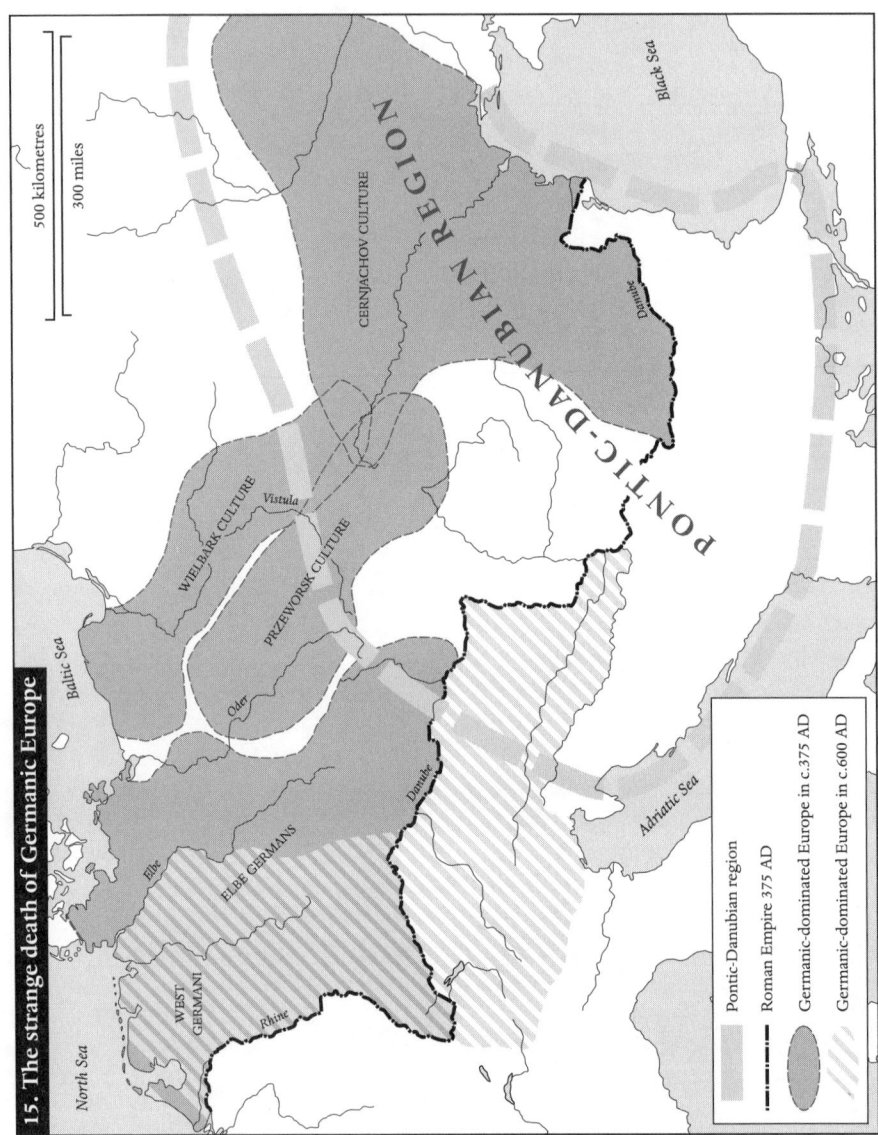

15. The strange death of Germanic Europe

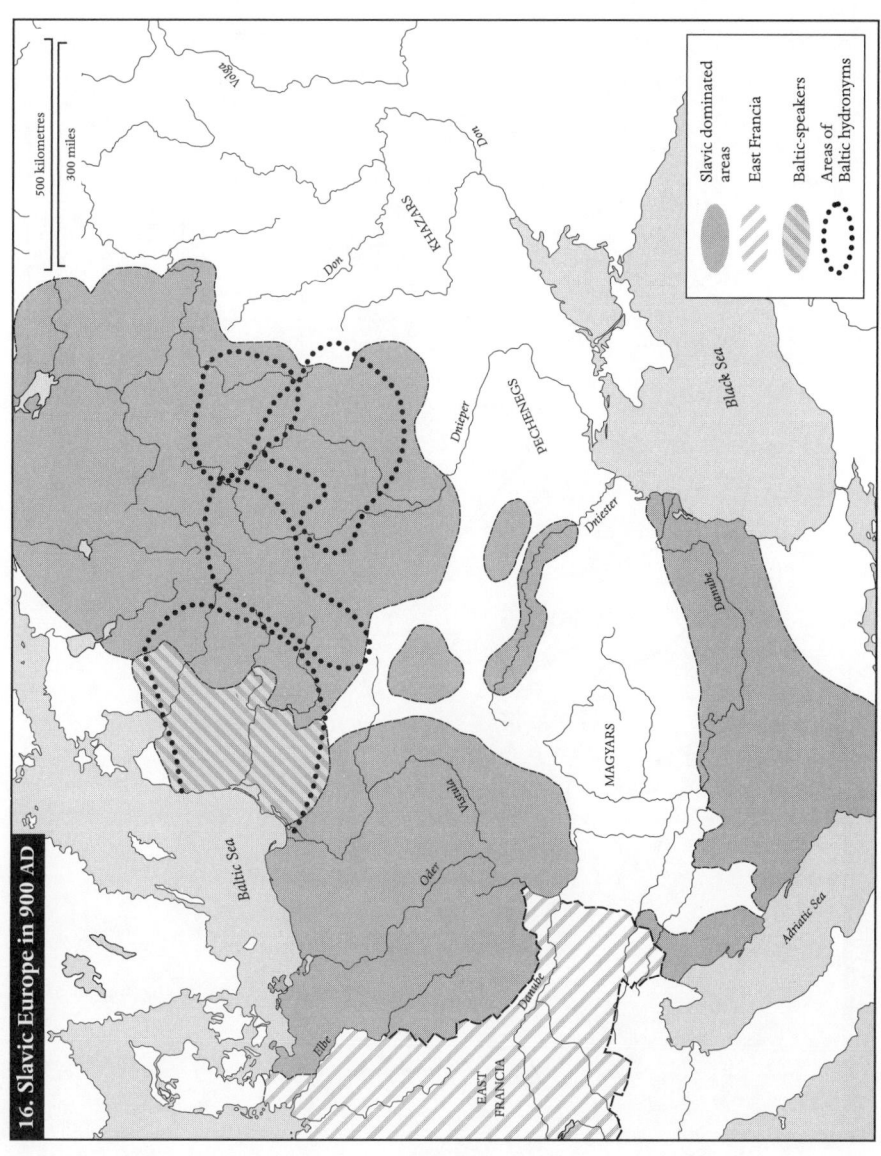

16. Slavic Europe in 900 AD

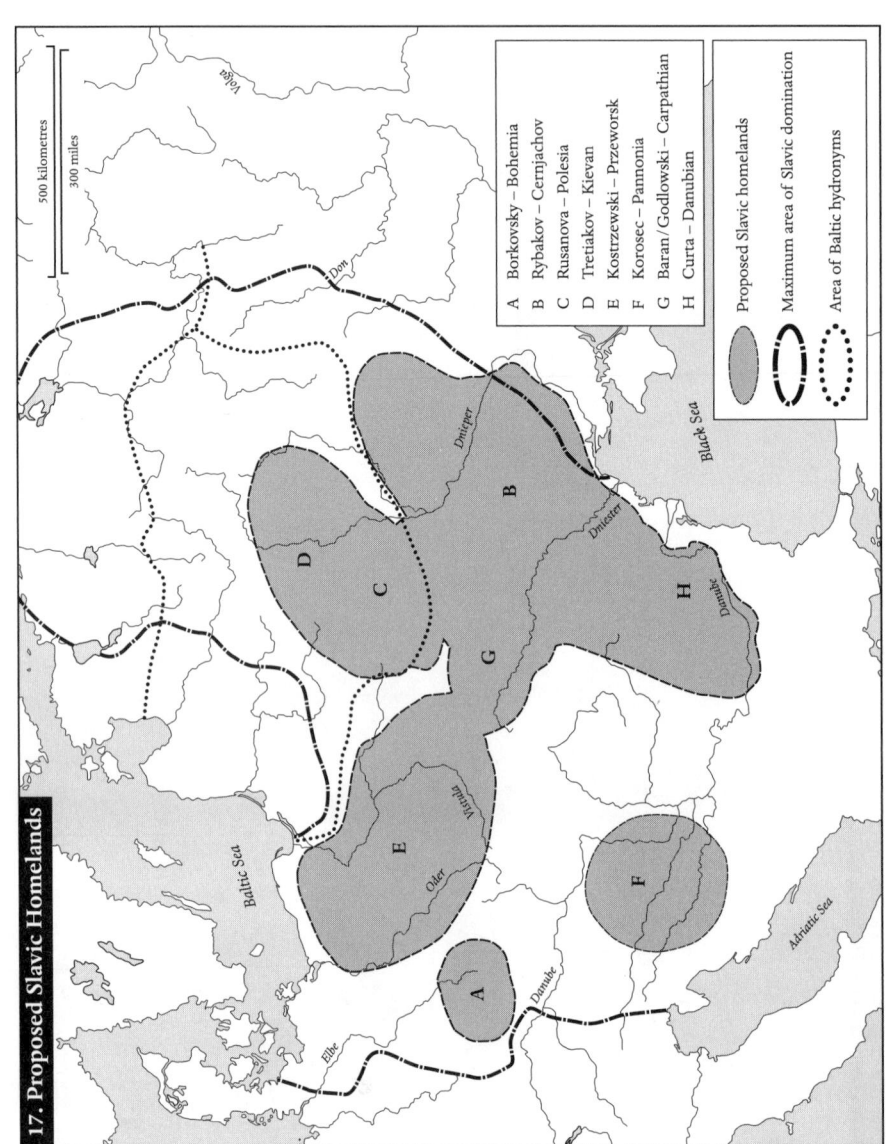

17. Proposed Slavic Homelands

A Borkovsky – Bohemia
B Rybakov – Cernjachov
C Rusanova – Polesia
D Tretiakov – Kievan
E Kostrzewski – Przeworsk
F Korosec – Pannonia
G Baran/Godlowski – Carpathian
H Curta – Danubian

Proposed Slavic homelands
Maximum area of Slavic domination
Area of Baltic hydronyms

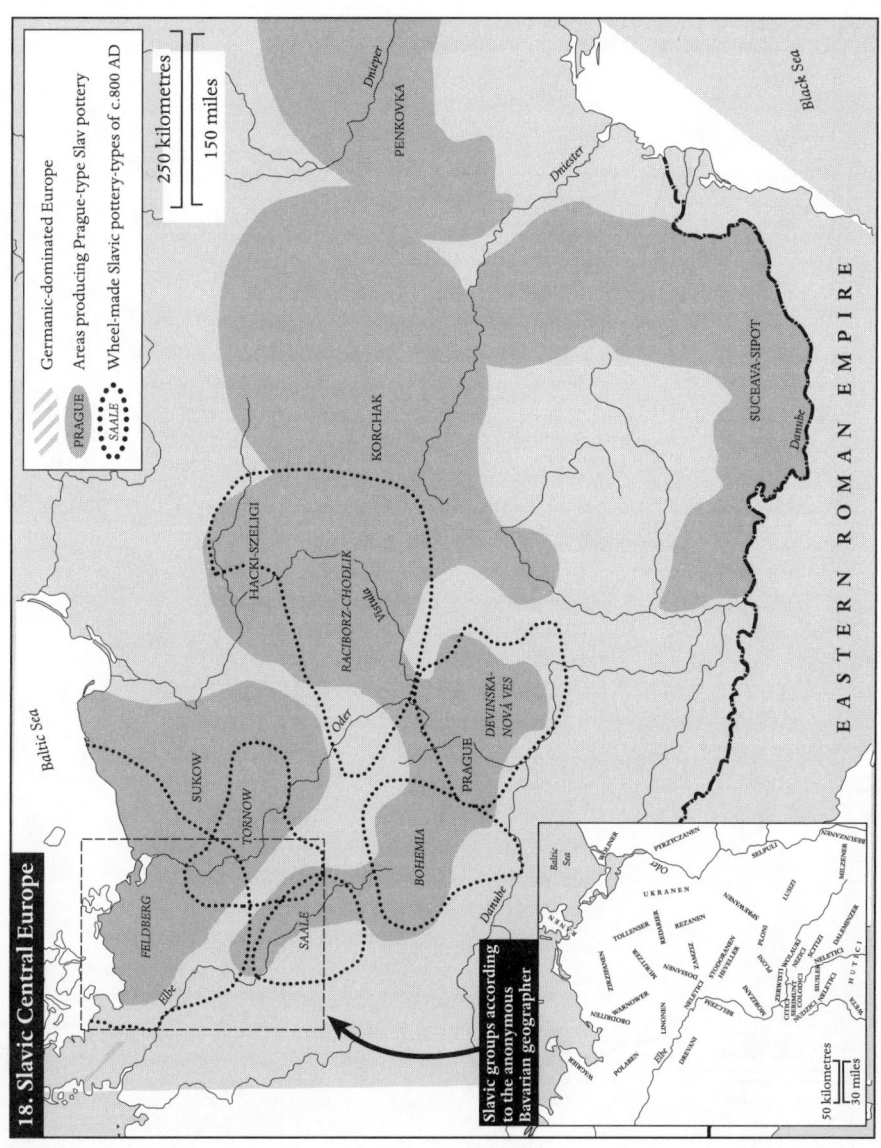

18. Slavic Central Europe

Germanic-dominated Europe
Areas producing Prague-type Slav pottery
Wheel-made Slavic pottery-types of c.800 AD

PRAGUE
SAALE

250 kilometres
150 miles

Baltic Sea

Black Sea

EASTERN ROMAN EMPIRE

PENKOVKA
KORCHAK
HACKI-SZELIGI
RACIBORZ-CHODLIK
DEVINSKA-NOVÁ VES
PRAGUE
BOHEMIA
SUKOW
TORNOW
FELDBERG
SAALE
SUCEAVA-SIPOT

Dnieper
Dniester
Danube
Vistula
Oder
Elbe

Slavic groups according to the anonymous Bavarian geographer

50 kilometres
30 miles

Baltic Sea

SMELDINGON
LINAA
BETHENICI
SMELDINGON
MORIZANI
HEHFELDI
SURBI
TALAMINZI
BECHELENZI
WELETABI
VUILCI
FRAGANEO
LUPIGLAA
OPOLINI
GOLENSIZI
SLEENZANE
LENDIZI
THADESI
MILZANE
BESUNZANE
VERIZANE
FRESITI
SERAVICI
LUCOLANE
UNGARE
VUISLANE
SLEENZANE
OSTERABTREZI
MISECO
PHESNUZI
THAFNEZI
ZUIREANI
BUSANI
UNLIZI
NERIUANI
ATTOROZI
EPTARADICI
VUILLEROZI
ZABROZI
ZNETALICI
ATURESANI
CHOZIROZI
LENDIZI

UKRANEN
WARNOWER
HEVELLI
OSTERABTREZI
MILOXI
MORIZANI
TOLLENSER
BEZUNZANE
PRISSANI
VUILOROZI
ZEUREANI
NABARENSI
FRESITI

Elbe
Oder

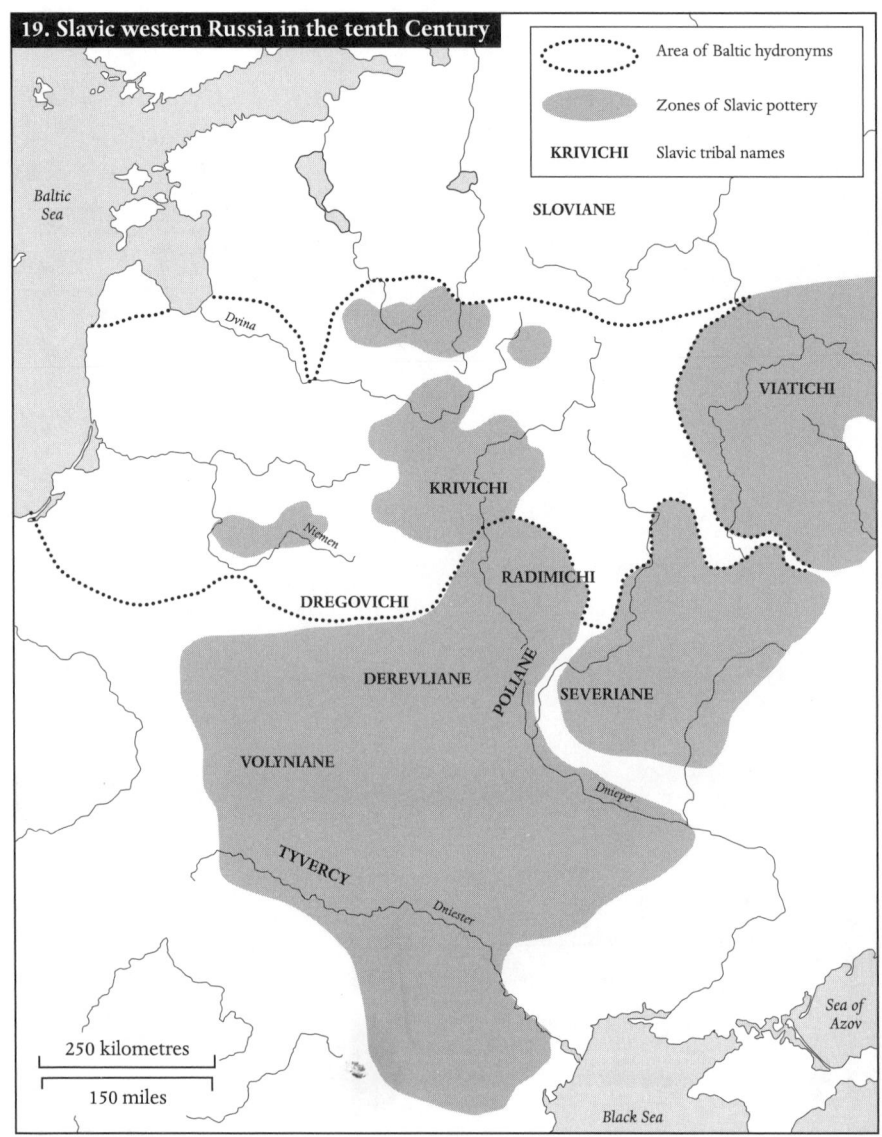

19. Slavic western Russia in the tenth Century

Area of Baltic hydronyms
Zones of Slavic pottery
KRIVICHI Slavic tribal names

Baltic Sea

SLOVIANE

Dvina

VIATICHI

KRIVICHI

Niemen

RADIMICHI

DREGOVICHI

DEREVLIANE

POLLIANE

SEVERIANE

VOLYNIANE

Dnieper

TYVERCY

Dniester

Sea of Azov

250 kilometres

150 miles

Black Sea

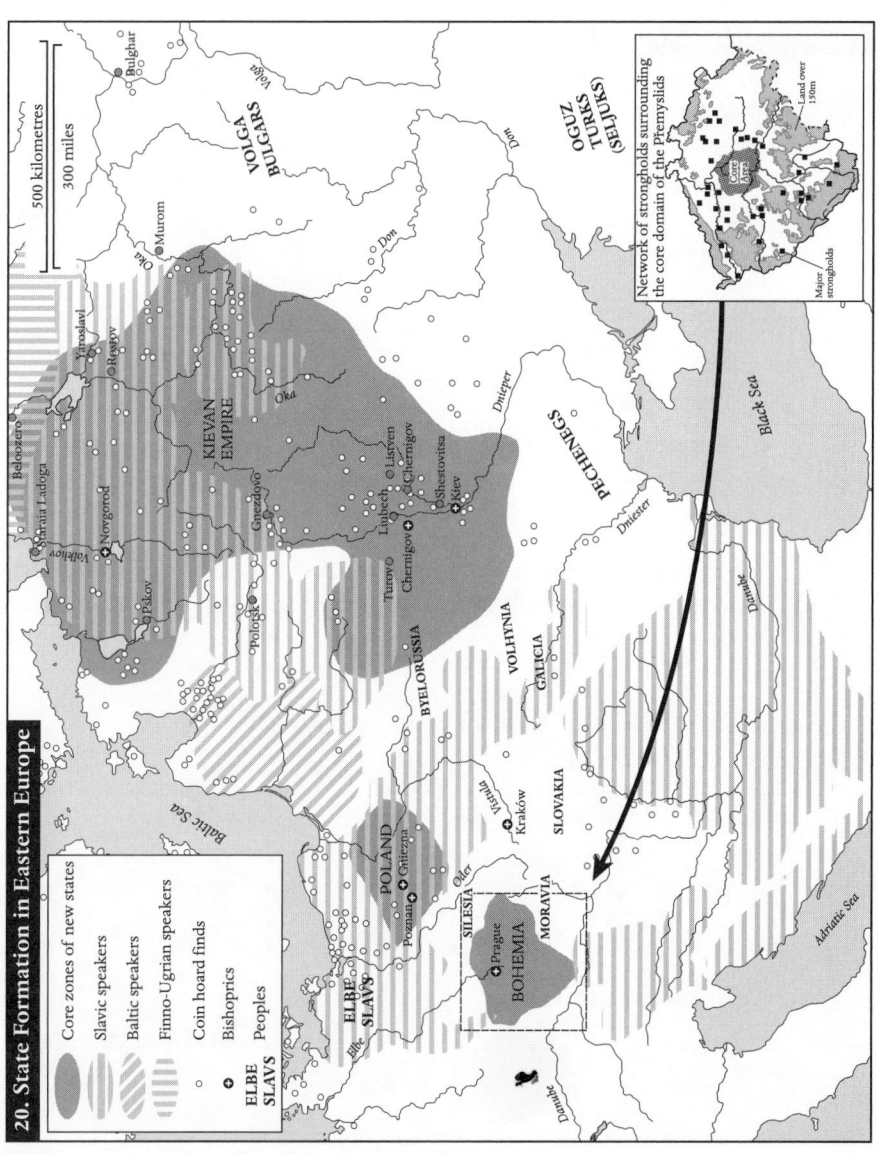

20. State Formation in Eastern Europe

Legend:
- Core zones of new states
- Slavic speakers
- Baltic speakers
- Finno-Ugrian speakers
- ○ Coin hoard finds
- ✚ Bishoprics
- ELBE SLAVS Peoples

Inset: Network of strongholds surrounding the core domain of the Přemyslids
- ■ Major strongholds
- Land over 150m

Scale: 500 kilometres / 300 miles

Labels: Bulghar, VOLGA BULGARS, Volga, Don, Murom, Oka, Yaroslavl, Rostov, Beloozero, Beloozero, Staraia Ladoga, KIEVAN EMPIRE, Oka, Volkhov, Novgorod, Gnezdovo, Lubech, Listven, Chernigov, Oshestovitsa, Kiev, Dnieper, OGUZ TURKS (SELJUKS), Black Sea, Pskov, Polotsk, Turov, Chernigov, PECHENEGS, Dniester, BYELORUSSIA, VOLHYNIA, GALICIA, Danube, Kraków, Vistula, SLOVAKIA, Oder, POLAND, Poznań, Gniezno, Baltic Sea, SILESIA, MORAVIA, BOHEMIA, Prague, ELBE SLAVS, Elbe, Adriatic Sea, Danube

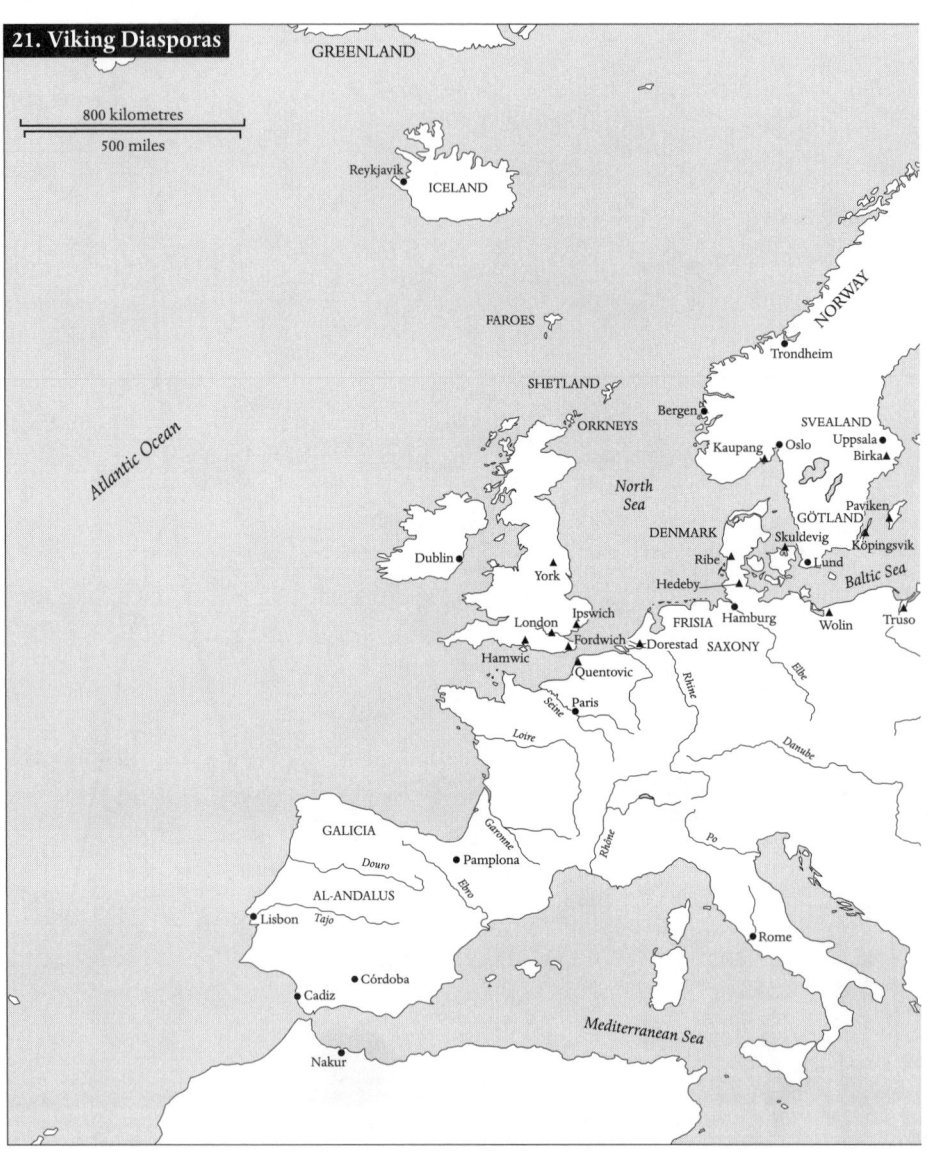

21. Viking Diasporas

800 kilometres

500 miles

GREENLAND

Reykjavik ICELAND

Atlantic Ocean

FAROES

SHETLAND

ORKNEYS

NORWAY

Trondheim

Bergen

North Sea

SVEALAND

Kaupang Oslo Uppsala
Birka

Paviken
GÖTLAND

DENMARK

Skuldevig Köpingsvik

Ribe Lund

Baltic Sea

Dublin

York

Hedeby Wolin Truso

Ipswich

London Hamburg

Fordwich FRISIA

Hamwic Dorestad SAXONY

Quentovic

Paris *Rhine*

Seine

Loire

Elbe

Danube

GALICIA

Douro Pamplona

Garonne

Rhône

Po

AL-ANDALUS

Lisbon *Tajo* *Ebro*

Rome

Córdoba

Cadiz

Mediterranean Sea

Nakur

NOTES

PROLOGUE

1 *Annal of Fulda* 882 for the incident with Poulik (1986) on the archaeology.

1. MIGRANTS AND BARBARIANS

1 Bohning (1978), 11.
2 For useful summaries of the modern evidence, see Salt and Clout (1976); King (1993); Collinson (1994), 1–7, 27–40; Holmes (1996); Cohen (1995), (1996), (1997), (2008); Vertovec and Cohen (1999). Canny (1994) provides an introduction to early modern migration evidence. 200,000 Germanic-speaking peasants: Kuhn (1963), (1973); Bartlett (1993), 144–5; and see, more generally, Phillips (1988), (1994).
3 For an introduction to the pre-Roman world of the Celts, see e.g. Cunliffe and Rowley (1976); Cunliffe (1997); James (1999). In fact, there is no one-to-one equation between Celts and the Oppida culture, and Roman conquest did advance just beyond its bounds: see Heather (2005), 49–58.
4 For useful introductions to the early Germanic world, see Hachmann (1971); Todd (1975), (1992); Krüger (1976), vol. 1; Pohl (2000). Note, though, that there is a strong tendency in some of this literature to avoid discussing Germanic groups around the Vistula and further east – a squeamishness resulting from the Nazi era, when the fact that ancient Germanic speakers had once dominated these lands was used as an excuse for territorial aggression.
5 For an excellent, recent overall introduction, see Batty (2007); on the broader cultural role played by Scythia in the formation of the Greek world view, see Braund (2005).
6 Khazanov (1984) provides an excellent introduction to the world of the steppe.
7 'The Veneti have taken': Tacitus, *Germania* 46.2 (cf. 46.4 on what lay beyond); see also Pliny, *Natural History* 4.97; Ptolemy, *Geography* 3.5.1 and 7. On the geography and ancient archaeological patterning of the society and economy of these regions, see Dolukhanov (1996). Within the Russian forest zone, many of the river names are actually Balt rather than Slavic in origin, even in areas where Slavs would be dominant by the year 1000 AD. It is thus unclear whether Tacitus' Veneti are likely to have been Slavic-speakers, Balt-speakers, or speakers of a tongue ancestral to them both (see Chapter 8).
8 Nomads too played their part: the Huns in the fall of the Roman Empire, the

Avars in the slavicization of central and eastern Europe, and the Magyars and Bulgars in laying the foundations of two substantial political entities whose lengthy histories underlie the existence of modern Hungary and Bulgaria.

9   The literature on the cultural significance of the rise of nationalism is now vast, but for introductions, see Gellner (1983); Anderson (1991); Geary (2002).

10 Early modern and modern accounts of Germanic migration consistently pictured migrants as family groups, while more contemporary Roman sources, when they said anything, also sometimes recorded the presence of women and children alongside the warriors. (I have simplified here, and the actual evidence will be surveyed in subsequent chapters.) Students of the collapse of the Roman Empire are broadly divided between viewing the Germanic invasions as its cause, and as its result. For useful overviews of the range of opinion, see Demandt (1984) and Ward Perkins (2005). With regard to the Slavs, one body of opinion has wanted to identify a very large, if submerged, population of Slavic-speakers throughout central and eastern Europe since the Bronze Age, but the evidence remains unconvincing (see Chapter 8). For a useful survey of traditional approaches to the Vikings, see Sawyer (1962), chapter 1. Nationalist conflicts also led to the downplaying of the so-called 'Normanist' view, that Vikings were responsible for the first Russian state: see Melnikova (1996), chapter 1 (and see also my Chapter 9).

11 Childe (1926), (1927).

12 See note 9 above. The general point is accepted even by those, such as Smith (1986), willing to conceive of relatively solid and sizeable group identities in at least some corners of the pre-nationalist past.

13 Leach (1954); 'evanescent situational construct': Barth (1969), 9. For more recent overviews, see e.g. Bentley (1987); Kivisto (1989); Bacall (1991).

14 That hypothesis was already marked in the work of Kossinna himself: see especially Kossinna (1928). It showed itself even more strongly in the equally influential work of Gordon Childe (see note 11 above), who generalized many of Kossinna's ideas, while dropping some of his assumptions about Nordic racial superiority. On Kossinna's legacy, see e.g. Chapman and Dolukhanov (1993), 1–5; Renfrew and Bahn (1991).

15 For an overview of these intellectual developments, see Shennan (1989); Renfrew and Bahn (1991); Chapman and Dolukhanov (1993), 6–25 (which includes an instructive difference in emphasis on the part of the two authors); Ucko (1995). The work of Ian Hodder – especially (1982) and (1991) – has been particularly important in rehabilitating the view that patterns of similarity and difference in material cultural items might sometimes reflect important aspects of human organization.

16 Clark (1966) represents a key turning point away from the invasion hypothesis. For accounts of the range of explanatory hypotheses that have been tried since, see e.g. Renfrew and Bahn (1991); Preucel and Hodder (1996); Hodder and Hutson (2003).

17 Halsall (1995b), 61; and see his further comment: '[The invasion hypothesis] is rarely given much credence in archaeological circles today. It is too simplistic, rather on a par with asserting that the change from neo-classical to neo-Gothic architecture or from classical to romantic art in the nineteenth century was the

result of an invasion' (p. 57). This 'before' and 'after' approach to migration is quite common. See, for a further example, the comments of Nicholas Higham in Hines (1997), 179, where a reinterpretation of a set of remains that had excluded migration from its discussion is lauded as 'more complex'. The discussion in question is in Hines (1984).

18 Wenskus (1961); cf., amongst others, Wolfram (1988) on the Goths, and Pohl (1988) on the Avars.

19 Geary (1985) and (1988) provide introductory essays composed from this perspective, Halsall (2007) a full-scale study of the fourth to sixth centuries. The migration topos features in Amory (1997) and Kulikowski (2002).

20 On the 'wave of advance' model, see, most famously, Renfrew (1987), chapters 1–2, 4 (summarizing previous approaches), and 6 (the model itself).

21 For a detailed case study of 'elite transfer', see my Chapter 6.

22 See note 13 above. Smith (1986) explores some historical applications of this more solid vision of group identity; Bentley (1987), 25–55 uses Bourdieu's concept of the *habitus* as the basis of a theorized approach towards how identity might be programmed into the individual by the society in which they grow up. When talking about the kinds of differences that prevent the individual from changing group identity so easily (religion, language, social values and so on) the 'primordialists' can sound as though they are still stuck in the intellectual world of pre-1945, making out checklists and ticking boxes. But in the primordialist view, it is not these 'things' themselves that decide identity, but the individual's reaction to them. In most of Europe, being a Catholic or Protestant is not a major determinant of group affiliation, but in Northern Ireland, for particular historical reasons, the same religious difference functions as a strong symbol of communal allegiance. It is not the item ticked in a box that decides group affiliation, but how individuals react to that item.

23 On the Greeks and Romans, see Sherwin-White (1973). Halsall (1999) objects to my earlier use of this analogy, but he doesn't seem to realize that Gastarbeiter and migrants without green cards don't enjoy remotely full citizenship rights in the societies in which they live, and ignores substantial evidence that even in the first millennium group identity was sometimes made the basis of differentiated rights in culturally complex contexts: see Chapter 5. He also takes the to my mind bizarre view that just anyone could turn up to claim a share when barbarian conquerors of different parts of the Roman west were handing out economic assets: see Heather (2008b).

24 Cf. Antony (1990), 895–9; Antony (1992) notes that these revised understandings render obsolete many older theoretical discussions that assumed much starker archaeological correlates of migration.

25 Härke (1998), 25–42, offers a fascinating insight into which contemporary archaeological traditions are more accepting of migration as a possible engine of change, and which more dismissive. British 'immobilism' – rejection of migration – finds parallels in the old Soviet Union and Denmark; the German tradition still incorporates migration as one of its basic paradigms.

26 Jerome (1926).

27 A recent five-hundred page book devoted to migratory activity around the fall of the Roman west, for instance, contents itself with drawing on a few

summaries of the literature drawn up for archaeologists rather than engaging with it at first hand: Halsall (2007), 417–22. By contrast, the same book devotes an entire chapter to the group-identity question, based on intense (and insightful) engagement with the specialist literature.

28 On Irish and Dutch migrants, see Bailyn (1994), 1–2. On overall patterns in modern instances, see Fielding (1993a); King (1993), 23–4; Rystad (1996), 560–1. On the historical parallels, see Canny (1994), especially 278–80 (with full references).

29 On the calculation of costs, see Rystad (1996), 560–1; Collinson (1994), 1–7 (both with useful further references.). On return migration, see e.g. Gould (1980); Kuhrt (1984).

30 For reviews of changing policies towards migrants in Western Europe, and their overall effects, see Cohen (1997); King (1993), 36–7; Fielding (1993b); Collinson (1994), chapter 4; Rystad (1996), 557–62; Cohen (2008). Obviously in recent years, EU enlargement has led to a huge influx of Eastern European migrants.

31 'Gives a shock': King and Oberg (1993), 2. For general discussions of a qualitative definition of mass migration, see e.g. King and Oberg (1993), 1–4; Fielding (1993a).

32 For discussion, though, of the high Middle Ages, see Phillips (1988), (1994); Bartlett (1993), 144–5.

33 In the 1990s there were discussions of how an end to Fordist mass-production techniques in industry were likely to affect future migration flows: Fielding (1993a). We now partly know the answer, with skilled labour being sucked into Western Europe, for instance, while the demand for mass labour in the Middle East continues to grow apace: Cohen (2008).

34 On Spanish migration to the new world, and British migration to Australia and New Zealand, see Sanchez-Albornoz (1994); Borrie (1994), 45ff. The convict ships to Australia were another kind of involuntary state-assisted scheme.

35 Bartlett (1993), 134–8.

36 Helpful general discussions of motivation include Fielding (1993a); Collinson (1994), especially 1–7; Voets et al. (1995), especially 1–10; Rystad (1996); Vertovec and Cohen (1999); Cohen (2008). Some case studies are provided by the essays of Atalik and Beeley, Cavaco, Montanari and Cortese, Oberg and Boubnova, in King (1993).

37 See e.g. Cohen (1996), (2008).

38 See e.g. Rystad (1996), 560–1; Bailyn (1994), 4–5.

2. GLOBALIZATION AND THE GERMANI

1 Ammianus 16.12.23–6. For attempts at localizing these early units, see Krüger (1976–83), vol. 1, 44–55, 202–19. For the view that little changed between the first and fourth century, see e.g. James (1989), 42, after Thompson (1965), 40.

2 The literature on Arminius and Maroboduus is enormous, but for introductions, see Krüger (1976–83), vol. 1, 374–412; Pohl (2000), 21–4. On early kingship, and its general lack, see Green (1998), chapter 7. On Maroboduus' lack of heirs, see Tacitus, Germania 42.

3 Chnodomarius, Serapio and Mederichus: Ammianus 16.12.23–6; Vadomarius and Vithicabius: Ammianus 27.10.3–4; Gundomadus: Ammianus 16.12.17. *Optimates*: Ammianus 16.12.23–6. This view of hereditary canton kingship would be accepted by the vast majority of scholars working in the field: see e.g. Pohl (2000), 29–30, 102ff.; Drinkwater (2007), 117ff. (with full references). Some of the old sub-group names within the Alamanni (Brisigavi, Bucinobantes, Lentienses) survive as modern place names (Breisgau, Buchengau, Linzgau).

4 On the first- and second-century leagues and alliances, see e.g. Tacitus, *Germania* 38–40 (on the Sueves). For more general commentary, see e.g. Hachmann (1971), 81ff.; Krüger (1976–83), vol. 1, 374–412; Pohl (2000), 65f. The revolt of Julius Civilis, for instance, combined elements from the Batavi, Frisians, Caninefates, Bructeri and Tencteri (Tacitus, *Histories* 4.18; 21) but no unity survived his fall.

5 'There fell in this battle': Ammianus 16.12.60; Julian's diplomacy is recounted at Ammianus 17.1, 17.6, 17.10 and 18.2. Vadomarius: Ammianus 21.3–4; Macrianus: Ammianus 28.5, 29.4, 30.3.

6 Early Medieval Ireland and England provide, respectively, more and less articulated examples: see e.g. Binchy (1970a) and the papers in Bassett (1989) for an introduction. I take here a very different view to the minimalist line in germanophone scholarship, a full introduction to which is provided by Humver (1998), and to Drinkwater (2007), 121ff., who argues that there was no urge to unification among fourth-century Alamanni, although he does admit that once Roman manipulation was removed in the fifth century, unification happened.

7 See Wolfram (1988), 62ff., with further arguments in Heather (1991), 97ff. against e.g. Thompson (1966), 43–55; cf. Thompson (1965), 29–41. The three generations are: Ariaricus (in power in 332), Ariaricus' anonymous son, and the son's son Athanaric. For this particular reconstruction of Gotho-Roman relations, which is again argued against Thompson (1966), see Heather (1991), 107–21. Others would reconstruct Gotho-Roman relations differently, but none doubts that the Tervingi survived heavy defeat at the hands of Constantine, or that the position of 'judge' survived.

8 Batavi: Tacitus, *Histories* 4.12; *Germania* 29. Chatti, Bructeri and Ampsivarii: Tacitus, *Annals* 58; *Germania* 33. Hermenduri: Tacitus, *Annals* 13.57.

9 On Ejsbøl Mose, see Ørsnes (1963). Sacrifices of the weapons of a defeated enemy are reported at Caesar, *Gallic War* 6.17; Tacitus, *Annals* 13.57.

10 Chnodomarius: Ammianus 16.12.60. Drinkwater (2007), 120–1 supposes that the king and his three friends had fifty followers each, rather than Chnodomarius having all two hundred, but if that were the case, it is hard to see why he was king. Tervingi: *Passion of St Saba*. On retinues more generally, see e.g. Hedeager (1987); Todd (1992), 29ff. (with references). The contrast with the public bodies of the early Roman period is very striking: see Thompson (1965), 29ff.

11 See Green (1998), chapter 7; cf. Wolfram (1997), chapter 1; Pohl (2000), 66ff. 'They chose kings': Tacitus, *Germania* 7 ('reges ex nobilitate duces ex virtute sumunt').

12 See Chapter 6 below on the rise of Clovis. Clovis operated on Roman soil, however, which meant that he could support a much larger retinue, whereas a Germanic economic context (see below) would have imposed tighter economic constraints and perhaps made this impossible.

13 On Chnodomarius' armour, see Ammianus 16.12.25; we will return to these swords on p. 78.

14 On Odry, see Kmiecinski (1968). In these eastern areas of Germania, the cemeteries were much more permanent than any settlements in the first two centuries AD, and are marked by large stone circles which contained few if any burials. It has been plausibly suggested that this reflects the fact that cemeteries rather than settlements provided the locus for social gatherings.

15 The fullest discussion is Haarnagel (1979).

16 On Wijster, see Van Es (1967). See more generally the relevant studies in Krüger (1976–83): compare vol. 1, chapter 11 with vol. 2, chapter 5; Myhre (1978); Steuer (1982), 258ff.; Hedeager (1988), (1992), 193ff.; Todd (1992), chapter 4. There is a useful discussion of the Roman side of the frontier in Carroll (2001), chapter 4.

17 Goffart (2006), 26–32 objects to old-style assumptions, based on the famous Jordanes, *Getica* 4.25, that Scandinavia in particular and Germania in general was a womb of nations, endlessly producing future invaders of the Roman Empire until it was overwhelmed. As a comment on old-fashioned historiography, this is fair enough, though his work does not engage with the detailed archaeological evidence.

18 See Urbanczyk (1997b).

19 On the Pietroasa treasure, see Harhoiu (1977). On *fibula* production at the Runder Berg (see note 24 below), see Christlein (1978), 43–7, 171. On pottery, see Heather and Matthews (1991), chapter 3 (Cernjachov); Drinkwater (2007), 89–93; cf., more generally, Krüger (1976–83), vol. 2, 123ff.

20 On glass, see Rau (1972). On combs, see Palade (1966).

21 The groundwork was laid by Steuer (1982).

22 For an introduction to the historiography, see Thompson (1965). I strongly suspect that measuring social status via artefacts will tend to place the basic erosion of human equality (to the extent that it ever existed) at far too late a date in the history of *Homo sapiens sapiens*.

23 For useful surveys, see Thompson (1965), chapters 1–2; Todd (1992), chapter 2; for more detailed discussions, see Gebuhr (1974); Hedeager (1987), (1988), (1992), chapters 2–3; Hedeager and Kristiansen (1981); Steuer (1982), 212ff.; Pearson (1989). For Odry, see note 14 above.

24 On Runder Berg, see Christlein (1978); Siegmund (1998); and cf. Brachmann (1993), 29–42; Drinkwater (2007), 93–106, which point out that there must have been other lowland Alamannic elite sites, none of which has yet been identified. On Feddersen Wierde, see Haarnagel (1979). On Gothic areas, see Heather (1996), 70ff. (with references). For more general discussion, see Krüger (1976–83), vol. 2, 81–90; Hedeager (1988), (1992), chapter 4; Todd (1992), chapter 6; Pohl (2000).

25 The two classic and highly influential general accounts are the solidly Marxist interpretation of Fried (1967), and the more optimistic line adopted by Service (1975). These studies set the agenda for more detailed subsequent studies of intermediate societies (between the very small and the more modern). The four areas I identify represent a distillation from the helpful collections of papers in Claessen and Skalnik (1978), (1981); Claessen and van de Velde (1987); Skalnik (1989); Earle (1991); Claessen and Oosten (1996).

26 This is true whether (see previous note) one adopts Service's view of the process (by which a wider range of functions is more efficiently fulfilled) or Fried's less optimistic Marxist view (whereby the growth of the bureaucracy entails the further rigidification of power structures).

27 The key term here is 'reciprocity', meaning that ruler and ruled exchange something that is of mutual value. This probably won't be (and certainly doesn't have to be) an equal exchange, but even the act of exchanging makes the interaction honourable. If it is one-sided, then it is demeaning.

28 Alamanni: Ammianus 16.12. Tervingi: Heather (1991), 109 (on pre-376 AD, based on Ammianus 20.8.1, 23.2.7, and 26.10.3), 146. Drinkwater (2007), 142–4 proposes that there were 15,000 Alamanni and allies at Strasbourg. He consistently downplays Alamannic numbers on the basis of his prior assumption that they posed no real threat to Roman frontier security, which is in my view a circular and unconvincing approach: see Heather (2008a). The evidence strongly suggests that these societies possessed slaves and that slaves were not normally liable for military service. We do not know the proportion of slaves, but they are likely enough to have been a significant portion of the population, so that merely to number fighting men will be to underestimate the total of young adult males in these societies.

29 For waterborne summits, see Ammianus 27.5.9 (cf. Themistius, *Orations* 10), 30.3.4–6. For Burgundian/Alamannic boundaries, see Ammianus 28.5.11.

30 For an introduction to the evidence, see Heather & Matthews (1991), chapter 5.

31 On the Gothic contingents, see note 28 above, with Heather (1991), 107ff. for the crucial link that military service was something imposed on the Goths by the Romans when they held the diplomatic upper hand. On the Alamannic contingents, see Heather (2001). On the loan word, see Green (1998), chapter 11.

32 Vannius: Tacitus, *Annals* 12.25. On Roman imports on elite Gothic sites, see Heather (1996), 70–2. On trade and diplomacy, see Heather (1991), 109. Of course, Chnodomarius may possibly just have been offering a share of war booty rather than cash up front.

33 On the 'wall' of Athanaric, see Ammianus 31.3.8, with Heather (1996), 100 for the identification. On Runder Berg and other sites, see note 24 above.

34 Based on a trawl through the literature cited in note 25 above. Not even the famously inert Irish kings of the Middle Ages – so wonderfully caricatured by the late Patrick Wormald as a 'priestly vegetable' – failed to exercise powers over dispute settlement. In the famous tract on Irish kingship, *Crith Gablach*, one day was reserved for this function: see Binchy (1970b); cf. Wormald (1986).

35 For an introduction to early Anglo-Saxon tax systems, see Campbell (2000); Blair (1994). These kinds of arrangement have also been found in areas of Britain that never fell under Roman rule: see Barrow (1973).

36 The mobility of Alamannic kings is suggested by the difficulty the Romans faced in trying to kidnap one of their number: see Ammianus 29.4.2ff. For an excellent introduction to the immense bibliography on itineration, see Charles-Edwards (1989).

37 See Thompson (1966); cf. Heather (1991), 177ff. (with full references). For Gundomadus, see note 3 above. Even if one accepts the hypothesis of Drinkwater (2007), 142–4 that there were 24 Alamannic canton kings, they would

have produced no more than 4,800 retinue warriors between them. On the range of material in burials, see e.g. Steuer (1982); Weski (1982); Harke (1992). On burials entirely empty of goods, see e.g. Heather and Matthews (1991), 62, for some examples from Gothic-dominated territories.

38 A quick read of the relevant law collections from the Visigothic, Frankish, Lombard, Burgundian, and Anglo-Saxon kingdoms brings out the importance of this group, who also feature in materials from the 'smaller' political entities, such from Thuringia, Bavaria, and Alamannia.

39 On the proportion of freemen to slaves, see Heather (1996), 324–5, after Procopius, *Wars* 3.8.12 (1 elite to 4 subordinates in one Gothic force); 8.26.12 (close to 1:1 in a Lombard force). On this warfare, see Heather (1996), especially Appendix I (collecting the evidence for two classes of warrior being mentioned in Roman narrative sources). For charter evidence, see Wickham (1992); (2005), part 3. Post-Roman society did not immediately fall under the sway of the much smaller landowning elite, who can be seen to be dominant from the Carolingian period of c.800 and beyond: see for example Chapter 6 above on the growth of landed estates, which was the basis of aristocratic/gentry domination in Anglo-Saxon England and northern Francia; and, for more general comment, Wickham (2005), part 2.

40 The village community at least attempted to protect the Christians in their midst: see *Passion of St Saba* 4.4; Heather and Matthews (1991), chapter 4.

41 See Ammianus 31.3.8.

42 The law codes again show that social value varied substantially with age, with women's value famously being highest during child-bearing years. But age was clearly important to men too: older men were buried with spurs but not weapons, for instance, suggesting that there was an age limit to military obligation: see Hedeager (1988). Children were likewise sometimes not buried in cemeteries: see e.g. Siegmund (1998), 179ff.

43 On the general importance of feasting as part of 'reciprocity' (see note 27 above), see Earle (1984), (1991). The first-century evidence is discussed in Thompson (1965). On Anglo-Saxon ideologies and realities, see Charles-Edwards (1989); Campbell (2000), chapter 8.

44 For the early Roman period, see Thompson (1965), 37ff. On Roman control of assemblies, see Dio 72.19.2; 73.2.1–4. On fourth-century village assemblies, see *Passion of St Saba*; cf. Heather and Matthews (1991), chapter 4. On the decision to cross into the Empire, see Ammianus 31.3.8: 'diuque deliberans' (see Chapter 4 below). Thompson (1965), (1966) emphasizes the absence of reference in fourth-century sources (which basically means Ammianus) to regular councils among the Goths and other Germani. While a correct observation, it does not mean they weren't happening.

45 The literature on sacral kingship is huge, but see e.g. Wenskus (1961) and Wolfram (1994). The terminology and concept of *heilag* is nevertheless clear: see Green (1998), chapter 7 for the linguistic evidence; and cf. Pohl (2000) and Moisl (1981) for a practical application. On the actual (as opposed to invented) history of the Amal dynasty, see Heather (1991), chapters 1–2, and part 3; Heather (1996), chapters 6, 8 and 9.

46 See Gregory of Tours, *Histories* 2.9; the Chatti are also mentioned at Ammianus

20.10. Salii: Ammianus 17.8; cf., amongst a huge range of possible secondary literature, James (1988), chapter 1, and the relevant papers in Wieczorek et al. (1997). The political processes behind the generation of the Alamanni may not have been totally dissimilar. No old names survived into the fourth century, but the confederation does seem to have built up gradually over time. In the third century, for instance, the Iuthingi (itself a new name) seem to have been a separate grouping, but by the fourth were operating as an integral part of the broader confederation: see Drinkwater (2007), 63ff.

47 For Gargilius' cow, see Boeles (1951), 130, plate 16, cited in Geary (1988), 3; the calculation of legionaries' demands is from Elton (1996).

48 Julian's treaties are discussed in more detail in Heather (2001). On the frontier and its operations, see generally Whittaker (1994); Elton (1996); Wells (1999), chapter 6; Carroll (2001), of which the two latter focus greater attention on the Roman side of the Rhine.

49 On loan words and trade, see Green (1998), 186f. and chapter 12. On iron production, see Urbanczyk (1997b); cf. more generally Krüger (1976–83), vol. 2, 157ff.

50 On the forced drafts of recruits, see Heather (2001).

51 See Green (1998), chapter 12.

52 Caesar, Gallic War 4.2; Tacitus, Germania 5 (who notes, however, that interior Germanic groups still did not value Roman silver coins); cf. Green (1998), chapter 12. On fourth-century coin concentrations, see Drinkwater (2007), 128–35; Heather and Matthews (1991), 91–3.

53 On the Tervingi and trade, see Themistius, Orations 10, with the commentary of Heather (1991), 107ff. For general orientation on Roman imports and their patterns, see Eggers (1951); Hedeager (1988); von Schnurbein (1995); Wells (1999), chapter 10; Drinkwater (2007), 34ff.

54 For Roman goods and social status, see Steuer (1982). For the amber causeways, see Urbanczyk (1997b). For tolls, see Green (1998).

55 See Caesar, Gallic War 6.17; Tacitus, Annals 13.57. For the bog deposits, see Orsnes (1963), (1968); Ilkjaer and Lonstrup (1983); Ilkjaer (1995); for more general comment, see e.g. Hedeager (1987); Steuer (1998); Muller-Wille (1999), 41–63.

56 For a thoughtful critique of the importance of trade, see Fulford (1985). On ninth- and tenth-century beneficiaries, see Chapter 10 above. For an introduction to 'agency', and its more particular problems, see Wilson (2008).

57 For a detailed report of the find, see Kunzl (1993); for an English summary, see Painter (1994).

58 For a more detailed account, with full references, see Heather (2001).

59 Ammianus 17.12–13, with Heather (2001). On the removal of potentially dangerous leaders, see Ammianus 21.4.1–5; 27.10.3; 29. 4.2 ff.; 29.6.5; 30.1.18–21.

60 On the rationale of hostage-taking, see Braund (1984). On subsidies, Klose (1934) collects the evidence from the early period, Heather (2001) for that of the later Empire.

61 'So eagerly did our forces': Ammianus 19.11. For further comment on the balance between resettlement and exclusion, see Chapter 3 above; and cf. e.g. Heather (1991), chapter 4, on standard Roman immigration policies; and Carroll

(2001), 29ff., on the amount of organized restructuring of adjacent populations that went on as Rome created its German frontier.

62 Valentinian's reduction of gifts: Ammianus 26.5; 27.1. For commentary, see Heather (2001); and Drinkwater (2007), chapter 8 (who seeks, in my view damagingly, to demonstrate that the Alamanni could never have represented any kind of threat).

63 On the Rhine–Weser, see Drinkwater (2007), 38–9. On fifth-century economic expansion in Alamannia, see ibid., 355–44.

64 See Wells (1999), chapters 10–11, following von Schnurbein (1995), who stresses the increase in imports of Roman weaponry into Germanic contexts after the mid-second century.

65 Athanaric: Ammianus 27.5; Macrianus: Ammianus 30.3. In both cases, though, the relevant emperor was being pressed by problems elsewhere – Valentinian in the Middle Danube, and Valens in Persia: see Heather and Matthews (1991), chapter 2.

3. ALL ROADS LEAD TO ROME?

1 'They were expecting . . .': Dio 32. 8–10.

2 For a good introduction, see Birley (1966), chapters 6–8, with Appendix III; see also Böhme (1975).

3 See Dio 72.20.1–2 (on the stationing of troops); 72.11–12, 72.20.2, 72.21 (on the movements of the Asdingi, Quadi, and Naristi respectively); 72.15, 72.16.1–2, 72.19.2, 73.3.1–2 (on trading privileges and neutral zones); 72.19.2, 73.2.1–4 (on assemblies).

4 'Not only were . . .': *Historia Augusta*: Marcus Aurelius 14.1; for an introduction to the trickeries of this text, see Syme (1968), (1971a), (1971b). For Roman aggression, see Drinkwater (2007), 28–32, who adds further thoughts on the possible impact of the plague, and Marcus' sense of duty, to the argument.

5 On Rhine frontier damage, see Carroll (2001), 138; and cf., on the legions and Marcus' self-monumentalization, Birley (as note 2 above). See also Chapter 2, note 28 above.

6 For the first-century homeland of the Langobardi, see Tacitus, *Germania* 40. That group of 6,000 clearly did not represent more than a subgroup, and they would be followed south by more Langobardi in the fifth century (see Chapter 5 below). These later Langobardi invaded the Middle Danube proper from intermediate settlements in Bohemia, but it is unknown whether this was true also of the second-century group. For references to permanent displacements, see note 3 above.

7 See Dio 72.3.1a.

8 See e.g. Barford (2001), introduction and chapter 1.

9 Of fundamental importance here is the work of the late Kazimierz Godlowski, especially his general treatment of north-central Europe in the Roman period: Godlowski (1970). Shchukin (1990) supplies a good general survey, building on Godlowski's pioneering work. The argument continues over details, and many more 'cultures', and phases within 'cultures', have acquired much more precise

and absolute dates. In pioneering days, only Roman coins provided any indication of absolute chronology. Since 1945, the chronological development of Roman wheel-turned pottery became better understood, both for fine wares (dinner services) and amphorae (storage jars for olive oil and wine). Two later techniques supply still more precise dates: carbon-14 (which produces a date-range) and dendrochronology, based on tree rings (which tells you precisely when a given tree was cut down). Combined with Godlowski's general method, these technical advances have generated a wealth of knowledge that would have astonished previous generations of scholarship.

10 In technical dating terms, the expansion occurred in Roman Iron Age periods B2, B2/C1a. These paragraphs distil information in two important collections of papers: *Peregrinatio Gothica* 1 and 2; and cf. Shchukin (2005).

11 For a fuller discussion, see Heather (1996), 35–8. There is a range of fragmentary references in classical sources indicating that Gothic groups were moving south and east: see Batty (2007), 384–7.

12 The relevant literature is huge. For a brief introduction, with full references, see Heather (2005), chapter 2.

13 For a recent comprehensive treatment, with full refererences, see Drinkwater (2007), chapter 2. (Note his important argument on pp. 43–5 that a group called the Alamanni clearly existed already in the 210s, a point to which we shall return.) On the brutal violence, see ibid., 78f. (with further examples); Carroll (2001), chapter 9.

14 On Alamannic origins, see Drinkwater (2007), 48f., 108–16 (with full references).

15 Argaith and Guntheric: Jordanes, *Getica* 16.91 (cf. *Historia Augusta*: Gordian 31.1 on 'Argunt', which probably represents a conflation of the two). Cniva: Zosimus 1.23; Jordanes, *Getica* 18.101–3; Zonaras, *Chronicle* 12.20.

16 The principal source is Zosimus 1.31–5. For additional sources and commentary, see Paschoud (1971–89), vol. 1, pp. 152ff., n. 59ff.

17 Zosimus 1.42–3, 46, with Paschoud (1971–1989), vol. 1, pp. 159ff. n. 70ff.

18 *Historia Augusta*: Aurelian 22.2. There is no evidence that he was related to the Cniva who had been operating in the same region a generation before: see note 15 above. On all these third-century attacks, see Batty (2007), 387–95.

19 Eutropius, *Breviarium* 8.2.

20 For first- and second-century references to the Goths, see Tacitus, *Germania* 43–4; Strabo, *Geography* 7.1.3 ('Butones'); Ptolemy *Geography* 3.5.8. Kulikowski (2007), chapters 3–4; cf. Jordanes, *Getica* 4.25–8 (on Filimer: see p. 122).

21 For more detail on the Tervingi, see Chapter 2 above. Jordanes' anachronisms were first demonstrated in Heather (1991), chapters 1–2 (where I show my own scepticism of Jordanes, *pace* Kulikowski).

22 On the first and second century, see Shchukin (1990); and cf. Batty (2007), 353ff. on Bastarnae, Sarmatians, and Dacians of various kinds (with full references, and noting the distorting political agendas that have sometimes been applied to these materials). For an introduction to Ulfila and his Bible, see Heather and Matthews (1991), chapters 5–7.

23 For the first- and second-century placement of Goths, see note 20 above. Rugi: Tacitus, *Germania* 44. Vandals: Courtois (1955), chapter 1. (Kulikowski does not discuss the broader range of evidence.)

24 Carpi: see note 38 below. On the 330s, see *Anonymous Valesianus* I.6.30.376, chapter 4. On the migration habit, see above, p. 30.

25 In technical terms, these transformations occurred in period B2–C1a/b. For fuller discussion, see Heather (1996), 43–50, drawing on the materials mentioned in note 9, and now supplemented by Shchukin (2005). Kulikowski (2007), 60ff. dismisses the importance of the archaeological evidence in very general terms without discussing the phenomenon of Wielbark expansion.

26 For introductions to this material, see Kazanski (1991); Shchukin (2005), with a fuller literature listed at Heather (1996), 47–50.

27 See Kazanski (1991); Heather (1996), 47–50; Shchukin (2005).

28 Jordanes, *Getica* 4.25–8.

29 Jordanes, *Getica* 16–17.90–100 records the third-century triumphs of the Amal King Ostrogotha. The king is entirely mythical, however, invented to explain why the Ostrogoths were so called, and his name has been added to known historical events: see Heather (1991), 22–3, 368.

30 For more detail, see Heather (1991), chapter 1 and 84–9.

31 Batavi: Tacitus, *Histories* 4.12, *Germania* 29. Chatti, Bructeri and Ampsivarii: Tacitus, *Annals* 58, *Germania* 33.

32 The element of fragmentation under Filimer is quoted on p. 122. Berig: *Getica* 4.25–6, with 17.94–5. Goffart (1988), 84ff. is reasonably concerned to undermine old assumptions that Gothic oral history suffuses the *Getica*, but is arguably a little too dismissive: see Heather (1991), 5–6, 57–8, 61–2.

33 See e.g. Borodzej et al. (1989); Kokowski (1995); Shchukin (2005).

34 See Drinkwater (2007), chapter 2 and 85–9 (with references).

35 See Ionita (1976).

36 On Heruli casualties, see George Syncellus, *Chronicle*, ed. Bonn, I.717. For other figures from the Aegean expedition (2,000 boats and 320,000 men), see *Historia Augusta*: Claudius 8.1. Cannabaudes' defeat is said to have cost 5,000 Gothic dead: *Historia Augusta*: Aurelius 22.2. Much of this material derives from the contemporary account of Dexippus. If the parallel with Viking activity is to be taken to the ultimate, one would suspect that relatively small groups made the initial moves, only for their very success to encourage larger entities to participate in the action. The state of the third-century evidence, however, does not make such a chronological progression certain. For further commentary on scale, see Batty (2007), 390ff.

37 Langobardi: Dio 72.1.9. Quadi: Dio 72.20.2 (explicitly *pandemei*, 'all the people').

38 For the protest of the Carpi, see Peter the Patrician fr. 8. For the exodus on to Roman soil, see Aurelius Victor, *Caesars* 39.43; *Consularia Constantinopolitana*, s.a. 295. See, more generally, Bichir (1976), chapter 14. A total of six campaigns were fought against the Carpi during the reign of the Emperor Galerius (293–311).

39 Naristi: see p. 98. Limigantes: see p. 85. On the Greek cities, the classic works of Minns (1913) and Rostovzeff (1922) remain essential. For an introduction to the archaeological evidence that has since become available, see Batty (2007), 284–9 (with references).

40 Drinkwater (2007), 43–5 rightly rejects the recent tendency to claim that Alamanni did not exist before the 290s, but then attempts to make all the action

of the third century, including the whole settlement of the Agri Decumates, into the result of warband activity. This argument fails to convince.

41 See above Chapter 9.

42 For female burial costume, see note 26 above. For an introduction to the Gothic Bible, see Heather and Matthews (1991), chapters 5–7. The contrast with the originally Norse Rurikid dynasty, who quickly took Slavic names (see above Chapter 10), is extremely striking. See also Chapter 6 below for discussion of the linguistic evidence from the Anglo-Saxon conquest of lowland Britain.

43 Quadi: Dio 72.20.2. Hasding Vandals: Dio 72. 12.1.

44 'Commonsense': Drinkwater (2007), 48. For Burgundian linguistic evidence, see Haubrichs (2003), (forthcoming).

45 For a qualitative definition of 'mass' migration, see pp. 31–2 above. If 'mass' sounds too redolent of the invasion hypothesis, then alternative terms might be found (perhaps 'significant'?), but there is surely virtue in bringing first-millennium usage into line with the norms prevailing in specialist migration studies.

46 *Panegyrici Latini* 3 [11].16–18.

47 See above Chapter 2.

48 On military inscriptions, see Speidel (1977), 716–18; cf. Batty (2007), 384–7. On shipping, see Zosimus 1.32.2–3.

49 I would therefore strongly argue that the 'interaction' theme that has been so marked a feature of frontier studies in recent years – e.g. Whittaker (1994); Elton (1996) – must be balanced with a proper appreciation of the frontier's equally real military function.

50 See pp. 43ff. Oddly, Drinkwater (2007), 48–50, while accepting the evidence for increased competition within the Germanic world, refuses to recognize that this would naturally lead to increased pressure on the Roman frontier, amongst other areas, as groups sought to escape the heightened dangers of their existence. Wells (1999), chapter 9 is similarly – and equally oddly – 'internalist' in interpretation, seeking to locate the causes of third-century disturbances within the frontier zone, and particular the Roman side of it.

51 Ammianus 26.5, 27.1; cf. Drinkwater (2007), chapter 8.

52 Tacitus, *Annals* 12.25.

53 See e.g. Anokhin (1980); Frolova (1983); Raev (1986).

54 Peter the Patrician fr. 8.

55 The rhythms of Roman frontier management perhaps aided the process. Thinning out the frontier zone periodically, as the Romans did, to reduce overcrowding and the potential for violence (see p. 85), can only have made it easier for more peripheral groups eventually to build up a sufficient manpower advantage to overthrow established Roman clients.

56 Cf. Chapter 2 above, p. 101. I would in any case strongly argue that freeman and retinue society were unlikely to be completely separate.

57 Jordanes, *Getica* 55.282 ('*ascitis certis ex satellitibus patris et ex populo amatores sibi clientesque consocians*').

58 For references, see note 10 above, with Kmiecinski (1968) on Odry. Descriptive terms like 'semi-nomad' are sometimes used, but to my mind misleadingly. What we're talking about here are mixed farming populations, who kept many animals, perhaps measured their wealth in cattle, but also engaged in extensive

arable agriculture, despite lacking the techniques to maintain the fertility of individual fields over the long term.

4. MIGRATION AND FRONTIER COLLAPSE

1 Before these tumultuous events of the late fourth century, the western border of Alanic territory lay on the River Don. This just about made them outer clients on Rome's Lower Danube frontier, especially since the Empire retained strong contacts with the southern Crimea. But they can only be classed as complete outsiders when it comes to the convulsion of 405–8, which affected the Middle Danubian frontier region.

2 The same basic vision of the crisis can be found, amongst other sources, in Ammianus 31; Eunapius frr. (and Zosimus 4.20.3 ff., which is largely but not completely dependent on Eunapius); Socrates, *Historia Ecclesiastica* 4.34; Sozomen, *Historia Ecclesiastica* 6.37. The total figure of 200,000 is provided by Eunapius fr. 42, whose account is generally vague and rhetorical, and therefore unconvincing by itself: see Paschoud (1971–89), vol. 2, 376 n. 143. The figure has, however, been accepted by some: see e.g. Lenski (2002), 354–5 (with references). On the 10,000 warriors, see Ammianus 31.12.3; these may have represented only the Tervingi: see Heather (1991), 139. On the wagon trains, see Ammianus 31.7, 31.11.4–5, 31.12.1ff. On social dependants, see e.g. Ammianus 31.4.1.ff.; Zosimus 4.20.6.

3 Matthews (1989) stresses Ammianus' literary artistry, where Barnes (1998) stresses his lack of candour. These two most recent studies disagree on many things but both stress that Ammianus is not a straightforward read. For further comment, see Drijvers & Hunt (1999); G. Kelly (2008).

4 For the 'more secret' archive, see Ammianus 14.9.1. For career documents, see Ammianus 28.1.30. For military dispatches, see Sabbah (1978).

5 On the migration topos, see Kulikowski (2002). On causation, see Halsall (2007), chapter 6.

6 For examples of the migration topos in action, see pp. 122 and 251 above. Ammianus on warbands: e.g. 14.4; 17.2; 27.2; 28.5. Ammianus on Strasbourg: 16.12.7; 31.8.3.

7 The recruitment of this extra mercenary support has sometimes been confused with the arrival of the Greuthungi alongside the Tervingi. This is a serious mistake: see Heather (1991), 144–5, and Appendix B.

8 On the split of the Tervingi, see Ammianus 31.3.8ff.; 31.4.13. The Greuthungi seem also to have fragmented, in that a leader called Farnobius and his followers, found alongside the main body as it crossed the Danube, then suffered an entirely different fate from the rest: see Ammianus 31.4.12; 31.9.3–4.

9 Only Kulikowski (2002) really dares to suggest that Ammianus might be completely misleading, and even he seems to backtrack substantially in Kulikowski (2007), 123ff., which, while wanting to minimize any unity among the Goths, still accepts that they formed a mixed group of humanity 'numbered at least in the tens of thousands, and perhaps considerably more' (p. 130). Among the other anti-migrationists, Halsall (2007) is willing to think in terms of over

10,000 warriors and a total mixed group of 40,000 people; while Goffart (1981), (2006) has never treated the events of 376 in any detail.

10 See Halsall (2007), 170ff., drawing particularly on the analysis of Socrates, *Historia Ecclesiastica* 4.33 in Lenski (1995).

11 I therefore remain entirely happy with the analysis of the 'Ammianus versus Socrates' issue I offered, with references, in Heather (1986). Halsall's desire to avoid a sequence of events that would put predatory migration at the heart of causation seems to provide the principal reason for rejecting the contemporary and more detailed Ammianus in favour of the later and less detailed Socrates, but he offers no good reasons based on the historical evidence, and in my view this line of argument allows preconception to justify unsound methodology.

12 Ammianus 31.3.8.

13 Zosimus 4.20.4–5.

14 Ammianus 31.3.2–8.

15 On the Caucasus raid, see Maenchen-Helfen (1973), 51–9 (who does think they came from the Danube). For other Goths north of the Danube in 383, see (Arimer) Achelis (1900); and (Odotheus) Zosimus 4.35.1, 4.37–9.

16 Some Hunnic groups did operate further west before 405–8, but the numbers were very small up to about 400: just the mercenaries who joined the Goths south of the Danube in autumn 377 (see note 7 above) and another Hunnic/Alanic warband found near Raetia in the 380s (Ambrose, *Epistolae* 25). Uldin's force from c.400 was clearly a bit larger, but even his command paled compared to the Hunnic forces that arrived in the Middle Danube after 405–8: see above Chapter 5. In general terms, all of this suggests to me that the action of 376 should be viewed rather along the lines of Caesar's description of the move of the Tenctheri and Usipetes west of the Rhine in the mid-first century BC. In that case, an extended series of smaller-scale raids and attacks, rather than one outright invasion, convinced them that they could no longer live securely east of the Rhine: Caesar, *Gallic War* 4.1.

17 On discussion, see Ammianus 31.3.8. On persuasion, see Heather (1991), 176–7,179–80.

18 On the archaeology and group identity, see above Chapter 1. The particular items within the Cernjachov culture that strike me as a priori promising for distinguishing the immigrant groups are its bone combs, particular *fibulae* and north European, Germanic longhouse types. Unfortunately, no detailed mapping of these items has yet been made.

19 For an introduction to Ulfila, see Heather and Matthews (1991), chapter 3. I suspect that the alternative view, of a swift social amalgamation, involves a degree of wishful thinking, largely inspired in reaction to the horrors of the Nazi era, that shrinks from accepting such unequal relationships between relatively bounded groups of human beings. The eastern expansions of the Goths and other Germanic groups in the late Roman period were enthusiastically seized upon by Hitler's propagandists to justify the poisonous activities of the Third Reich: see Wolfram (1988), chapter 1. But a laudable determination to condemn Nazi atrocities becomes muddled thinking if we try to make the past conform to our wish rather than to the reasonable probability of its evidence.

20 Different grades of warrior are not specifically mentioned in the Hadrianople campaign, but they do feature in the evidence for Radagaisus' Gothic force of 405 (Olympiodorus fr 9.) and Theoderic the Amal's Ostrogoths (see Chapter 5), as well as in later Visigothic laws. Moreover, the *Historia Augusta*'s vivid account of third-century mass Gothic migratory bands, complete with families and slaves, may well be based on fourth-century events (*Historia Augusta*, Claudius 6. 6, 8.2; cf. Chapter 3 above), and I strongly suspect it was those of lower status that the hard-pressed Tervingi were selling into Roman slavery in return for food on the banks of the Danube: Ammianus 31.4.11.

21 On Carpo-Dacai, see Zosimus 4.34.6. On Cernjachov continuity, see Kazanski (1991).

22 For the Carpi, see Chapter 3 above. On the Sarmatian move, see *Anonymous Valesianus* 6.31.

23 For Goths in the fourth century, see Chapter 2. The point about information is also applicable to the minority under Athanaric who moved into Sarmatian territory: this is what the Tervingi as a whole had tried in 332, only to be frustrated by Roman counteraction (see previous note).

24 Noel Lenski (2002), 182ff., 325f. seeks to locate the reason for Valens' aggression towards Persia in the Goths' arrival, and thoughts of the extra recruits he could muster from them. I find the argument unconvincing, and remain confident that the Gothic crisis left Valens with very little room for manoeuvre: see Heather (1991), 128ff.

25 Kulikowski (2007), 123ff. implies that the Tervingi and Greuthungi came to the Danube and requested asylum on separate occasions, so that Valens had sequential decisions to make, but this is not what Ammianus' account suggests (31.4.12–13; 31.5.2–3).

26 Ammianus 31.10; cf. more general frontier studies such as Whittaker (1994).

27 Ammianus 31.5.3–4. This might possibly be Roman paranoia; I don't think it is.

28 For example, the Goths of Sueridas and Colias (Ammianus 31.6.1); perhaps also the Alamannic unit under Hortarius (Ammianus 29.4.7).

29 I suspect, but am unable to prove, that this would have been particularly true of indigenous groups who merely paid some tribute to the Goths and were otherwise left substantially alone. For a similar range of relationships between the Huns and their different subjects, see Chapter 5.

30 For full references, see *PLRE* 2, 934.

31 For Vandals in Raetia, see Claudian, *Gothic War* 278–81, 363–5, 400–4, 414–29. On the identity of the Sueves, see most recently Goffart (2006), 82–3, who adopts the most plausible Marcomanni/Quadi approach. The Rhine crossing is generally dated 31 December 406 on the basis of Prosper, *Chronicle* AP 379; for the argument that the chronicler might have meant 31 December 405, see Kulikowski (2000a), 328–9. Following the counterargument of Birley (2005), 455–60, however, Kulikowski (2007), 217 n. 37 appears less sure.

32 For Uldin, see Sozomen, *Historia Ecclesiastica* 9.25.1–7; *Codex Theodosianus* 5.6.3. On Burgundians, see Demougeot (1979), 432; 491–3.

33 On Olympiodorus, see, above all, Matthews (1970), with the further thoughts of Blockley (1981), (1983).

34 For Vandal losses, see Gregory of Tours 2.29. The 1:5 ratio was customarily

employed, e.g. by Schmidt (1933), 286, 293. For Vandal/Alanic numbers, see Procopius, *Wars* 3.5.18–19; Victor of Vita, *History of the Persecution* 1.2. For Burgundian numbers, see Orosius 7.32.11. On Radagaisus' following, see note 30 above.

35 Jerome, *Chronicle* 2389 (= 371 AD).

36 On the '*distributio*' and its significance, see Jones (1964), Appendix III. Jones's argument is unaffected by Kulikowski (2000b) since it works from the comparison of two well-dated sections of the *Notitia*: the eastern field army of c.395, and its western counterpart of c.420. For the thirty '*numeri*', see Zosimus 5.26.4. For the 12,000 followers of Radagaisus, see Olympiodorus fr. 9.

37 Victor of Vita, *History of the Persecution* 1.2. I therefore take a more optimistic view of Victor than does Goffart (1980), Appendix A.

38 Halsall (2007), 206, for example, has Radagaisus leading 'a large force', characterizing the Rhine crossers as a 'huge force' (p. 211). It is really only Drinkwater (2007), especially 323–4, who thinks that warbands will adequately explain the action.

39 Zosimus 5.26.3 has Radagaisus engaged in widespread recruitment prior to attacking Italy (although I remain slightly worried that he is here confusing Radagaisus and the Rhine crossing). *Codex Theodosianus* 5.6.3 makes clear that Uldin's following was a mixture of Huns and Sciri, and therefore a new, post-376 alliance.

40 On Radagaisus' followers: Zosimus 5.35.5–6. On Alans in Gaul: Paulinus of Pella, *Eucharisticon* 377–9. On Vandals and Alans in North Africa: Victor of Vita, *History of the Persecution* 1.2. For Burgundians, see notes 34 and 35 above, together, of course, with the fact that this group were able to preserve their east German dialect throughout these moves: see Chapter 3 above. No one doubts Ammianus' report that the Goths of 376 also came with women and children in tow (31.3–4), so the basic principle that Germanic and Alanic armed forces might have moved with their dependants seems well enough established. Against this, the assertions of Drinkwater (1998), especially 273, that it is commonsense that only warriors took part in the action are underwhelming. Cf. Drinkwater (2007), 323–4.

41 Both points – i.e. the Middle Danubian origins of the crisis, and the subsequent appearance there of the Huns – were first argued by Heather (1995a), and are now been generally accepted: see for example Goffart (2006), chapter 5; Halsall (2007), 206ff. The crucial passage of Claudian which has been misunderstood to refer to Huns on the Danube is *Against Rufinus* ii.26ff. (especially 36ff.)

42 Heather (1995a).

43 Goffart (2006), chapter 5, especially 75–8 (Huns appear in Middle Danube shortly after the crisis); 78–80 (Radagaisus); 94–5 (summarizing the knock-on effect among the expectations of other groups of the fact that the Goths had survived their arrival on Roman soil with their coherence more or less intact).

44 Halsall (2007), 195–212; cf. Halsall (2005), particularly on the disruptive effects of ending subsidies.

45 On Tribigild, see Heather (1988); Synesius, *De Regno* 19–21.

46 The first practical help from the east consisted of 4,000 soldiers who arrived in Ravenna in 409/10: Zosimus 6.8.

47 For slaves, see Orosius 7.37.13ff.

48 Either Constantine III or Flavius Constantius has usually been considered responsible for the transfer: see Chastagnol (1973); cf. Kulikowski (2000a). Halsall (2007), 209 raises doubts, but offers no specific evidence in their support.

49 On Constantine III, see Zosimus 6.1, which specifically identifies British, Gallic and Spanish military forces as sufficiently united behind him to drive the Vandals, Alans and Suevi into Spain, and to take the usurper to the brink of Empire: see Matthews (1975), 312ff. On the general role of subsidies in Roman diplomacy, see Heather (2001).

50 The relevant sources are, above all, Ammianus 17.12–13; *Anonymous Valesianus* 6.31–2. For a recent discussion of the Vandals in the fourth century, see Goffart (2006), 82–7, who convincingly concludes that the evidence places them in Silesia and on the Upper Tisza.

51 For Vandals in Raetia, see note 31 above. For their fourth-century placement, see previous note.

52 For fourth-century Goths, see for example Heather (1991), chapter 3. For Alans, see Goffart (2006), 89–90, with Ammianus 31.3.1, who records that the western-most group of Alans in c.375 were called 'The Don People' (Tanaites).

53 For the Alans in 377, see Ammianus 31.8.4ff., with Heather (1991), 144–5 and Appendix B; and in 378, see Ammianus 31.11.16. For their drafting into the Roman army, see Zosimus 4.35.2.

54 The identity of Uldin's followers emerges from Sozomen, *Historia Ecclesiastica* 9.25.1–7 and *Codex Theodosianus* 5.6.3.

55 Ammianus 17.12–13 (Constantius' arrangements in 358); 19.11.1–3 (the return of the Limigantes in 359).

56 On the differences between the cyclical movements inherent in a nomad economy and 'real' migration, see pp. 208–12).

57 Ammianus 31.4.13; I take it these are the Sarmatians defeated by Theodosius prior to his elevation: Theodoret, *Historia Ecclesiastica* 5.5; *Panegyrici Latini* 12(2).12.9–10.

58 The nomadic character of the Alans' economy makes one expect a priori that they had a different social structure from agricultural Germani such as the Vandals or Goths, and this is strongly implied, if in a rather general way, by Ammianus 31.2.25.

59 On the Sciri, see *Codex Theodosianus* 5.6.3. The contrasting fates of the 'better' among Radagaisus' following who were drafted into the Roman army, versus the many others sold into slavery, might suggest that the latter had had little choice over whether to participate in the action.

60 On Uldin's force, see Sozomen, *Historia Ecclesiastica* 9.25.1–7; and *Codex Theodosianus* 5.6.3. Radagaisus: Olympiodorus fr. 9 for the 'best', as against Orosius 7.37.13ff., who records the miserable fate of the mass of the rank and file sold into slavery. 'Best' translates Olympiodorus' *optimates*, which has sometimes been translated as 'nobles', but to reckon so many nobles is absurd, so the word can only make sense as a reference to the higher-grade caste of warriors: see Chapter 2 above. The elites of both the Rhine crossers and the Burgundians, neither of whom of course were faced with as powerful and immediate a

Roman counterattack as Radagaisus or Uldin, showed no obviously similar propensity to abandon the migrant mass.

61 The only group for whom any case can be made for an invitation is the Tervingi, by Valens, in 376, but in my view even here Valens had no real choice: see p. 169 above.

62 Cf. Heather (1991), chapter 5, and Appendix B. I don't believe that the Emperor Gratian made a separate peace agreement with the Greuthungi in the summer of 380. That this adjustment in traditional Roman policy affected only these particular Goths is well understood: see Stallknecht (1969). Kulikowski (2002) and Halsall (2007), 180ff. have recently tried to argue that nothing out of the ordinary was granted in 382, but the case does not stand up to scrutiny: see Heather (forthcoming).

63 It may well be, then, that local Roman landowners cut a deal with the invaders to prevent less organized and hence inherently more damaging assaults upon their property. Cf. Hydatius, *Chronicle* 41[49]: the settlement saw particular groups of invaders settle in particular provinces, so it is possible that the Spanish provincial councils were responsible for the Roman provincial side of the negotiations.

64 See for example Kulikowski (2002); Halsall (2007), chapters 7–8.

65 Claudian, *De Bell. Get.* 166ff., 610ff. (dating to 402); Synesius, *De Regno* 19–21 (dating to 399), with Heather (1988). Neither Kulikowski (2002) nor Halsall (2007), 189–94 offers any explanation of the fundamental distortion they suppose these authors to be incorporating.

66 Zosimus 5.5.4. Privileging the short, non-contemporary and confused Zosimus over the more contemporary sources is the basic approach adopted (even if leading to slightly different interpretations of Alaric's career) in Liebeschuetz (1992); Kulikowski (2002); Halsall (2007), 191–4. Amongst other problems, Zosimus conflates Stilicho's two campaigns against Alaric (in 395 and 397) and wipes out ten years of the history of Alaric's Goths in making the join between his two main sources here: Eunapius and Olympiodorus (at Zosimus 5.26.1: see Heather (1991), 210). To say that Zosimus had no real grasp of Alaric's career, therefore, is an understatement.

67 The activities of Gainas are well covered, if certainly with hostility, in Synesius, *De Providentia*; cf. Cameron and Long (1993).

68 To my mind, this is why Liebeschuetz (1992) cannot be correct in viewing Alaric as leading no more than a regiment or two of Gothic auxiliaries in 395. Halsall (2007), 192–3 tries to wriggle round this problem by continuing to deny the overlap with the Goths of 382 while accepting that Alaric's armed following must have been large, mostly Gothic and from the Balkans. Having accepted these points, he is in fact most of the way to the conclusion that Alaric led the 382 Goths in revolt. He resists this conclusion because he doesn't believe there was a peace deal in 382 which licensed Gothic autonomy, but see the next note.

69 Themistius, *Orations* 16.211. Continuing Gothic autonomy up to and beyond c.390 is signalled, beyond Themistius, in sources both sympathetic to Theodosius and his treaty such as Pacatus, *Panegryici Latini* 12.(2).22.3–5 (where the Goths are one of a series of foreign peoples serving Theodosius), and hostile to them:

Synesius, *De Regno* 19–21, with the commentary of Heather (1988). Halsall (2007), 180–4 oddly argues that there is no evidence that any continued Gothic autonomy was licensed in 382; he appears not to have read the closing words of Themistius' speech closely enough. Cf. Kulikowski (2002).

70 For Roman policies towards leaders, see Heather (2001). Neither the original leaderships (Athanaric, the dynasty of Ermenaric) nor their immediate successors (Fritigern, Alatheus and Saphrax) survived the struggles of 376–82: for more detail, see Heather (1991), 188–92.

71 Fritigern: Ammianus 31.12.8–9, with Heather (1991), 175–6, 179–80. The best example of the post-382 jockeying is provided by the quarrel between Fravitta and Eriulph. Both led factions and both held different views over the proper ordering of Gotho-Roman relations: see Eunapius fr. 59, dated by the summary of it at Zosimus 4.56. For further discussion, see Heather (1991), 190–1. That Theodosius should have held such a banquet undermines the contention of Halsall (2007), 188–9 that Alaric couldn't have been the leader of the Goths of 382 in revolt, because there is no evidence that their sociopolitical hierarchies had continued in place after that date (Halsall does not discuss the incident). For Sarus and Sergeric in more detail, see Heather (1991), 197–8. Neither Kulikowski (2002) nor Halsall (2007) bothers to discuss its potential significance. In my view, extracting Roman recognition of his leadership was also precisely the significance of the generalship that Alaric periodically demanded of compliant Roman regimes – probably along with the financial package for his followers that came with it. But note that the generalship was an optional extra that he was willing to drop to make a deal: see Heather (1991), chapter 6.

72 Kulikowski (2002) – largely followed by Halsall (2007), 187–9 – denies large-scale Gothic military service in the years between the 382 treaty and Alaric's revolt in 395, but this involves too much special pleading to be convincing. *Panegryrici Latini* 12.(2)32.3–5 strongly implies that the main Gothic contingent was recruited only for the campaign against Maximus (especially Pacatus' explicit comment that it would have been dangerous to leave the Goths behind), while Eunapius fr. 55 and Zosimus 4.45.3 note Maximus' attempts to undermine the recruited Goths' loyalties, which again implies that this was something unusual. A range of sources note the participation of large numbers of Goths in the campaign against Eugenius (Zosimus 4.58; John of Antioch fr. 187; Orosius 7.37.19), and Theodosius' banquet for the Gothic leaders (see note 71 above) was held precisely when Theodosius was mulling over his answer to Eugenius' envoys (Zosimus 4.56). In my view, the banquet was probably a first move towards securing Gothic participation.

73 For Maximus' revolt, see Eunapius fr. 55; Zosimus 4.45.3, 48–9. Alaric of course led the revolt after the Eugenius campaign. The arguments of Kulikowski (2002) and Halsall (2007), 187–93 comment neither on the Maximus revolt nor on the significance suggested by the exact chronology of the banquet quarrel.

74 Orosius 7.35.19 (casualties confirmed at Zosimus 4.58). Neither Kulikowski (2002) nor Halsall (2007), 187–93 discusses this backdrop to the Gothic revolt.

75 In recent times, we have seen one successful example of this kind of diplomatic strategy in the Good Friday Agreement in Northern Ireland, and one so far unsuccessful example in the Oslo Accords on the Middle East.

76 Zosimus 5.5.5ff.

77 Themistius, *Orations* 16.211.c–d.

78 Whether this made any practical difference to Alaric's position in Illyricum in the short term is unclear; having been commanding general there, he was perhaps in a position to retain control of the levers of power.

79 For this argument in more detail, see Heather (forthcoming).

80 See in more detail Heather (1991), chapter 6.

81 Fravittas, Sarus and Modares: *PLRE* 1, 605; 372–3. On the 402 battle, see Claudian, *VI cons. Hon.* 229ff.; cf. Cameron (1970), 186–7.

82 It disappeared to the extent that of the sources that discuss the events of 376, only Ammianus knew there were originally two separate groups of Goths. In my view, both Greuthungi and Tervingi were settled under the treaty of 382 and Alaric's revolt in 395 involved and definitively united both. An alternative view sees the unification happening when Alaric summoned his brother-in-law Athaulf from Pannonia in 408: Zosimus 5.37.1ff.

83 Zosimus 5.35.5–6. *Pace* Kulikowski (2002), it is hard to see who this large body of barbarian soldiery in Roman service was, if not mainly the 12,000 followers of Radagaisus drafted by Stilicho: Olympiodorus fr. 9.

84 Heather (1991), 151ff. looks to unravel Zosimus' confusions.

85 Gothic subgroups were destroyed by Frigeidus (Ammianus 31.9), Sebastianus (Ammianus 31.11) and Modares (Zosimus 4.25), and there is no reason to think this a comprehensive list. On this process in general, see Heather (1991), 213–14, 223–4, 314ff.

86 Zosimus 5.45.3; cf. Liebeschuetz (1990), 75ff.; Kulikowski (2002).

87 Exactly how much larger this Gothic force was involves a huge amount of guesswork, but if it is right to calculate the military manpower of fourth-century Gothic units at around 10,000, then the Visigoths who formed around Alaric could certainly field at least twice this number of soldiers, and possibly between three and four times as many.

88 Victor of Vita, *History of the Persecutions* 1.2.

89 Hydatius, *Chronicle* 77 [86].

90 On the mid-410s: Hydatius, *Chronicle* 59–60 [67–8]; on the 420s: ibid. 69 [77]; on the 440s and 460s: Heather (2005), 289ff. and 390ff.

91 Suevi: Hydatius, *Chronicle* 63 [71]. Alans: see the convenient listings of Bachrach (1973).

5. HUNS ON THE RUN

1 Jordanes, *Getica* 50.261–2.

2 Uldin's Huns and Sciri: Sozomen, *Historia Ecclesiastica* 9.5; *Codex Theodosianus* 5.6.3. Huns' Gothic subjects in 427: Theophanes AM 5931; cf. Procopius, *Wars* 3.2.39–40, with Croke (1977). The date could be either 421 or 427. The best general survey of Attila's subject peoples is Pohl (1980).

3 See Maenchen-Helfen (1973), chapters 8–9, who also notes that leaders like Attila could easily have had 'proper' Hunnic names as well as Germanic nicknames; Attila means 'little father' in Germanic.

4 Ammianus 31.2.1–2; Zosimus 4.20.3–5 (cf. Eunapius fr. 42); Jordanes, *Getica* 24.121–2.

5 Ammianus 31.2.3ff.

6 The Alanic digression: Ammianus 31.2; the Saracen digression: Ammianus 14.4. In treating this material, the approach of Maenchen-Helfen (1945) was much more critical than that of Thompson (1995), even though it was Maenchen-Helfen who noticed the meat being placed under saddles. For further comment, and recent bibliography, see G. Kelly (2008), chapter 2.

7 For some orientation on nomadism, particularly of the Eurasian-steppe variety, see Cribb (1991); Khazanov (1984); Krader (1963); Sinor (1977), (1990).

8 Bury (1928).

9 Avars: Pohl (1988). Magyars: Bakony (1999).

10 For general accounts, see Thompson (1995); Maenchen-Helfen (1973); cf. Heather (1995a) on relations with Aetius.

11 Attila's more or less complete indifference to additional territorial gains emerges with striking clarity from the surviving fragments of Priscus' history.

12 Huns up to 376: Ammianus 31.3. Huns and Alans in 377: Ammianus 31.8.4ff. Huns and Carpo-Dacians: Zosimus 4.35.6.

13 'Improvised leaders': Ammianus 31.2.7. Jordanes does place a Hunnic king called Balamber in this era, but these are really events of c.450 and Balamber is in fact the Gothic king Valamer: see Heather (1989), and p. 234 above.

14 Uldin: Sozomen, *Historia Ecclesiastica* 9.5, with further comment and full references in Heather (1995a). Analogous phenomena occurred in the Viking era, when leaders thrown up in the first generation of small-scale expansion were quickly subdued as larger numbers under more important leaders joined in the flow: see Chapter 9.

15 Olympiodorus fr. 19; cf. Priscus fr. 11.2, p. 259 (on the Akatziri).

16 On the Huns' bow, see Heather (2005), 154–8, with further references.

17 For some calculations based on grazing room on the Great Hungarian Plain, see Lindner (1981). On the great raid of 395, see Maenchen-Helfen (1973), 51–9.

18 On nomad devolution in general, see the literature cited in note 7 above.

19 See Heather (2005), 325ff., with full references.

20 Procopius, *Wars* 8.5 seems to preserve by an indirect route the story originally told by the contemporary Eunapius, which was cast after Herodotus 5.9 (on the Sigynnae).

21 On the third-century Heruli, see Chapter 3. Sciri: Zosimus 4.35.6. Rugi: Tacitus, *Germania* 43. On likely placements within the Middle Danubian region, see Pohl (1980).

22 We will return to the Amal-led contingent in more detail. Bigelis: Jordanes, *Romana* 336. For the third group under Dengizich's control, see Priscus fr. 49.

23 Dengizich: *PLRE* 2, 354–5. Hernac: Jordanes, *Getica* 50.266, with *PLRE* 2, 400–1. Hormidac: *PLRE* 2, 571. Bigelis: see previous note.

24 Jordanes, *Getica* 50.264. Pohl (1980) suggests – in a compromise – that the Amal-led Goths may have moved at this point only from Transylvania. Much ink has been spilled on the relationship between the surviving Gothic history of Jordanes and the Gothic history of one Cassiodorus, which we know to have been written down at Theoderic's court in Italy. In my view, the textual evidence indeed

suggests that Jordanes worked using Cassiodorus' text (as he claims) and I find the various conspiracy theories that have been offered against this unconvincing: see Heather (1991), chapter 2; Heather (1993). The archaeological evidence for such a late Gothic move to the Middle Danube is indecisive. Kazanski (1991) has placed the end of the Cernjachov culture as late as c.450, but this is not the usual view, and the argument is essentially circular since it is based on Jordanes' report that there were still Goths east of the Carpathians at this date.

25 Odovacar: *PLRE* 2, 791–3. On the Balkan adventures of the Amal-led Goths, see Heather (1991), part 3.

26 For full references, see *PLRE* 2, 457 and 484–5.

27 The Lombards should strictly be called Langobards. The narrative source is Paul the Deacon, *History of the Lombards* 1.19. For modern commentary, see e.g. Christie (1995); Jarnut (2003); Pohl and Erhart (2005).

28 Procopius, *Wars* 6.14–15. Cf. Pohl (1980): the archaeological evidence suggests that the Gepids were slowly expanding south into Transylvania at this point.

29 An earlier exception would be the Goths rescued from Hunnic domination in 427: see note 2 above. These may also be the same as the Thracian Goths, as we shall shortly see. The Gepids too engaged in expansion within the region: see previous note.

30 'He was a Greek trader . . .': Priscus fr. 11.2.422–35. Edeco: *PLRE* 2, 385–6.

31 You can tell how people were dressed by where they wore their safety pins, which is all that tends to survive of clothing in most graves. Possible reasons for archaeological invisibility can range from the dramatic (where bodies are left exposed to the elements and wild animals) to the prosaic (cremation followed by scattering of ashes), or the generation of customs where bodies are buried without any chronologically identificatory gravegoods (something which often makes medieval cemeteries undatable in northern Europe once populations convert to Christianity). The horizons of the Hunnic Middle Danube are differentiated from one another by slight changes in the manner in which broadly similar sets of gravegoods were decorated. In chronological order (and there are overlaps between them), the sequence starts with the Villafontana horizon, succeeded in turn by those of Untersiebenbrunn and Domolospuszta / Bacsordas. For introductions to this material, see Bierbrauer (1980), (1989); Kazanski (1991); Tejral (1999). There are excellent illustrations in Wolfram (1985).

32 Many of the Germanic groups of central Europe had practised cremation in the first to the third century, but inhumation was already spreading more widely before the arrival of the Huns.

33 Historical sources do occasionally supply enough information, however, which can be used in conjunction with the archaeological evidence approximately to identify some particular groups.

34 See most recently Halsall (2007), 474–5; for a similar view of the Avar Empire, see Pohl (1988), with Pohl (2003).

35 Priscus fr. 14.

36 Priscus fr. 11.2, p. 259. See Chapter 4 for the Tervingi and Greuthungi; cf. the junior status and grimmer treatment handed out to the Sciri after Uldin's defeat: see above, note 2.

37 For references, see note 2 above; and see p. 248.

38 Priscus fr. 2, p. 225.

39 Priscus fr. 2, p. 227.

40 Priscus fr. 2, p. 227.

41 Priscus fr. 49.

42 The Romans provided Attila with a succession of secretaries, and a prisoner called Rusticius wrote the odd letter (Priscus fr. 14, p. 289). This governmental machine could keep lists of renegade princes who had fled to the Romans, and could keep track of the supplies required from subject groups, but little more. Akatziri: Priscus fr. 11.2, p. 259. Goths: Priscus fr. 49.

43 Jordanes, *Getica* 48.246–51, with Heather (1989), (1996), 113–17, 125–6.

44 Gepids: Jordanes, *Getica* 50.260–1.

45 Franks: Priscus: fr. 20.3. Akatziri: see note 42 above. Of the subject groups in between, the most dominated were apparently the Goths who appear in Priscus fr. 49, part of which is quoted above; the least dominated were the Gepids, who led the revolt against Attila's sons (see previous note). In between were the Pannonian Goths of Valamer: see note 43 above.

46 *Miracles of St Demetrius* II.5.

47 See e.g. Agadshanow (1994).

48 For further discussion, see Heather (2005), 324ff., with references.

49 As we have seen, modern anthropological evidence indicates that the most you will sometimes find in such circumstances is that a very few particular items have significance for signalling group identity, but that does not mean that the group identity is in any sense unreal: above, p. 26, after, in particular, Hodder (1982).

50 While Attila could extract annual subsidies measured in thousands of kilos of gold, the most that even a successful Hunnic successor group like the Amal-led Goths could manage was three hundred: Priscus fr. 37.

51 Jordanes, *Getica* 50.265–6. Jordanes himself came from this Balkan military milieu, and there is every reason to suppose this catalogue correct. When exactly in the 450s or 460s these settlements occurred is not clear: that of Hernac is firmly dated to the later 460s, however, and it may be that they all belong to the post-465 meltdown of Hunnic power that also saw moves into Roman territory by Bigelis and Hormidac. Hernac's willingness to have his power base broken up might explain why he was treated more favourably than Dengizich (see note 23 above).

52 Jordanes, *Getica* 53.272; cf. Agathias 2.13.1ff.

53 Paul the Deacon, *History of the Lombards* 2.26ff.; cf. Jarnut (2003). It is a consistent theme both within Paul's narrative and some of the other early Lombard texts that victory led to the inclusion of warriors in the group, but not always on terms of equality: see e.g. *Origo Gentis Langobardorum* 2 (as *aldii*: 'half-free'); *History of the Lombards* 1.20, 1.27, 5.29.

54 Goths: Heather (1996), Appendix 1. Lombards: ibid., and see previous note. See also Chapter 2 above.

55 Lombards: see e.g. Jarnut (2003), who argues that kingship among the Lombards may have been a temporary phenomenon restricted to the leading of expeditions. Goths: Heather (1989), (1996), chapters 8–9.

56 For the Rugi joining Theoderic in 487: John of Antioch fr. 214.7; on their still

being identifiable in 541: Procopius, *Wars* 7.2.1ff. (they had swapped sides twice during the Gothic conquest of Italy). Heruli: Procopius, *Wars* 6.14–15.

57 The account of the Heruli is doubted by Goffart (1988), 84ff.; that of the Rugi by Halsall (1999). See Chapter 1 above for general comments on modern understandings of group identity.

58 The Gundilas papyrus (translated by him as Appendix 1) is central to Amory (1997). But see also Heather (1996), chapter 9, and Appendix 1, (2003).

59 Malchus fr. 20, p. 446.215ff. (the 6,000 men), p. 440.83ff. (non-combatants and baggage). Cf. Jordanes, *Getica* 55.281–2 (Theoderic had earlier also used 6,000 men in the expedition that proved his manhood following his return from being a hostage in Constantinople). For further commentary, see Heather (1991), chapter 7.

60 See Amory (1997); but see also, in addition to Procopius, *Wars* 5.1.6ff., Ennodius, *Panegyric on Theoderic* 26–7 and *Life of Epiphanius* 118–19 (cf. 111—12).

61 The east Romans captured 2,000 wagons in a surprise attack (Malchus fr. 20), but there is nothing to suggest that this was the total baggage train. The Goths were offered 'unoccupied' land, which strongly implies that they were to do their own farming, as do all the negotiations between Theoderic and Constantinopolitan representatives: Malchus fr. 18.3, p. 430.5ff.; fr. 20, p. 438.55ff., p. 446.199ff.; cf. Heather (1991), 244ff.

62 For fuller discussion and complete references, see Heather (1991), 259–63; for Bigelis, see note 22 above.

63 For pay and rations for 13,000, and 910 kilos of gold per annum, see Malchus fr. 18.4, p. 434.12ff. and fr. 2, p. 408.22ff. For full discussion and references, see Heather (1991), 253–6.

64 For Strabo's death, and Recitach's assassination, see John of Antioch fr. 211.4 and fr. 214.3. For Theoderic's forces in Italy, see Hannestad (1960). For full discussion, see Heather (1991), 300–3.

65 For references, see notes 22 and 23 above.

66 On Herule numbers in 549: Procopius, *Wars* 7.34.42–3. It is generally tempting to think that the Heruli were smaller than the Amal-led Goths because the latter are portrayed as so victorious in the post-Attilan competition on the Middle Danubian plain. Our only source for this, however, is Jordanes, and it may be that Theoderic's following only acquired superpower status when he added the Thracian Goths to his following.

67 The migration topos entirely suffuses Paul the Deacon's *History of the Lombards*: the brothers Ibor and Agio lead the first move from Scandinavia, Agilmund the second into Bohemia, Godo takes them into Rugiland, Tato fights the Heruli, and Wacho leads the annexation of part of Pannonia. For modern secondary comment, see the works cited in note 27 above.

68 See especially Jarnut (2003), with references, and for the thought – as note 55 above – that early Lombard kings may fundamentally have been expedition leaders; cf. Christie (1995), 14–20.

69 See Curta (2001), 190–204, with his figure 18.

70 On various occasions, groups of Ostrogoths, Heruli, Huns, Rugi, and Lombards all fall into this category of mass migration. Lombard migration may well have taken the form of an initial flow that had to reorganize itself in mass form when

it was necessary to fight the Heruli head-on. In this, it resembles the third-century Goths: see Chapter 3.

71 Vidimer: Jordanes, *Getica* 56.283–4. Procopius, *Wars* 1.8.3 explicitly names Bessas and Godigisclus among the Thracian Goths who didn't follow Theoderic; see Heather (1991), 302 for some other contenders.

72 The Amal-led Goths were receiving 136 kilos of gold per annum in the 460s (Priscus fr. 37), while the Thracian *foederati* pulled in 910 (see note 63 above). On Theoderic and the wealth of Italy, see Heather (1995b).

73 For Hun-generated wealth, see note 31 above. It is possible, however, given their seemingly non-centralized political structures, that the further spread of Lombard groups south of the Danube into old Roman Pannonia may have again taken the form of a variegated flow rather than a single directed movement.

74 *Life of Severinus* 6.6.

75 For references, see note 56 above. Alternatively, it may be , given Theoderic's subsequent success, that they had no real choice in the matter.

76 For markets, see Priscus fr. 46. For other references, see note 23 above.

77 On Theoderic's spell as a hostage, and its ending, see Heather (1991), 264–5. On his mention of Italy in 479, see Malchus fr. 20.

78 Rodulf of the Rani: Jordanes, *Getica* 3.24.

79 For the route of the 473 trek: Jordanes, *Getica* 56.285–6.

6. FRANKS AND ANGLO-SAXONS: ELITE TRANSFER OR *VÖLKERWANDERUNG?*

1 Campbell (1982), chapter 2.

2 The old maximalist tradition runs from scholars such as Freeman (1888) to Stenton (1971). It never went unchallenged, but scholars such as Higham (1992) and Halsall (2007), especially 357–68, are representative of the more substantial minimizing tradition of recent years. Recent scholars thinking in terms of large-scale migration include Campbell (1982), Härke (1992), Welch (1992). Hills (2003) is representative of an ultra-minimalist position adopted by some younger archaeologists. A good introduction to the variety of opinion is Ward Perkins (2000).

3 See Woolf (2003).

4 H. R. Loyn, quoted in P. Sawyer (1978). The best introduction to late Roman Britain remains Esmonde-Cleary (1989).

5 For an overview of Anglo-Saxon settlement and the development of place names, see Hooke (1998).

6 See Heather (1994).

7 Esmonde-Cleary (1989) is very balanced on the end of Roman Britain, as is Halsall (2007), 79–81, 357ff. For an introduction to the literature on systems collapse, see amongst others Faulkner (2000); Jones (1996); Higham (1992). Dark (2002) stands against this position.

8 One recent example is Halsall (2007), 519ff., with references to some of the alternatives.

9 For an introduction to such materials, see Dumville (1977).

10 Campbell (1982), chapter 2 provides a clear introduction to the *Chronicle*.

11 For useful introductions to this material, see Campbell (1982), chapter 2; Arnold (1997); Welch (1992).

12 This much is accepted even by such a general anti-migrationist as Halsall (1995b), (2007), 357ff.

13 See e.g. Arnold (1997), 21ff.

14 *Ine's Law* 24.2 (cf. 23.3); cf. Arnold (1997), 26ff., with discussion of Warperton.

15 Compare, for example, Weale et al. (2002) with Thomas et al. (2006). The sample was of men whose pre-Industrial Revolution ancestors can be shown to have been living in the same area as the modern descendant.

16 One other line of thought has therefore taken a more indirect route, attempting to identify and analyse so-called 'epi-genetic' features of the skeletal remains unearthed from the inhumation cemeteries of the fifth to seventh centuries. Such factors reflect the impact of inherited genes rather than diet or environmental factors. The work was able to establish that the element of the population buried with weapons was noticeably taller than those buried without. The argument continues as to whether the height differences should be explained genetically – i.e. as a sign that the weapons-bearers were an intrusive population – or by something else, such as differences in diet, and no firm conclusions have yet emerged: see Härke (1989), (1990).

17 For 446 AD, see Bede, *Ecclesiastical History* 2.14; 5.23, 24 (after Gildas, *Ruin of Britain* 20). For 450 AD, see *The Greater Chronicle*, year of the world 4410; cf. *Anglo-Saxon Chronicle* (449 AD) for the arrival of the Kentish dynastic founders, Hengist and Horsa.

18 Everything is reasonably clear up to about 409, when Zosimus 6.5 records a British revolt. Controversy really begins with Zosimus 6.10, which is traditionally interpreted as Honorius telling the British provincials to look after their own defence, although the text is corrupt. For an introduction to these events and the historiography, see Salway (2001).

19 Gildas, *Ruin of Britain* 23–6.

20 On the Saxon attacks in c.410, see *Gallic Chronicle* of 452 (though this chronicle does not always date events to individual years). For the first datable remains, see Welch (1992), chapter 8. The appendix to Halsall (2007) attempts to extend the generally accepted sequence still further, arguing that Gildas' unnamed tyrant, who issued the invitation, usually thought of as a post-Roman figure, was in fact the usurping Emperor Maximus (383–7), and that it was Maximus who brought the first Saxon mercenaries to Britain. This is not an impossible reading, but neither is it the most obvious, so the jury is still out. The further arguments which Halsall erects on the back of this first hypothesis are unconvincing: see notes 44 and 46 below.

21 Gildas' report that Roman Britain's final appeal to the central imperial authorities came when Aetius was (or had been) consul for the third time (446 or after) might provide some further confirmation that the 440s were a period of

particular disaster. The British leader on the Loire was Riotamus: see *PLRE* 2, 945.

22 See e.g. Campbell (1982), chapter 2; Higham (1994); Halsall (2007), Appendix.

23 See Dumville (1977).

24 Gregory of Tours, *Histories* 4.42; cf. Paul the Deacon, *History of the Lombards* 2.6ff.

25 Bede, *Ecclesiastical History* 1.15 (Angles, Saxons and Jutes); 5.9 (the others).

26 Higham (1992), 180–1.

27 Gregory of Tours, *Histories* 5.26, 10.9; Procopius, *Wars* 8.20.8–10; cf. especially Woolf (2003).

28 Famously, the supposed Gothic migration from Scandinavia is also said to have taken place in three ships: Jordanes, *Getica* 1.25, 17.95.

29 See Chapter 4 above.

30 For Norse DNA evidence, see Chapter 9. For language change, see p. 296 above.

31 Gildas, *Ruin of Britain* 23–6.

32 On the Saxon attack on Gaul: Ammianus 28.5. For introductions to the 'Saxon Shore', see Johnston (1977); Rudkin (1986).

33 On coastal inundation, see the excellent discussion of Halsall (2007), 383ff. On Frankish pressure: Gregory of Tours, *Histories* 4.10, 14.

34 Carausius: *PLRE* 1, 180. On parallel phenomena in the Viking period, see Chapter 9.

35 Gildas, *Ruin of Britain* 20. The archaeological evidence for Pictish and especially Scottish (= Irish) intrusion into western Britain is irrefutable, even if there is little in the way of historical evidence. A good recent account is Charles-Edwards (2003), Introduction and chapter 1.

36 See Woolf (2003), 345f.

37 On the nautical evidence, see Jones (1996), though his discussion includes neither a consideration of Roman ships nor the extended nature of the Anglo-Saxon migration flow. For Goths and the Black Sea, and Vandals and North Africa, see above Chapter 4.

38 For further exploration of these issues, see e.g. Dark (2002); Woolf (2003).

39 See e.g. Higham (1992); Halsall (1995a), (2007), 357ff.

40 An excellent general survey is Hooke (1998).

41 Relevant general surveys include Hooke (1998); Williams (1991). An excellent case study is Baxter (2007), chapter 7. On the demotion of the peasantry, see Faith (1997), chapter 8.

42 For introductions to this issue, see Hooke (1997); Powlesland (1997).

43 See Esmonde-Cleary (1989), 144–54; cf. Loseby (2000) and Halsall (2007), 358f., both with references, on attempts to generate a substantial post-Roman urbanism.

44 On the peasants' revolt, see e.g. Jones (1996); cf. Halsall (2007), 360ff.

45 Constantius, *Life of St Germanus* 13–18, 25–7. For the Romance-speaking elite, see note 3 above. The famous Llandaff charters may provide further confirmation of essential sub-Roman continuity, although this has been disputed: see Davies (1978).

46 'Is simply to dispose . . .': Halsall (1995), 61. This 'before' and 'after' approach to migration is quite common. For another example, see the comments of Higham in Hines (1997), 179, where a reinterpretation of a set of remains – by Hines

(1984) – is praised as 'more complex' because it ejected migration from its usual role in their discussion. See pp. 160 and 192 for two instances where the determination to minimize the importance of migration has led scholars, including Halsall again, to make methodologically problematic choices in their handling of the evidence.

47 There are many parallel examples, but for a recent overview of the decline of Roman structures in the Balkans, see Heather (2007).

48 See the review of the literature in Woolf (2007), 123ff., which draws on, amongst others, Denison (1993) and Hall (1983), which have effectively countered the attempts of Preussler (1956) and Proussa (1990) to detect deeper Celtic influences on Old English. On later medieval cases of language change, see Bartlett (1993), 111ff.

49 See further Chapter 2 above.

50 This emerges with huge clarity from all the literary sources – everything from critiques of individual kings in historical narratives to the value systems underlying heroic poetry. Introductions to the mix of land and cash expected over the course of an individual's lifetime are provided by Charles-Edwards (1989); Campbell (2000), chapter 10.

51 For a general introduction to the pre-Viking great powers, see Campbell (1982), chapters 3–4.

52 See Hooke (1998), chapter 3; Powlesland (1997); Esmonde-Cleary (1989).

53 For weapons burials, see Härke (1989), (1990). For continental parallels, see Chapter 2 above.

54 Ward Perkins (2000).

55 See e.g. Kapelle (1979).

56 Woolf (2007), 127ff.

57 Even the land-grabbing that followed the Norman Conquest was not under William's control, despite his relatively great authority, and the need for clarity as to who now held what was one of the reasons for the great survey that underlay the *Doomsday Book*: *Anglo-Saxon Chronicle* 1085 AD. I would imagine that the land-grabbing process of the fifth and sixth centuries was infinitely more chaotic, given that central authority among the incoming Anglo-Saxons was so much weaker than among the eleventh-century Normans, and that the land-grabbing followed piecemeal in the wake of many small victories rather than a single decisive one like the battle of Hastings.

58 This is also the conclusion of Woolf (2007), as, of course, of all of those writing in the tradition that does not seek to minimize the importance of Anglo-Saxon migration.

59 Julian and the Franks: Ammianus 17.8.3–5. This is an isolated incident, however, and it is therefore impossible to say whether the Franks had the same kind of functioning confederative structure as the Alamanni, although it is certainly possible. For introductions to early Frankish history and archaeology, see Zollner (1970); Perin (1987); James (1988); Ament (1996); Reichmann (1996); R. Kaiser (1997).

60 For materials and commentary on Childeric's grave, see Perin and Kazanski (1996); Halsall (2001).

61 On Childeric's career, see *PLRE* 2, 285–6, with referencess. The 'Roman' Clovis

has been argued for by Halsall (2001), (2007), 269–71, 303–6. On Gundobad, see *PLRE* 2, 524–5.

62 James (1988) started the controversy, and it has drawn counter-arguments from Perin (1996); MacGeorge (2002).

63 Gregory of Tours, *Histories* 2.40–2.

64 On the dating controversies surrounding Clovis' conversion, see Shanzer and Wood (2002), with more general commentary on his career in Wood (1985); Halsall (2001).

65 On the parallel rise of Theoderic and the Amal family, see Chapter 5.

66 See Halsall (2007), 346f. On the rise of Marseilles, see Loseby (1992), (1998).

67 For a general introduction, see Halsall (2007), 347ff. Halsall (1995a) looks at the re-emergence of the landed basis of real aristocracy in northern Gaul in the seventh century.

68 On the inscriptions at Trier, see Handley (2001), (2003). On Remigius, see Castellanos (2000). On the broader cultural changes, specifically on language, see Haubrichs (1996). On the disruption to ecclesiastical structures, especially bishoprics, across northern France, see Theuws and Hiddinck (1996), 66f.

69 The most recent general surveys are Perin and Feffer (1987), vol. 2; Wieczorek et al. (1997).

70 For introductions to rural settlement, see Van Ossel (1992); Van Ossel and Ouzoulias (2000); Lewitt (1991).

71 Werner (1950); Böhme (1974). The grander of these Frankish officers met with in historical sources are Fraomarius, Erocus, Silvanus, Mallobaudes, Bauto and Arbogast.

72 Halsall (2007), 152–61, with references; the argument was first made in Halsall (1992). Reichmann (1996), 61–4 discusses the funerary habits of Frankish groups before the rise of the Merovingians.

73 On 'tutulus' brooches, see Halsall (2007), 157–9; cf. Böhme (1974). Furthermore, as Halsall observes, even if it were established that the brooches signified Germanic origin, the appearance of weapons would still remain to be explained, since this would then become a 'new' habit adopted on Roman soil.

74 See Chapter 5, and p. 271 above.

75 On the Armorican revolt: Zosimus 6.5. On the 410s: Exsuperantius, *PLRE* 2, 448. On the subsequent history, see MacGeorge (2002).

76 Slightly different views of the phenomenon of Bagaudae have been adopted by Van Dam (1985), 16–20, 25f.; Drinkwater (1989), (1992); Minor (1996), all of whom nevertheless step back from the old Marxist, class-warfare analysis to think instead in terms of local self-help in the face of Roman central control's fragmentation.

77 For policies towards Alamannic overkings, see Chapter 2.

78 For general surveys, see note 69 above.

79 The historiography of the study is discussed by James (1988).

80 Pirling (1966) and Pirling and Siepen (2003) summarize the ongoing investigations.

81 James (1988), 25–8 surveys the tradition, which runs through Werner (1935); Bohner (1958); Perin (1980). Frénouville was the work of Buchet: see James (1988), 110f. For a useful overview, see Perin (1987), 138ff.

82 See p. 311 above.

83 For the 'social-stress' interpretation, see Halsall (2007), 350ff.

84 For the traditional argument, see e.g. Perin (1996) or Wieczorek (1996); for its critique, see Halsall (2007), 269f.

85 For references, see note 69 above.

86 For the ending of cremation in England, see e.g. Welch (1992).

87 For an excellent recent survey, see Haubrichs (1996). For the earlier emergence of structured estates, see Halsall (1995a).

88 I will return to this broader issue in the following chapter.

89 Ammianus 17.8.3–5.

90 See Holt (1987).

7. A NEW EUROPE

1 The radical wing on identity and the supposed migration topos is led by Amory (1996), and Kulikowski (2002), (2007), but the germanophone tradition had long been thinking in terms of very fluid group identities: see Wenskus (1961); Wolfram (1988). The idea that the fall of the western Empire was a surprisingly peaceful process is particularly associated with Goffart (1980), (1981), (2006). In a different combination of these trends, Halsall (2007) sees the Empire coming apart from the edges because of its own internal divisions, particularly that between east and west, with barbarian invasion as consequence rather than cause. In various combinations, these ideas have been exercising a huge influence over the scholarship of the last twenty years or so, on which see the excellent Ward Perkins (2005).

2 We have already encountered these examples in Chapters 4, 5 and 6. The idea that the western Empire was actually going to end really only dawned on much of the Roman west after the defeated attempt to conquer North Africa in 468: see Heather (2005), chapter 9.

3 Heather (2005), 375–84.

4 See Chapter 4 above.

5 See Halsall (2007), chapter 7, with Chapter 4 above.

6 By my reckoning, the eleven campaigns comprise: Ad Salices (377); Hadrianople (378); the Macedonian defeat of Theodosius (381); Frigidus (393); the Macedonian campaign (395); Epirus (397); Verona and Pollentia (402); the defeat of Radagaisus (406); the sack of Rome (409/10); the assault of Flavius Constantius (413/15); and the savaging of the Vandals (416–18).

7 See Chapter 5.

8 The same was broadly true, with different date ranges, of the smaller migrant' groups the Suevi and Burgundians. A third stage, where the migrantsí new states engaged in competition with one another, was also carried forward largely by violence, and occasionally generated further migration, such as that of the Visigoths to Spain – see Heather (1996), chapter 9 – but that is beyond the limits of the story being explored here.

9 For further discussion of this model, see Heather (1995). I should note that this account of how the Empire unravelled does not significantly differ from that,

for example, of Goffart (1981), from the point at which barbarian groups were already on Roman soil.

10 See Jones (1964), vol. 3, Appendix III.

11 For more detailed discussion, see Heather (2005), chapter 6.

12 For lowland Britain, see Chapter 6.

13 In the 430s, the Burgundians first suffered a heavy military defeat at the hands of the Huns, and were then resettled on Roman soil: for an introduction, see Favrod (1997).

14 On the creation of these new and larger groups, see pp. 189ff.

15 See Chapter 5 above.

16 On the developing understanding of group identity, see Chapter 1 above.

17 It is well documented that the Goths' motivation for moving in 376 was political and negative. In my view, the same can be deduced with a high degree of probability for the migrants of 405–8, although the lack of explicit evidence means that other views of causation are possible (see above Chapter 5). These other views have no effect on my account of the amount of migration under way in the middle of the first decade of the fifth century.

18 There is no evidence for the Sueves, but the fact that so many other groups moved with women and children must make it a reasonable probability.

19 See Chapter 6.

20 See Chapter 4 note 20 above for the suggestion that the *Historia Augusta's* description of migrant Goths with large numbers of slaves, which ostensibly relates to the third-century migrations, actually relates to events after 376.

21 The distinction – but simultaneous intersection – between local Roman life and the structures of the state is often missed, but is a deeply important historical phenomenon. Just to give one example, the effects of wrenching the Church out of its Roman context show up vividly in the highly fragmented western Christendom examined in Brown (1996) and Markus (1997).

22 Goffart (1980) made the initial case, responding to the some of the critiques in Goffart (2006), chapter 6, which are well summarized, with much extra value besides, by Halsall (2007), 422–47.

23 Among Theoderic's Ostrogoths in Italy, who are better documented than most other intrusive groups, there is good evidence for the existence of intermediate leaders who stood in between the king and their own personal followings among the rank and file. These leaders would presumably have been responsible for distributions of booty and property which affected their own men. See Heather (1995a).

24 Victor of Vita, *History of the Persecution* 1.13, with Moderan (2002).

25 For an outline of this view, together with the settlement evidence, see Heather (1996), chapter 8. It should be noted in passing that Goffart (2006), chapter 6 has commented that there is no evidence of public rather than private land ever being recycled to barbarians. However, this ignores *Novels of Valentinian 34*, which records that the Roman state precisely compensated displaced landowners from Proconsularis with incomes from publicly owned land.

26 On the Burgundian settlement, see Wood (1990); cf. Halsall (2007), 438ff. on the Visigothic material.

27 For the conflict in Spain, see p. 204 above. For the conquest of North Africa, see Heather (2005), chapter 6.

28 For Ostrogoths, see Chapter 5 above. Wood (1990) gives some thought to the trauma suffered by the Burgundians.

29 To this extent, we are seeing here an extension of third-century patterns, but with the added negative stimulus imparted by the Huns (see Chapter 3).

30 Or, at least, the capacity to create large enough forces, combined with the ability to exploit further movement: see p. 189ff.

31 For the Ostrogoths and the Franks, see pp. 248 and 309 respectively.

32 As we have seen, Priscus' evidence makes it entirely clear that Attila's chief aim in attacking the Roman Empire was to siphon off some of its wealth.

33 'You [Anastasius] are the fairest . . .': Cassiodorus, *Variae* 1.1., trans. Hodgkin (1886); cf. more generally Heather (1996), chapter 8, with full references.

34 See Chapter 2, but note that the third-century migrations had partly collapsed the outer periphery into the inner zone.

35 On Theudebert in particular, see Collins (1983); cf., more generally on the growth of Merovingian power, Wood (1994), chapters 3–4.

36 Justinian's decision to attack in the west was a highly contingent one: see Brown (1971). On the collapse of Theoderic's quasi-imperial edifice, see Heather (1996), 248ff., with Wood (1994), chapters 3–4 on Frankish expansion.

37 As Rome's capacity to conquer most of the known world on the back of its Mediterranean assets makes clear.

38 See Wood (1994), chapters 4 and 10.

39 In my view, this process can broadly be characterized as its transformation from an outright conquest state to something more like a community of provincial communities, built on consent: see Heather (2005), chapters 1 and 3.

40 For a useful outline narrative, see Wood (1994), chapters 13 and 15.

41 For more detailed discussion, especially on the nature of Charles Martel's rule and the strategies used to cement his control, see Wood (1994), chapter 16; Fouracre (2000).

42 Famously, seizing the crown involved an appeal to the Pope, Zacharias, who replied that kingship should reside in the hands of a man wielding real power, not in a figurehead. Less famously, but probably more important at the time, the change of dynasty was also sanctioned by a major Frankish assembly. For a good account, see McKitterick (1983).

43 On all this, see most recently Collins (1998).

44 For an excellent introduction, see Dunbabin (2000).

45 For useful introductions to the Ottonian Empire, see Leyser (1989) and Reuter (1991).

46 See Reuter (1985), (1990).

47 The phenomenon of culture collapse, with its more precise chronology, was identified by Kazimierz Godlowski: see e.g. Godlowski (1970), (1980), (1983). One anomaly within the overall pattern is the so-called 'Olsztyn group'. Established in Mazovia on the south-east shores of the Baltic east of the Vistula, and beyond the long-standing limits of Germanic domination, the material culture of this group is characterized by the presence of some of the traditional

Germanic items and also a fair quantity of Mediterranean imports, both of which date this group firmly to the sixth century. What remains unclear, of course, is whether the remains were deposited by a group of newly arrived Germanic immigrants to the area, or else represent some locals (perhaps Baltic-speakers) who adopted a new kind of material culture. Either way, the group was relatively short-lived, since no Olsztyn remains can be dated to the seventh century: see Barford (2001), 33, with references.

48 See e.g. Koch and Koch (1996); Wieczorek (1996); Hummer (1998).

49 Those Saxons who were never completely conquered by the Merovingians, although brought under Frankish hegemony, insulated the Scandinavian world from any explicit Frankish interference.

50 For primary references, see note 47 above, taken further by Parczewski (1993), 120ff., (1997).

51 Historical sources also provide a possible analogy to explain the Olsztyn group. As we have seen (above, Chapter 5), one fragment of the Heruli, defeated by the Lombards in 508, moved north from the Middle Danube region, eventually establishing themselves in Scandinavia. It is perfectly conceivable, therefore, that other Germanic-speakers, taking a similar option, might have ended up further east.

52 See e.g. Urbanczyk (1997b), (2005).

53 The case of Frankish migrants into northern France also deserves comment, though it is unclear whether these came from the areas that suffered from culture collapse, so I have omitted them from this thought experiment.

54 See above Chapter 4

55 See Batty (2007), 39–42. For Greater Poland, which fell within the areas of the Przeworsk and Wielbark culture collapse, an extensive field-walking and survey-ing project prompts the parallel conclusion that its population density after culture collapse was still around 1 person per square kilometre: see Barford (2001), 89–91 (with references). This again suggests that the departure of half a million people might well be significant, but, representing a maximum one-third of the population, would not have generally emptied the landscape, while many of the more southern areas affected are likely to have had larger populations.

56 See pp. 64ff above.

57 See Chapter 8 above.

58 For a good recent account, see Kennedy (2007).

59 Why the long-established habit of limited warfare between the two empires should have given way to such a mutually destructive conflict thus becomes a central question.

60 This emerges very clearly from Sartre (1982).

61 For useful introductions, see Whittow (1996); Haldon (1990).

62 The great expansion of the tenth century came when the Abbasid Caliphate had fragmented as a political entity, and itself fell apart when the Seljuks restored a measure of Islamic unity in the eleventh.

63 For an introduction, see Kennedy (2004).

64 Not at least since the second century, when formal relations between Vandals and Empire are recorded during the Marcomannic war: see Chapter 2 above.

8. THE CREATION OF SLAVIC EUROPE

1 These remains were originally labelled 'Prague' by Borkovsky, who first identified them in what is now the Czech Republic in a study of 1940. On the change
of name, see note 9.

2 Especially since much of the third zone of Europe, beyond Rome's outer
periphery, had been inhabited by people living this kind of life. Hence it is
unsafe to assume the kind of exclusive, one-to-one association between 'the'
Slavs and Korchak remains that would have been posited in the old world of
culture history (see Chapter 1).

3 The map is after Barford (2001), 326. The collapse of the Iron Curtain has made
it possible to discuss these matters with much greater candour. For good
introductions in English to the politicized history of Slavic studies, see e.g.
Barford (2001), especially the introduction and chapter 13; Curta (2001), chapter
1.

4 Kostrzewski (1969) provides a good summary of his position, drawn up at the
end of his highly eventful life. Having studied with Kossinna from 1910, he
spent the Second World War in hiding from the Gestapo since his visions of an
early, utterly Slavic Poland were considered unacceptable.

5 Shchukin (1975), (1977). In Poland, the work of Godlowski on the Przeworsk
system and early Slavic cultures was crucial; its results are most easily accessible
to English-speakers through Godlowski (1970). Thanks to the work of him and
his pupils, the Wielbark and Przeworsk systems have come to be understood
as thoroughly dominated by Germanic-speakers, with earlier archaeological
'proofs' that the latter comprised just a very few migrants from southern
Scandinavia being overturned. Godlowski was also responsible for demonstrating
how huge an archaeological upheaval separated the Germanic-dominated Poland
of the Roman period from the Slavic-dominated Poland of the early Middle
Ages.

6 Procopius, *Wars* 8.40.5 mentions that attacks began in the time of Justin. Slavic
raids of different kinds feature regularly in Procopius' narrative of Justinian's
reign: Curta (2001), chapter 3 offers a good recent analysis.

7 See Barford (2001), 41f.; Curta (2001), 228–46.

8 Jordanes, *Getica* 5.34–5; cf. Tacitus, *Germania* 46.2 (on the Venedi) and 46.4 (on
what lay beyond). For further references to the Venedi, see Pliny, *Natural History*
4.97; Ptolemy, *Geography* 3.5.1 and 7.

9 The 'tree argument' was first made by the Polish botanist Rostafinski in 1908:
Curta (2001), 7–8. Rusanova published entirely in Russian; for discussion of her
work with full references, see Curta (2001), 230ff.

10 See Curta (1999), (2001), especially 39–43 (Jordanes); 230ff. (Rusanova); chapters
3 and 6 (the Slavs' dynamic transformation via contact with eastern Rome).

11 Godlowski (1983); Parczewski (1993), (1997: an English summary); Kazanski
(1999), chapter 2; cf. Barford (2001), 41ff. (who remains open-minded).

12 Jordanes, *Getica* 48.247 (Boz and the Antae), with Heather (1989) establishing
the chronology (see p. 234 above); 50.265–6 (Hunnic and other settlements on
the Danube: see p. 223 above).

13 Dolukhnaov (1996) is good on the background of the long-term development of the simple farming cultures of eastern Europe.

14 For useful introductions to the linguistic evidence, see Birnbaum (1993); Nichols (1998).

15 Procopius, *Wars* 7.29.1–3 (547 AD); 7.38 (548 AD); 7.40 (550 AD). Procopius elsewhere reports that the raids were annual: *Secret History* 18.20; cf. Curta (2001), 75–89.

16 Turris: Procopius, *Wars* 7.14.32–5. On forts more generally, see Curta (2001), 150ff.

17 On the Avars, see e.g. Pohl (1988), (2003); Whitby (1988); with Daim (2003) for an introduction in English to the archaeological materials of the Avar Empire.

18 See Whitby (1988), especially 156ff.

19 On the Persian war, see Chapter 7 above. On the disasters of the 610s: John of Nikiu, *Chronicle* 109; *Miracles of St Demetrius* I.12, 13–15; II.1, 2. The siege of Constantinople is recounted in *Chronicon Paschale* AD a. 626.

20 *Miracles of St Demetrius* II.4, 5. Miracle II.4 names the Runchine, Strymon and Sagoudatae Slavs as attacking Thessalonica at this point; Miracle II.1 adds the names of the Baiounitae and Buzetae. For the transplanting, see Theophanes, *Chronicle* AM 6180 (687/8 AD). Justinian later tried to use them to fight the Arabs, but they changed sides at the crucial moment in the battle of Sebastopol in 692: Theophanes, *Chronicle* AM 6184 (691/2 AD), where the figure of 30,000 appears. For archaeological materials from the north and west Balkans, see Kazanski (1999), 85–6, 137; Barford (2001), 58–62, 67ff.

21 The seven Slavic tribes: Theophanes, *Chronicle* AM 6171 (678/9 AD). For the developing archaeological picture, see Kazanski (1999), 138; Barford (2001), 62ff., with references. For an introduction to the Bulgars, see Gyuzelev (1979).

22 *Miracles of St Demetrius* II.4, with *De Administrando Imperio* 49–50 on Patras. For the archaeology, see Kazanski (1999), 85f., 137; Barford (2001), 67f.; and in particular the correctly critical account of Curta (2001), 233–4, responding in part to overly enthusiastic past attempts to use these materials to 'prove' the *Chronicle of Monemvasia*'s account of an early and massive Slavicization of the Peloponnese: see e.g. Charanis (1950).

23 *De Administrando Imperio* 30 and 31 (respectively Croat and Byzantine versions of the arrival of the Croats); 32 (the Serbs). Samo: Fredegar, *Chronicle* 4.48; cf. 4.72 (on the Bulgars). For further comment, see Pohl (2003). Scholarly opinion divides on how much credence to give the *De Administrando*'s account.

24 For further comment, see Barford (2001), 73–5; Curta (2001), 64–6, with references. An Iranian origin to some of the names recorded of Antae leaders has also been argued for, but the etymologies continue to be contested.

25 For references, see note 21 above.

26 The Geographer's information underlies all accounts of ninth-century Slavic central Europe, and discussion of the preceding centuries is always framed with this outcome in mind. Ninth-century Carolingian diplomatic manoeuvring concentrated on groups within this area: the Elbe Slavs, the Bohemians, and the Moravians.

27 For the tenth century, see Chapter 10. For the Roman era, see Map 1.

28 512 AD: Procopius, *Wars* 6.15.1–2. Hildegesius: Procopius, *Wars* 7.35.16–22; cf. Curta (2001), 82, with full references to other secondary literature, on Slovakia as his likely recruiting ground. Samo: Fredegar, *Chronicle* 4.48, 68.

29 The literature is enormous, but for recent general accounts see Brachmann (1997); Parczewski (1997); Kazanski (1999), 83–96; Barford (2001), 39–44; Brather (2001). These draw on and update such earlier accounts as Donat and Fischer (1994); Szydlowski (1980); Brachmann (1978); Herrmann (1968)

30 On the new wheel-turned potteries, see Barford (2001), 63ff., 76–9, 104–12; Brather (2001); cf. Brather (1996). For the older view of a second migration, see Brachmann (1978), with references.

31 For a general discussion, see Godlowski (1980), (1983), with pp. 371ff above. The departure of the Lombards for Italy in 568 greatly changed the complexion of archaeological patterns in the Middle Danube region.

32 Barford (2001), 53–4, 65–6, with references.

33 For the basic information, see Kobylinski (1997); Barford (2001), 65–7, 76–7. For an introduction to older views, see Herrmann (1983). Sukow-Dziedzice burial customs are not known; they must have have consisted of some archaeologically invisible rite such as surface disposal or cremation of the body without any additional, identifying objects.

34 See Kobylinski (1997).

35 For references, see note 33 above.

36 For useful introductions, see Franklin & Shepard (1996), 71ff.; Goehrke (1992), 34–43.

37 For the linguistic evidence, see note 14 above.

38 For the evidence, see Goehrke (1992), 14–19; Parczewski (1993); Kazanski (1999), 96–120; Barford (2001), 55–6, 82–5, 96–8. The term 'Slavic-dominated' is a carefully chosen formulation to remind the reader that the old assumptions of culture-historical interpretation may be as misleading in the Slavic era as in its Germanic predecessor: see Chapter 1 above.

39 For an outline and further information, see Goehrke (1992), 20–33; Barford (2001), 85–9, 96–9.

40 The different possible answers are nicely defined by two recent books on early Slavic history. Kazanski (1999), especially 120–42, argues that overall similarities in lifestyle between the Prague-Korchak, Penkovka, and Kolochin cultures suggests that if the first two were Slavic, then so was the third. In his view, much of the East European Plain, the territory covered by the Kolochin culture, was already Slavic-speaking in c.500 AD (cf. Map 16). Korchak/Penkovka expansion from the seventh century onwards represented a political but not a linguistic revolution. Barford (2001) would identify the generation of Prague–Korchak itself as a moment of primary Slavicization, when Balts and Slavs really came to distinguish themselves from one another. For him, therefore, the spread north and east of Prague-Korchak in the seventh century, followed by the generation of the Luka Raikovetskaia, Volyntsevo, and Romny-Borshevo traditions, represents not just a political revolution, but the moment when Slavs first came to dominate the landscape, albeit while absorbing much of the indigenous population into their new social structures.

41 The mixed group of 1,600 Huns, Antae and Sclavenes: Procopius, *Wars* 5.27.1; the 3,000 Slavs: Procopius, *Wars* 7.38. Hildegesius: Procopius, *Wars* 7.35.16–22. The 5,000 Slavs at Thessalonica: *Miracles of St Demetrius* I.12.

42 Possibly also consistent with some kind of 'wave of advance' model is the fact that the same names seem to have been used by different Slavic groups who found themselves in very different places at the end of the migration process. The usual explanation adopted for this phenomenon is that originally unified groups split into fragments, which moved in different directions as Slavic migration progressed. Such a process might also explain why Prague-Korchak, Penkovka, and even some Kolochin materials have been found intermixed with one another in the Balkans (see note 40 above). The problem remains, however, that the best-documented examples of multiply appearing names refer to Serbs and Croats, who appear to have been military specialists (see pp. 424–5 above), rather than the small conservative type of social grouping that carried Korchak culture in its complete form across the European landscape.

43 *Strategicon of Maurice* 11.4. Given the relatively small size of the groups in which they operated, this preference presumably reflected a desire for additional protection, rather than an inherent love of difficult terrain. On the Indo-European wave of advance, see Renfrew (1987).

44 For references, see notes 20 and 22 above. The political context also provides good reasons why the Balkan settlements would have been undertaken by larger units. In the case of the Peloponnese, likewise, the named Slavic groups were distinct from a local Greek-speaking population, so, once again, the named units would appear to have been properly Slavic, as opposed to the result of any reorganization among native and immigrant populations.

45 Musocius: Theophylact 6.8.13–6.9.15. Ardagastes: Theophylact 1.7.5, 6.7.1–5, 6.9.1–6. Perigastes: Theophylact 7.4.8 ff. Dabritas: Menander fr.21. The quarrel over the prisoners: Theophylact 6.11.4–21. On the sociopolitical transformation of the Slavs nearest the east Roman frontier, see Curta (2001), especially chapter 7. To keep matters in proportion, a total group population of c.10,000 individuals could not have fielded more than one or two thousand fighting men, and was much smaller – by as much as a factor of ten – than some of the migrant groups attested among the Germani of the Hunnic era (see Chapter 4).

46 For references, see notes 23 and 24 above; for the 5,000 'elite' Slavs at Thessalonica, see note 41 above.

47 For general references, see note 39 above. For Novotroistkoe, see Liapushkin (1958).

48 Maurice, *Strategicon* 11.4.

49 For Bohemia, see Godja (1988); cf., more generally, Kolendo (1997). For pollen studies, see Brachmann (1978), 31–2; Herrmann (1983), 87–9. Discontinuity is also the theme of Henning (1991). On Germanic culture collapse, see also pp. 371ff.

50 Fredegar, *Chronicle* 4.48. On agriculture and its expansion, see Barford (2001), chapter 8, (2005), with full references. Really good information on population expansion is limited to only a few areas, but the field-walking and surveying project in Greater Poland has established that population densities increased from less than 1 person per square kilometre in c.500 AD to 3 per square kilometre by 900 AD, to 7 per square kilometre by 1200 AD: see Barford (2001),

89–91, with references. Indications from agricultural technology tell the same story in qualitative terms. For example, ploughs only came into use at all in the more northerly reaches of the Russian forest zone with the spread of Slavic dominance there in the second half of the first millennium: see Levaskova (1994).

51 For references, see note 33 above.

52 See Halsall (2007), 383ff.

53 On the *Chronicle of Monemvasia*, see Charanis (1950). For Patras and Ragusa, see *De Administrando Imperio* 49–50; cf. (on the Salona evacuation) Whitby (1988), 189–90, with references.

54 See Chapter 4 above.

55 Urbanczyk (1997b), (2005). There is no explicit historical evidence to support this view of an exploited Germanic peasantry, but, as a kind of parallel, highly exploited Roman peasantry certainly sometimes sought refuge in (perhaps relative) tax havens beyond the frontier. One aspect of the Emperor Constantius' activities north of the Danube in 358, as we have seen, was to 'liberate' peasantry who had cleared off north of the frontier: see Chapter 3.

56 Fredegar, *Chronicle* 4.48; cf. Urbanczyk (2002).

57 This might also explain how Slavs came to take over some Germanic river and place names, the island of Rügen and Silesia, for example, seemingly named after the Rugi and the Siling Vandals respectively.

58 See Henning (1991), correcting and exposing the political bias of the DDR era in Herrman (1984), (1985), 33ff.

59 Topirus: Procopius, *Wars* 7.39. The events of 594: Theophylact, 7.2.1–10.

60 This provides an alternative explanation – and a much more convincing one – to the ideologically generated nationalist models of 'submerged' Slavs living under the rule of just a small Germanic-speaking elite.

61 See Chapter 10.

62 For an excellent overview, see Barford (2001), chapters 3–8.

63 'When [Vinitharius] attacked . . .': Jordanes, *Getica* 48.247, with note 12 above.

64 *Chronicon Paschale* (626 AD); cf. the more general accounts of Avar–Slav relations in Whitby (1988), 80ff.; Curta (2001), 90ff.

65 Fredegar, *Chronicle* 4.48.

66 The Mogilany group predates the arrival of the Avars, but they may have given added momentum to the generation of the Sukow-Dziedzice system, although, as we have seen, the internal chronology is as yet too unclear to allow too much emphasis to be given to this point: for references, see note 33 above.

67 For useful introductions to the history and archaeology of the Avar Empire, see Pohl (2003); Daim (2003).

68 See p. 203 above.

69 On Dulcinea, see Curta (2006), 56–7.

70 The Slavs' dugouts: *Chronicon Paschale* (626 AD); *Miracles of St Demetrius* II.1.

71 Buko (2005), chapter 3.

9. VIKING DIASPORAS

1 'From Hernar in Norway one should keep sailing west to reach Hvarf in Greenland and then you are sailing north of Shetland, so that it can only be seen if visibility is very good; but south of the Faroes, so that the sea appears halfway up their mountain slopes; but so far south of Iceland that one only becomes aware of birds and whales from it': from the fourteenth-century *Hauksbok*, quoted in Bill (1997), 198.

2 There is a strong tendency from a British perspective to distinguish two ages of major Viking invasion: one in the ninth century, and another right at the end of the tenth and the beginning of the eleventh. The latter, however, was substantially different in character, being organized by a centralized Danish monarchy and involving little in the way of actual migration; it will therefore be considered in Chapter 10.

3 On the logistics of sailing these northern waters, see Crawford (1987), chapter 1.

4 There is an almost infinite bibliography on the Viking raids in the west, but, between them, Nelson (1997), Keynes (1997) and O Corrain (1997) provide an excellent introduction, usefully supplemented by the appropriate chapters in Forte et al. (2005) and Loyn (1995).

5 See Crawford (1987), chapter 4 (place names); 136ff. (types of settlement); cf. Ritchie (1993). Hints of what must have been happening in the north emerge from the action unfolding in Ireland (see following note).

6 See for example the *Chronicle of Ireland* for the years 807, 811, 812 and 813; the record of attacks becomes pretty much annual from 821, suggesting that the assault on Ireland intensified just a little before that on England and the continent. The *Chronicle of Ireland* (848 AD) calls the Viking leader Tomrair a *tanaise rig*, in Irish terms an heir apparent or second in command to a king (Charles-Edwards (2006), vol. 2, 11). He may well have been an 'earl' (Old Norse, *Jarl*), therefore, rather than a 'king': see further below. The action in England and on the continent is well covered in Nelson (1997); Keynes (1997).

7 For further detail, see Nelson (1997); Keynes (1997); Coupland (1995), 190–7.

8 See O Corrain (1997), with the very helpful commentary of Charles-Edwards (2006) in the notes to his translation of the *Chronicle of Ireland*. On the two kings, see the ground-breaking work of Smyth (1977). Something more of the actual death of Reginharius is reported in the *Translatio* of St Germanus: see Nelson (1997).

9 For useful summaries, see Coupland (1995), 197–201; Keynes (1997). The narrative of the *Anglo-Saxon Chronicle* is itself excellent (in early medieval terms!) for these years.

10 On the nature of the Great Armies, see especially Brooks (1979); cf. Smyth (1977). The switching of manpower back and forth between England and the continent can be followed in more detail in Nelson (1997) and Keynes (1997).

11 Studies of Alfred's reforms abound, but Brooks (1979) is an extremely helpful introduction. It can be supplemented in greater detail by e.g. Smyth (1995); Abels (1998). Many of the relevant texts are conveniently collected and translated in Keynes and Lapidge (1983).

12 For further detail, see Nelson (1997); O Corrain (1997).

13 For Brittany, see J. Smith (1992), 196–200; Searle (1988), 29–33. For (somewhat) contrasting introductions to the history of Normandy, see Bates (1982); Searle (1988), especially chapters 5 and 8.

14 For Orkney, see Crawford (1987), 51ff., with Rafnsson (1997) on Iceland and the Atlantic diaspora.

15 For a broad summary, see O Corrain (1997); for a much more detailed, indeed slightly controversial treatment, see Smyth (1979).

16 Various materials have come down to us in excerpts made in the Middle Ages from Ibn Rusteh – see Wiet (1957) – Ibn Jaqub – see Miquel (1966) – and Ibn Fadlan – see Canard (1973); cf. Melnikova (1996), 52–4.

17 For the 'rapids', see De Administrando Imperio, chapter 9. For trade treaties, see Russian Primary Chronicle (911 and 944 AD). For an introduction to the debate, with full sources, see Franklin and Shepard (1996), 27–50; Melnikova (1996), 47–9; Duczko (2004), 3ff.

18 Ibn Fadlan also makes the Rus sound like Nordic stereotypes: tall and fair, with reddish complexions.

19 Russian Primary Chronicle (860–2 AD). For the textual tradition, see the introduction to the translation of Cross and Sherbowitz-Wetzor (1953), with the further thoughts of Franklin and Shepard (1996), 27ff.; Melnikova (1996), chapters 7–8.

20 Noonan (1997) provides an excellent introduction; the comprehensive treatment is now Duczko (2004).

21 Ibn Jaqub, Relation (see note 16).

22 Life of St Anskar 30. On Ottar, see Lund (1984); cf. Melnikova (1996), 49–52.

23 As we shall see in Chapter 10, there is good reason to suppose that many of the slaves were also being acquired from indigenous intermediaries.

24 Ibn Jaqub, Relation (see note 16). For the winter circuit of the Rus in the first half of the tenth century, see De Administrando Imperio 9.

25 Russian Primary Chronicle (911 and 944 AD); for detailed comment, see Franklin and Shepard (1996), 106ff. and 118ff.: comparison shows, amongst other things, an increase in the numbers of Rus trading with Constantinople.

26 De Administrando Imperio 9.

27 For an excellent introduction, see Noonan (1997).

28 On the archaeological evidence for this earliest phase of Scandinavian activity in northern Russia, see Duczko (2004), chapter 2. For the comparison between Constantinople and the Caliphates as potential markets, see Chapter 7.

29 Swedish Vikings: Annals of St Bertin AD a 839. (It must be questioned whether this was the first time that the Dnieper route was actually tried.) The death of Sviatoslav: Russian Primary Chronicle (972 AD).

30 The relevant boats will presumably have been Slavic 'monoxyla', hollowed from single tree trunks, however, rather than the longships so prominently deployed in the west.

31 On the Abaskos attack and its aftermath, see Franklin and Shepard (1996), 50ff.; Duczko (2004), chapter 1.

32 On the anarchy at Samarra, see Kennedy (2004).

33 On these coin flows, see Noonan (1997).

34 On the archaeological evidence for Scandinavian settlers from this era, see Franklin and Shepard (1996), 91ff.; Duczko (2004), chapters. 3–5.

35 See Franklin and Shepard (1996), chapter 3; Melnikova (1996), 54–60; Duczko (2004), chapter 6.

36 See Likhachev (1970); Melnikova (1996), 105–9.

37 Quoted in O Corrain (1997), 94.

38 See Sawyer (1962).

39 *Chronicle of Ireland* ad a. 848.

40 Healfdan: *Anglo-Saxon Chronicle* (878 AD). The argument for the scale of the Great Armies was made in full by Brooks (1979); cf. Smyth (1977) on their structure, with the identifications of Olaf and Ingvar. (The data has since been accepted by Sawyer.) On the continent, likewise, when the Franks won their great victory at the Dyle, they killed two Viking kings and captured sixteen royal standards.

41 Settlement entries: *Anglo-Saxon Chronicle* (876, 877 and 880 AD). Estimates based on *Doomsday Book* suggest that the total population of England in 1086 was perhaps a million and a half, and the settlements did not affect the whole country.

42 *Anglo-Saxon Chronicle* (896 AD).

43 See Vince (2001); Leahy and Paterson (2001); cf. the more general studies of Hart (1992); and for a good survey of the place-name evidence, with full references, see Fellows-Jenson (2001).

44 See O Corrain (1997), (1998); Smyth (1979).

45 For a useful introduction, see Ritchie (1993), 25–7.

46 *Chronicle of Ireland* (856, 857, 858 AD); cf. O Corrain (1998), 326–7; Charles-Edwards (2006), vol. 2, 4–5.

47 For the DNA evidence, see Helgason et al. (2000), (2001), (2003); Goodacre et al. (2005).

48 For a useful introduction, see Rafnsson (1997).

49 The Great Army: *Anglo-Saxon Chronicle* (892 AD). For the DNA evidence, see note 47. As we have seen in the case of the Anglo-Saxon evidence (Chapter 6), there is a substantial margin for error in reading modern proportions as a direct reflection of those in the past.

50 See note 28 above.

51 And to suggest that the Norse had in fact arrived here before any Slavic immigrants: see note 34 above.

52 For references, see note 35 above.

53 Such commemorative runestones were put up in far larger numbers in the late Viking period: see B. Sawyer (1991).

54 See Wormald (1982).

55 For introductions to post-Viking history in England and Scotland respectively, see Campbell (1982); Broun et al. (1998). Davies (1990), chapter 4 tackles the same issue for Wales.

56 For references, see notes 13 and 14 above.

57 See e.g. P. Sawyer (1982), (1997a); Sawyer and Sawyer (1993), chapter 2, with references.

58 On the dirhems, see Noonan (1997), 145 (and cf. the comments of Arab

travellers on the astonishing silver wealth of Rus merchants: see note 16 above).
On Francia, see Nelson (1997), 37.

59 For the background of the Roman period, see Chapter 2.

60 For useful introductions, see Crawford (1987), chapter 1; Bill (1997); Rafnsson (1997).

61 See Melnikova (1996), 3–18 (for the mixture of practical information, amidst the learned and biblical material, in medieval Scandinavian geography), and 31–44 (for a brilliantly evocative account of the main Russian river routes and their interconnections).

62 The Anglo-Saxon expansion into England nevertheless bears some comparison: see Chapter 6.

63 Asser Saxe: Aarhus runestone no. 6; cf. Roesdahl (1991), 58. *Sodalitates*: *Annals of St Bertin* (862 AD); cf. Nelson (1997), 36.

64 For Asser Saxe, see previous note. On ship types, see Bill (1997). For *monoxyla*, see note 26 above; cf. Melnikova (1996), 33. Because of the rapids, shallows and sandbanks of the River Volkhov, ships had to be changed for rivercraft precisely at Ladoga.

65 For a useful introduction, see Bill (1997).

66 For an excellent recent survey, see Wickham (2005), 680–90, 809–11, with full references to the excavations, most of which have happened within the last scholarly generation.

67 'Emporia' were centres of movable wealth, so it is at least possible that initial accident turned into eventual design here; that would certainly be my best guess.

68 See Wormald (1982). For an introduction to this 'first' Danish state, see Roesdahl (1982), especially chapters 5 and 8; Hedeager (1992); Lund (1995), 202–12.

69 Anoundas: *Life of St Anskar* 19. Reginharius: see note 8 above.

70 This phenomenon has generated the concept of 'New Medievalism' in international-relations theory, a conceptualization of the fact that Third World states in particular find that they have in practice no monopoly of power or authority within their notional territorial space. For an introduction to these debates, see Friedrichs (2004), chapter 7.

10. THE FIRST EUROPEAN UNION

1 Thietmar, *Chronicle* 4.45–6.

2 The classic treatment in English of all three of these kingdoms remains Dvornik (1949); cf. Dvornik (1956). The available literature is now immense, and the references below are largely restricted to those written in western European languages, in which most of the main players in the Slavic world have anyway always tended (and now increasingly tend) to write, at least at regular intervals. Important supplements to Dvornik are provided, firstly, by useful general works such as Barford (2001); Curta (2006); and, secondly, by collections of papers dealing with a range of Slavic states: Manteuffel and Gieysztor (1968); Settimane (1983); Brachmann (1995); Urbancyzk (1997a), (2001); Curta (2005); Garipzanov et al. (2008). Then, thirdly, there are the studies devoted to individual kingdoms,

as follows. Poland: Manteuffel (1982); Urbancyzk (2004). Bohemia: Wegener (1959); Graus and Ludat (1967); Turek (1974); Sasse (1982); Prinz (1984); Godja (1988), (1991). Moravia: Dittrich (1962); Bosl (1966); Graus and Dostál (1966); Poulik et al. (1986); Bowlus (1995); M. Eggers (1995). These works provide the basis of my understanding of state formation in eastern Europe in the late first millennium; I will only footnote very specific points in the rest of this chapter, but this literature will always be implicit.

3 Dvornik was well aware of this point, although he did not treat Denmark in the same detail as his Slavic kingdoms. Papers relevant to Scandinavian state formation appear in some of the secondary sources listed in the previous note. In addition, the following studies, from an immense literature, of the early Danish state's emergence from the Viking period, are particularly useful: Randsborg (1980); Roesdahl (1982); Hedeager (1992); Sawyer and Sawyer (1993); Rumble (1994); Lund (1997).

4 Dvornik treated the history of the first Rus state in some detail (see note 2 above, with the supplementary works detailed there). For further information, see e.g. Kaiser and Marker (1994); Franklin and Shepard (1996); Melnikova (1996); Duczko (2004); cf., on legal structures, D. Kaiser (1980), (1992).

5 The first homegrown historian of the Danish state, Saxo Grammaticus, worked about seventy-five years later than his Slavic counterparts.

6 For an introduction to the new cultural patterns of the Carolingian era and beyond, see McKitterick (1989), (1994).

7 For a useful commentary on Thietmar's *Chronicle*, see Schröder (1977). Dvornik was immensely interested in the conversion of these states to Christianity: with the works cited in note 2 above, see especially Dvornik (1969). Many of the studies cited in notes 2–4 deal with Slavic conversion, but useful additional information and analysis can be found in e.g. Wolfram (1979), (1995); Kantor (1990); Urbancyzk (1997b); Wood (2001).

8 In addition to the studies cited in note 3 above, there are several useful papers in Scragg (1991) and Cooper (1993), especially those of Sawyer (1993) and Lund (1993). See, too, in Rumble (1994), the papers of Sawyer (1994) and Lund (1994). The rehabilitation of the reputation of Aethelred – see e.g. Keynes (1987) – only emphasizes the military capacity of the Danish monarchy.

9 On the fortifications of Dux Rastiz of Moravia: *Annals of Fulda* 869 (cf. ibid 855). For Slavic and Danish fortifications, see notes 2–4 above, but particularly helpful are Kurnatowska (1997a); Dulinicz (1997); Petrov (2005). For the Alamanni, see Chapter 2. Admittedly, the political unity of the Tervingi collapsed when their leader induced them to build fortifications, but only when they were simultaneously faced with Hunnic assault: Ammianus 31.3.8, and see Chapter 4 above.

10 See Ibn Fadlan, *Relation*, with Thietmar, *Chronicle* 4.46 (quoted on p. 515). Polish forces in 1003: Thietmar, *Chronicle* 5.36–7. For other sources of revenue, see pp. 563ff above.

11 One Byzantine source reports that the Rus force assisting Basil II numbered 6,000 men: Franklin and Shepard (1996), 161–3. Territorial contingents: *Russian Primary Chronicle* (1015 AD, 1068 AD). Retinues appear regularly in the Bohemian sources translated by Kantor (1990).

12 *Encomium of Queen Emma* II.4. Lund (1986), (1993) argues firmly for a solely

mercenary army, but the descriptions in the *Encomium* sound more like a mixed force, and it is worth noting that the much less powerful earls of Orkney had imposed carefully defined military obligations on their populations from an early date: Crawford (1987), 86–91. I think it is in the nature of the Danish state, as with its Slavic peers, that there are likely to have been substantial differences between separate parts of the kingdom: see pp. 526ff above.

13 On the Tithe Church, see Franklin and Shephard (1996), 164–5. For other references, see notes 2–4 above, with Kurnatowska (1997a); Shepard (2005); Font (2005).

14 For Danish transport infrastructure, see Randsborg (1980), 75ff.; Roesdahl (1982), chapter 3.

15 For an introduction, see Dvornik (1949), 105–10, with Appendix 5, though the details of the territories it defines are much disputed.

16 On Moravia and Bohemia, see Jirecek (1867) and Friedrich (1907) for some of the texts, with Kantor (1983), (1990) for commentary. On Russia, see D. Kaiser (1980), (1992).

17 The pattern does not apply to Moravia, however, whose capacity to operate as state centre was destroyed by the rise of Magyar power in the 890s, after which it became one of the territories to be swapped.

18 For a more detailed narrative in English, see Dvornik (1949). Randsborg (1980), 75ff. is excellent on the principles on itineration. For Bohemian documents, see note 16 above. For excellent commentaries on the Polish information, see Lowmianski (1960); Gorecki (1992).

19 *Russian Primary Chronicle* (945–55 AD).

20 See e.g. Roesdahl (1982), 147–55. One-quarter of the buildings in Fyrkat were residential, for instance, and one-third for storage.

21 Sobibor would later die in Prague in 1004 fighting the expelled Premyslid Jaromir, having returned to Prague with a Polish army: Thietmar, *Chronicle* 6.12. For a general account, see Urbanczyk (1997c).

22 *Annals of Fulda* (845, 872, 895 AD), with secondary references as note 2 above.

23 For commentaries, see Wolfram (1995); cf. the different geographical reconstructions of Bowlus (1995); M. Eggers (1995). Wherever it is placed, however, the basic political process stays the same.

24 The traditional picture was of 30 small 'tribes' in the seventh century, eventually evolving into 8 greater ones, according to Marxist principles, in the ninth. This was mostly guesswork based on extrapolation from the *Anonymous Bavarian Geographer*, which didn't cover lands beyond the Oder (see Chapter 8), and by analogy with Bohemia: cf. Barford (2001), chapter 12. The pattern may not be so far from the historical reality, except that we must reckon with a much more violent finale: see especially Kurnatowska (1997a); Dulinicz (1994), (1997).

25 On the emergence of the Rus state, see Franklin and Shepard (1996), chapter 3.

26 *Russian Primary Chronicle* (974 AD: Sveinald); 978 AD: Rogvolod/Ragnvaldr and Tury; 993 AD: the concubines).

27 For fuller discussion of the Jelling dynasty, see references in note 3 above. On the fate of the ninth-century state, see p. 511 above.

28 On the kings associated with the ninth-century Great Armies, see Chapter 9. For more detail on tenth-century patterns, see the literature cited in note 3 above.

29 For general references, see previous note. For the 'mark' in Denmark, see Lund (1984), 21–2; cf. Lund (1997).

30 See Chapter 8 above, with Curta (2001), chapter 7.

31 Miesco: Ibn Jaqub. Bohemia: *Annals of Fulda* (845 AD), with the texts translated in Kantor (1990). Moravia: *Annals of Fulda* (894 AD). Russia: Ibn Fadlan, *Relation*; *Russian Primary Chronicle* (945–6 AD). The fourth-century Germani: see Chapter 2 above. The sixth-century Slavs: Curta (2001), chapter 7.

32 For further discussion, see Chapter 8, and Chapter 1 (on the early Germanic world).

33 For fuller discussions, see the literature cited in notes 2–4 above. On the Germanic world, see pp. 64ff above.

34 For post-Avar leaders, see the literature cited in note 23 above. Wiztrach and his son ruled their own *civitas* in Bohemia: see *Annals of Fulda* (857 AD). On Moravia, see the studies cited in note 2 above. For excellent introductions to the changing patterns of hillforts, see Godja (1991), chapter 3; Kurnatowska (1997a), with full references. Nothing similar has been found in the early Slavic world to the Runder Berg and other *Herrenhöfe* of the leaders of the fourth-century Alamanni: see Chapter 2.

35 See, in particular, Roesdahl (1982); Hedeager (1992); Sawyer and Sawyer (1993).

36 For pollen, see Donat (1983); cf. Barford (2001), 153–9, both with full references.

37 In addition to the literature cited in note 2 above, see most recently, on agricultural expansion, Henning (2005); Barford (2005). These studies show that full manorialization followed rather than preceded state formation (as Marxist orthodoxy required). Agricultural expansion did, however, take other forms between the sixth and tenth centuries.

38 For the reasons we have previously encountered, the availability of food is one of the most basic limiting factors on possible population sizes.

39 For similar processes among the Germani, see Chapter 2.

40 Hedeby: *Royal Frankish Annals* (808 AD), with Roesdahl (1982), 70–6. Prague: Ibn Rusteh. Kiev: *De Administrando Imperio*, chapter 9; cf. Thietmar, *Chronicle* 8.2. Poland's participation in these networks is clear from the silver dirham distribution map: see Map 16.

41 *Russian Primary Chronicle* (911 and 945 AD). As far back as 808, Godfrid had moved the merchants to Hedeby because he wanted the toll revenue: see previous note.

42 For literature on the destruction of tribal castles, see note 24 above. On Vladimir's transfers, see *Russian Primary Chronicle* (1000 AD). On service villages and the organization of the heartlands of Bohemia and Poland, see respectively Godja (1991), chapters 3–4; Kurnatowska (1997a).

43 Oleg's army: *Russian Primary Chronicle* (880–2 AD). Sviatoslav: *Russian Primary Chronicle* (971–2 AD). For Vladimir, see previous note, with general commentary in Franklin and Shepard (1996), chapter 4.

44 For an introduction, see Bartlett (1993), chapter 5. A top estimate is that some 200,000 German peasants were eventually attracted east of the Elbe by the excellent terms on offer.

45 On Carolingian expansion and its structural importance, see Reuter (1985), (1990).

46 On the feuds, see Leyser (1989). On the burgwards, see Reuter (1991).

47 For an introduction to the Elbe Slavs, and a convenient collection of the relevant materials, see Lübke (1984–88), with Lübke (1994), (1997) for further analysis.

48 Dvornik (1949) provides a useful narrative. For the Northern Crusades, see e.g. Christiansen (1980).

49 Gero: Widukind of Corvey 2.20, with Heather (1997) more generally on the Abodrites. Zwentibald: *Annals of Fulda* (870–2 AD).

50 On the Christianization of Moravia, see the references in note 2. Werinhar's mutilation: *Annals of Fulda* (882 AD). Violence and plunder are regular features in all the warfare of this period, as recorded in Thietmar's *Chronicle*, the *Russian Primary Chronicle*, Adam of Bremen's *History of the Bishops of Hamburg* and Helmold's *Chronicle of the Slavs*, the two latter both having much to report on the plunderings and wars between the Empire and the Elbe Slavs.

51 On Saxon military evolution, see Leyser (1982), essays 1 and 2. For the Capitulary of Thionville, see Boretius (1883), 44.7.

52 *Miracles of St Demetrius* II.5; cf. the swift appearance of powerful leaders such as Liudewit: for references, see note 23 above.

53 For further discussion, and references, see Chapter 9.

54 On the slave raids of the Rus and Western Slavs: Ibn Jaqub; cf. McCormick (2001), on the general importance of these new connections.

55 Ibn Fadlan, *Relation*; cf. *Russian Primary Chronicle* (993 AD), on Vladimir. If the trade was essentially in women, the Rus presumably had to carry their own boats round the Dnieper rapids, but this may just be the literature of shock. Certainly the western slave trades – overland and by sea – involved males as well as females; cf. Verlinden (1955), the source of the map in question.

56 For introductory references, see note 7 above. The same tendency of trying to avoid taking your Christianity from a near imperial neighbour is also visible in the case of the Bulgarians, who did the same, trying to avoid a Byzantine connection: see e.g. Browning (1975), for an introduction. The Bulgarians equally failed to avoid the imperial connection, but, like the Poles, were eventually granted their own archbishop.

57 The availability of Bede's extraordinary narrative and a host of other sources from the early conversion period in England means that the Anglo-Saxon case study has often been a vehicle for exploring these ideas. For an excellent introduction, see Mayr Harting (1972); cf. Mayr-Harting (1994) for a comparison with Bulgaria.

58 There is little sign that conversion to Christianity changed the nature of immediate political competition in the Slavic context, any more than it did in the Anglo-Saxon, on which see the wonderful paper of Wormald (1978). On the administrative front, the last of the Anglo-Saxon kingdoms was brought to Christianity in 681, when Wessex conquered the Isle of Wight, but the beginnings of an administrative system that was working convincingly via literacy are, to my mind, visible only two or three generations later, and it is really only in the ninth and tenth centuries that the evidence multiplies.

59 Carolingian imperial rule had established tithing as a norm from the later eighth century: see McKitterick (1977). This made the later Slavic conversions different examples, like that of the Anglo-Saxons, where religious taxation was not yet so firmly established.

60 For the Tervingi, see Chapter 2. For the 983 revolt, see Reuter (1991); Lübke (1994), commenting particularly on the narrative of Thietmar, *Chronicle* 3.17ff. The *Russian Primary Chronicle* is the basic source of information on the Russian case, upon which Shepard (2005), with full references to the earlier literature, provides an excellent recent treatment.

61 Vladimir: *Russian Primary Chronicle* (978–80 AD); with Shepard (2005) for commentary. For the Elbe Slavs, see the references in note 47 above.

62 For an introduction to the concept of 'peer polity interaction', and some case studies, see Renfrew and Cherry (1986).

63 For a detailed discussion of coin flows, see Noonan (1997), (1998).

64 For comparative case studies, both ancient and modern, see Gottmann (1980); Rowlands et al. (1987); Bilde et al. (1993); Champion (1995). It is extremely important, however, to factor in a generalized concept of agency: cf. Wilson (2008).

11. THE END OF MIGRATION AND THE BIRTH OF EUROPE

1 *Annals of Fulda* (900 AD).

2 See Faith (1997) on the extent to which the Normans rewrote the rules by which peasant life was governed.

3 See in particular Chapters 6 and 9.

4 As we saw in Chapter 6, the 50–75 per cent spread of possibly 'Anglo-Saxon' Y chromosomes recorded in samples of the modern English population can be accounted for by an invading group that was anywhere between 50–75 per cent of the fifth-/sixth-century population, or only 10 per cent if you give them even a marginal breeding advantage.

5 These key cases reported by Ammianus and Procopius are explored in detail in Chapters 4 and 5.

6 In these cases, the evidence is currently not good that the groups actually crossed the frontier, but reliable contemporaries describe at least their subsequent moves, to Spain and North Africa respectively, and to join Alaric, in precisely such terms.

7 The Gothic and Lombard migration flows are discussed in Chapters 3 and 5.

8 The ideas set out here, and through earlier chapters, are discussed in more detail in Heather (2008a).

9 For Anglo-Saxon return migration, see Chapter 6 above.

10 Autumn 376, although this has been challenged, is also the likeliest time for the move of the Gothic Tervingi: see Heather (2005), 153.

11 This presumably limited the amount of fundamental social (as opposed to political) change that was generated by the flows of new wealth into Scandinavia, since the wealth was bound initially to fall largely into the hands of those who where already reasonably wealthy.

12 And possibly also by non-militarized slaves: see Chapter 4 above.

13 Of course, Jordanes' migration topos gave more than an excuse to do so: see Chapter 3.

14 The one successor state not founded by a new coalition created on the march was that of the Burgundians, but there is a crucial lack of narrative evidence to help us understand fifth-century Burgundian history, which was certainly traumatic.

15 Hunnic imperial history confirms the point, since the huge supraregional power created by Attila and his predecessors was entirely dependent upon large-scale flows of Mediterranean wealth for its continued existence: see Chapter 5.

16 Just one surviving vignette illustrates Gotho-Slav interaction: Jordanes, *Getica* 48.247, with p. 234 above.

17 The sources suggest, however, that some Slavic groups had already developed a considerable degree of political and military organization on the back of the new wealth flows of the sixth century: see Chapter 8.

18 Bartlett (1993), especially chapters 2 and 5, provides an excellent introduction to these new patterns.

19 Tacitus, *Germania* 46.4.

PRIMARY SOURCES

Following normal conventions, specific editions and translations of standard classical works are not listed in the bibliography, though all those works cited in this book appear below, and most are translated in either or both of the Loeb and Penguin Classics series. All Christian authors are available, if sometimes in outdated form, in *Patrologia Latina* or *Patrologia Graeca* editions. More recent (sometimes competing) editions of most of the texts cited in the introductions and notes can be found in *GCS* (*Die Griechischen Christlichen Schriftsteller der ersten Jahrhunderte*), *CSEL* (*Corpus Scriptorum Ecclesiasticorum Latinorum*), *CC* (*Corpus Christianorum*), and *SC* (*Sources Chrétiennes*). Many are translated in the *Nicene and Post-Nicene Fathers*, and *Library of the Fathers* collections. Otherwise, the following editions and translations of late Roman and early medieval sources have been used.

Adam of Bremen, *History of the Bishops of Hamburg*, ed. Schmeidler (1917); trans. Tschan (1959)

Agathias, *History*, ed. Keydell (1967); trans. Frendo (1975)

Ammianus Marcellinus, ed. and trans. Rolfe (1935–39)

Anglo-Saxon Chronicle, ed. and trans. Whitelock et al. (1961)

Annals of Fulda, ed. Pertz and Kurze (1891); trans. Reuter (1992)

Annals of St Bertin, ed. Waitz (1883); trans. Nelson (1991)

Anonymous Bavarian Geographer, ed. Bielowski (1946)

Anonymous Valesianus, ed. and trans. in Rolfe (1935–39), vol. 3

Aurelius Victor, *Caesars*, ed. Pichlmayr (1911); trans. Bird (1994)

Bede, *Ecclesiastical History*, ed. and trans. Colgrave and Mynors (1969)

Caesar, *Gallic War*

Cassiodorus, *Variae*, ed. Mommsen (1894b); trans. Hodgkin (1886); Barnish (1992)

Chronicle of Ireland, trans. Charles-Edwards (2006)

Chronicle of Monemvasia, ed. and trans. Charanis (1950)

Chronicon Paschale, ed. Dindorf (1832); trans. Whitby and Whitby (1989)

Claudian, *Works*, ed. and trans. Platnauer (1922)

Codex Theodosianus, ed. Mommsen and Kreuger (1905); trans. Pharr (1952)

Constantius, *Life of St Germanus*, ed. Noble and Head (1995), 75–106

Consularia Constantinopolitana, ed. Mommsen (1892)

Cosmas of Prague, *Chronicle of Bohemia*, ed. Bretholz and Weinberger (1923)

Crith Gablach: ed. Binchy (1970).

De Administrando Imperio, ed. and trans. Moravcsik and Jenkins (1967)

Dio Cassius, *Roman History*, ed. and trans. Cary (1914–27)

Encomium of Queen Emma, ed. and trans. Campbell (1949)

Ennodius, *Works*, ed. Vogel (1885)

Eugippius, *Life of Severinus*, ed. Noll and Vetter (1963); trans. Bieler (1965)

Eunapius, *Histories*, ed. and trans. Blockley (1982)

Eutropius, *Breviarium*, ed. Santini (1979); trans. Bird (1993)

Fredegar, *Chronicle*, ed. and trans. Wallace-Hadrill (1960)

Gallic Chronicle of 452, ed. Mommsen (1892)

Gallus Anonymous, *Chronicle*, ed. Maleczynski (1952); trans. Knoll and Schaer (2003)

Gildas, *Ruin of Britain*, ed. and trans. Winterbottom (1978)

Gregory of Tours, *Histories*, ed. Krusch and Levison (1951); trans. Thorpe (1974)

Helmold, *Chronicle of the Slavs*, ed. Lappenberg and Schmeidler (1909); trans. Tschan (1966)

Herodotus, *Histories*

Historia Augusta, ed. and trans. Magie (1932)

Hydatius, *Chronicle*, ed. Mommsen (1894); trans. Burgess (1993)

Ibn Fadlan, ed. and trans. Canard (1973)

Ibn Jaqub, ed. and trans. Miquel (1966)

Ibn Rusteh, ed. and trans. Wiet (1957)

Jerome, *Chronicle*: an online edition and English translation may be found at http://www.tertullian.org./fathers/jerome_chronicle_oo_eintro.htm

John of Antioch, ed. Mueller (1851–70); trans. Gordon (1966)

John of Nikiu, *Chronicle*, trans. Charles (1916) (from the Ethiopian)

Jordanes, *Romana* and *Getica*, ed. Mommsen (1882); *Getica*, trans. Mierow (1915)

Laws of Ine, ed. Liebermann (1903–16); trans. Whitelock (1955)

Life of Anskar, ed. Trillmich et al. (1978); trans. Robinson (1921)

Malchus, ed. and trans. Blockley (1982)

Menander Protector, ed. and trans. Blockley (1985)

Miracles of St Demetrius, ed. Lemerle (1979–81)

Notitia Dignitatum, ed. Seeck (1962)

Novels of Valentinian III, ed. Mommsen and Kreuger (1905); trans. Pharr (1952)

Origo Gentis Langobardorum: see Paul the Deacon.

Olympiodorus of Thebes, ed. and trans. Blockley (1982)

Orosius, *Against the Pagans*, ed. Arnaud-Lindet (1990–91); trans. Defarri (1964)

Panegyrici Latini, ed. and trans. Nixon and Rogers (1994)

Passion of St Saba, ed. Delehaye (1912); trans. Heather and Matthews (1991)

Paul the Deacon, *History of the Lombards*, ed. Bethmann and Waitz (1878); trans. Foulke (1974)

Paulinus of Pella, *Eucharisticon*, ed. and trans. Evelyn White (1961), vol. 2

Peter the Patrician, ed. Mueller (1851–70)

Pliny, *Natural History*

Priscus, ed. and trans. Blockley (1982)

Procopius, *Works*, ed. and trans. Dewing (1914–40)

Prosper Tiro, *Chronicle*, ed. Mommsen (1892)

Ptolemy, *Geography*

Royal Frankish Annals, ed. Kurze (1895); trans. Scholz (1972)

Russian Primary Chronicle, trans. Cross and Sherbowitz-Wetzor (1953)

Saxo Grammaticus, *History of the Danes*, ed. Knabe et al. (1931–57); trans. Fisher (1996) (Books I–IX); Christiansen (1980–81) (Books X–XVI)

Socrates, *Historia Ecclesiastica*, ed. Hansen (1995); trans. in *Nicene and Post-Nicene Fathers*, vol. 2

Sozomen, *Historia Ecclesiastica*, ed. Bidez and Hansen (1995); trans. in *Nicene and Post-Nicene Fathers*, vol. 2

Strabo, *Geography*

Strategicon of Maurice, ed. Dennis (1981); trans. Dennis (1984)

Synesius of Cyrene, ed. Garzya (1989)

Tacitus, *Annals*; *Histories*; *Germania*

Themistius, *Orations*, ed. Schenkl et al. (1965–74); trans. Heather and Matthews (1991) (*Orations* 8 and 10); Heather & Moncur (2001) (*Orations* 14–16 and 34)

Theodoret, *Historia Ecclesiastica*, ed. Parmentier and Hansen (1998); trans. in *Nicene and Post-Nicene Fathers*, vol. 3

Theophanes, *Chronographia*, ed. Niebuhr (1839–41); trans. Mango and Scott (1997)

Theophylact Simocatta, *History*, ed. De Boor and Wirth (1972); trans. Whitby and Whitby (1986)

Thietmar of Merseburg, *Chronicle*, ed. Holtzmann (1935); trans. Warner (2001)

Victor of Vita, *History of the Persecution in Africa*, ed. Petschenig (1881); trans. Moorhead (1992)

Widukind of Corvey, ed. Lohmann and Hirsch (1935)

Zonaras, *Chronicle*, ed. Weber (1897)

Zosimus, *History*, ed. Paschoud (1971–89); trans. Ridley (1982)

BIBLIOGRAPHY

Abels, R. (1998). *Alfred the Great: War, Kingship and Culture in Anglo-Saxon England* (London)

Achelis, H. (1900). 'Der älteste deutsche Kalender', *Zeitschrift für die neutestamentliche Wissenschaft* 1, 308–35

Adamson, S. et al., eds. (1990) *Papers from the 5th International Conference on English Historical Linguistics* (Amsterdam)

Agadshanow, S. G. (1994). *Der Staat der Seldschukiden und Mittelasien im 11.–12. Jahrhundert* (Berlin)

Ament, H. (1996). 'Frühe Funde und archäologische Erforschung der Franken in Rheinland', in *Die Franken: Wegbereiter Europas: vor 1500 Jahren, König Chlodwig und seine Erben* (Mainz), 22–34

Amory, P. (1997). *People and Identity in Ostrogothic Italy, 489–554* (Cambridge)

Anderson, B. (1991). *Imagined Communities: Reflections on the Origin and Spread of Nationalism* (London)

Anokhin, V. A. (1980). *The Coinage of Chersonesus* (Oxford)

Antony, D. W. (1990). 'Migration in Archaeology: the Baby and the Bathwater', *American Anthropologist* 92, 895–914

———— (1992). 'The Bath Refilled: Migration in Archaeology Again', *American Anthropologist* 94, 174–6

Arnaud-Lindet, M-P., ed. and trans. (1990–91). *Orose: Histoires contre les païens* (Paris)

Arnold, C. J. (1997). *An Archaeology of the Early Anglo-Saxon Kingdoms*, 2nd ed. (London)

Arnold, E. (1973). *The Kingdom of the Scots: Government, Church and Society from the Eleventh to the Fourteenth Century* (London)

Bacall, A. (1991). *Ethnicity in the Social Sciences: A View and a Review of the Literature on Ethnicity* (Coventry)

Bachrach, B. S. (1973). *The Alans in the West* (Minneapolis)

Bailyn, B. (1994). 'Europeans on the Move, 1500–1800', in Canny (1994), 1–5

Bakony, K. (1999). 'Hungary', in Reuter (1999), 536–52

Barford, P. (2001). *The Early Slavs* (London)

———— (2005). 'Silent Centuries: The Society and Economy of the Northwestern Slavs', in Curta (2005), 60–102

Barnes, T. D. (1998). *Ammianus Marcellinus and the Representation of Historical Reality* (Ithaca; London)

Barnish, S. J. B. (1992). *The Variae of Magnus Aurelius Cassiodorus Senator* (Liverpool)

Barrett, J. C. et al., eds. (1989). *Barbarians and Romans in North-west Europe from the Later Republic to Late Antiquity* (Oxford)

Barrow, G. W. S. (1973). *The Kingdom of the Scots: Government, Church and Society from the Eleventh to the Fourteenth Century* (London)

Barth, F., ed. (1969). *Ethnic Groups and Boundaries: The Social Organization of Ethnic Difference* (Boston)

Bartlett, R. (1993). *The Making of Europe: Conquest, Colonization and Cultural Change 950–1350* (London)

Bassett, S., ed. (1989). *The Origins of Anglo-Saxon Kingdoms* (London)

Bates, D. (1982). *Normandy before 1066* (London)

Batty, R. (2007). *Rome and the Nomads: The Pontic-Danubian Realm in Antiquity* (Oxford)

Baxter, S. (2007). *The Earls of Mercia: Lordship and Power in late Anglo-Saxon England* (Oxford)

Bentley, G. C. (1987). 'Ethnicity and Practice', *Comparative Studies in Society and History 29*, 25–55.

Bethmann, L. and Waitz, G., eds. (1878). *Pauli Historia Langobardorum*, Monumenta Germaniae Historica: Scriptores rerum Germanicarum (Hanover)

Bichir, G. (1976). *The Archaeology and History of the Carpi* (Oxford)

Bidez, J. and Hansen, G. C., eds. (1995). *Sozomenus Kirchengeschichte* (Berlin)

Bieler, L. (1965). *Eugippius: The Life of Saint Severinus* (Washington)

Bielowski, A., ed. (1946). *Descriptio civitatum et regionum ad septentrionalem plagam Danubii, Monumenta Poloniae Historica* 1, 10–11

Bierbrauer, V. (1980). 'Zur chronologischen, soziologischen und regionalen Gliederung des ostgermanischen Fundstoffs des 5. Jahrhunderts in Sudosteuropa', in Wolfram and Daim (1980), 131–42

—— (1989). 'Ostgermanische Oberschichtsgräber der römischen Kaiserzeit und des frühen Mittelalters', in *Peregrinatio Gothica* 2, *Archaeologia Baltica* 8 (Lodz), 40–106

Bilde, P. et al. (1993). *Centre and Periphery in the Hellenistic World* (Aarhus)

Bill, J. (1997). 'The Ships', in P. Sawyer (1997a), 182–201

Binchy, D. A. (1970a). *Celtic and Anglo-Saxon Kingship: the O'Donnell Lectures for 1967–8 delivered in the University of Oxford on 23 and 24 May 1968* (Oxford)

—— , ed. (1970b). *Críth gablach* (Dublin)

Bird, H. W. (1993). *Eutropius Breviarium* (Liverpool)

—— (1994). *Liber de Caesaribus of Sextus Aurelius Victor* (Liverpool)

Birley, A. R. (1966). *Marcus Aurelius* (London)

—— (2005). *The Roman Government of Britain* (Oxford)

Birnbaum, H. (1993). 'On the Ethnogenesis and Protohome of the Slavs: the Linguistic evidence', *Journal of Slavic Linguistics* 1.2, 352–74

Blair, J. (1994). *Anglo-Saxon Oxfordshire* (Oxford)

Blench, R. and Spriggs, M., eds. (1998). *Archaeology and Language II: Correlating Archaeological and Linguistic Hypotheses* (New York)

Blockley, R. C. (1981). *The Fragmentary Classicizing Historians of the Later Roman Empire: Eunapius, Olympiodorus, Priscus and Malchus*, vol. 1 (Liverpool)

—— (1983). *The Fragmentary Classicizing Historians of the Later Roman Empire: Eunapius, Olympiodorus, Priscus and Malchus*, vol. 2 (Liverpool)

—— (1985). *The History of Menander the Guardsman* (Liverpool)

Boeles, P. C. J. A. (1951). *Friesland tot de elfde eeuw* (Gravenhage)

Böhme, H.-W. (1974). *Germanische Grabfünde des 4 bis 5 Jahrhunderts zwischen untere Elbe und Loire. Studien zur Chronologie und Bevölkerungsgeschichte* (Munich)

———— (1975). 'Archäologische Zeugnisse zur Geschichte der Markomannenkriege (166–80 n. Chr.)', *Jahrbuch des römischgermanischen Zentralmuseums Mainz* 22, 155–217

Bohner, K. (1958). *Die frankischen Altertumer des Trierer Landes*, 2 vols. (Berlin)

Bohning, W. R. (1978). 'International Migration and the Western World: Past, Present and Future', *International Migration* 16, 1–15.

Boretius, A., ed. (1883). *Capitularia regum Francorum* (Hanover)

Borodzej, T. et al. (1989). *Période romaine tardive et début de la période des migrations des peuples (groupe de Masłomecz)* (Warsaw)

Borrie, W. D. (1994). *The European Peopling of Australasia: A Demographic History, 1788–1988* (Canberra)

Bosl, K. (1966). *Das Grossmährische Reich in der politischen Welt des 9. Jahrhunderts* (Munich)

Bowlus, C. (1995). *Franks, Moravians, and Magyars: the Struggle for the Middle Danube, 788–907* (Philadelphia)

Brachmann, H. (1978). *Slawische Stämme an Elbe und Saale: zu ihrer Geschichte und Kultur im 6.–10.* (Berlin)

———— (1993). *Der frühmittelalterliche Befestigungsbau in Mitteleuropa: Untersuchungen zu einer Entwicklung und Function im germanisch-deutschen Gebiet* (Berlin)

————, ed. (1995). *Burg – Burgstadt – Stadt: zur Genese mittelalterlicher nichtagrarischer Zentren in Ostmitteleuropa* (Berlin)

———— (1997). 'Tribal Organizations in Central Europe in the 6th–10th Centuries AD: Reflections on the Ethnic and Political Development in the Second Half of the First Millennium', in Urbanczyk (1997a), 23–38

Brather, S. (1996). *Feldberger Keramik und frühe Slawen: Studien zur nordwestslawischen Keramik der Karolingerzeit* (Bonn)

———— (2001). *Archäologie der westlichen Slawen: Siedlung, Wirtschaft und Gesellschaft im früh- und hochmittelalterlichen Ostmitteleuropa* (Berlin)

Braund, D. C. (1984). *Rome and the Friendly King: The Character of Client Kingship* (London)

———— (2005). *Scythians and Greeks: Cultural Interactions in Scythia, Athens and the early Roman Empire (Sixth Century BC–First Century AD)* (Exeter)

Bretholz, B. and Weinberger, W., eds. (1923). *Die Chronik der Böhmen des Cosmas von Prag*, Monumenta Germaniae Historica: Scriptores rerum Germanicarum, n.s. 2 (Berlin)

Brooks, N. (1979). 'England in the Ninth Century: The Crucible of Defeat', *Transactions of the Royal Historical Society*, 5th series, 29, 1–20

Broun, D. et al., eds. (1998). *Image and Identity: The Making and Re-making of Scotland through the Ages* (Edinburgh)

Brown, P. (1971). *The World of Late Antiquity: from Marcus Aurelius to Muhammad* (London)

———— (1996). *The Rise of Western Christendom: Triumph and Diversity, AD 200–1000* (Oxford)

Browning, R. (1975). *Byzantium and Bulgaria: A Comparative Study across the Early Medieval Frontier* (London)

Buko, A. (2005). *Archeologia Polski* (Warsaw)

Burgess, R. (1993). *The Chronicle of Hydatius and the Consularia Constantinopolitana* (Oxford)

Bury, J. B. (1928). *The Invasion of Europe by the Barbarians* (London)

Cameron, A. D. E. (1970). *Claudian: Poetry and Politics at the Court of Honorius* (Oxford)

Cameron, A. D. E. and Long, J., with a contribution by Sherry, L. (1993). *Barbarians and Politics at the Court of Arcadius* (Berkeley)

Campbell, A. (1949). *Encomium Emmae Reginae* (London)

Campbell, J., ed. (1982). *The Anglo-Saxons* (Oxford)

———— (2000). *The Anglo-Saxon State* (London)

Canard, M. (1973). *Miscellanea Orientalia* (London)

Canny, M., ed. (1994). *Europeans on the Move: Studies on European Migration 1500–1800* (Oxford)

Carroll, M. (2001). *Romans, Celts and Germans: The German Provinces of Rome* (Stroud)

Cary, E. (1914–27). *Dio's Roman History* (London)

Castellanos, S. (2000). 'Propiedad de la tierra y relaciones de dependencia en la Galia del siglo VI. El Testamentum Remigii', *Antiquité Tardif* 8, 223–7

Chadwick-Hawkes, S., ed. (1989). *Weapons and Warfare in Anglo-Saxon England* (Oxford)

Champion, T. C. (1995). *Centre and Periphery: Comparative Studies in Archaeology* (London)

Champion, T. C. and Megaw, J. V. S., eds. (1985). *Settlements and Society: Aspects of West European Prehistory in the First Millennium BC* (Leicester)

Chapman, R. and Dolukhanov, P., eds. (1993). *Cultural Transformations and Interactions in Eastern Europe* (Aldershot)

Charanis, P. (1950). 'The Chronicle of Monemvasia and the Question of the Slavic Settlements of Greece', *Dumbarton Oaks Papers* 5, 141–66

Charles, R. H. (1916). *The Chronicle of John, Bishop of Nikiu* (London)

Charles-Edwards, T. C. E. (1989). 'Early Medieval Kingships in the British Isles', in Bassett (1989), 28–39.

————, ed. (2003). *After Rome* (Oxford)

———— (2006). *The Chronicle of Ireland* (Liverpool)

Chastagnol, A. (1973). 'Le repli sur Arles des services administratifs gaulois en l'an 407 de notre ère', *Revue Historique* 97, 23–40

Childe, V. G. (1926). *The Aryans: A Study of Indo-European Origins* (London)

———— (1927). *The Dawn of European Civilization* (London)

Christiansen, E. (1980). *The Northern Crusades: The Baltic and the Catholic Frontier, 1100–1525* (London)

———— (1980–81). *Saxo Grammaticus: Danorum Regum heroumque historia: Books X–XVI* (text and translation), 3 vols. (Oxford)

———— (2002). *The Norsemen in the Viking Age* (Oxford)

Christie, N. (1995). *The Lombards: The Ancient Longobards* (Oxford)

Christlein, R. (1978). *Die Alamannen* (Stuttgart)

Claessen, H. J. M. and Skalnik, P. (1978). *The Early State* (The Hague)

———— (1981). *The Study of the State* (The Hague)

Claessen, H. J. M. and van de Velde, P. (1987). *Early State Dynamics* (Leiden)

Claessen, H. J. M. and Oosten, J. G. (1996). *Ideology and the Formation of Early States* (Leiden)

Clark, G. (1966). 'The Invasion Hypothesis in British Archaeology', *Antiquity* 40, 172–89

Cohen, R., ed. (1995). *The Cambridge Survey of World Migration* (Cambridge)

———, ed. (1996). *Theories of Migration* (Cheltenham)

———, ed. (1997). *The Politics of Migration* (Cheltenham)

——— (2008). *Global Diasporas: An Introduction*, 2nd ed. (Abingdon)

Colgrave, B. and Mynors, R. A. B., eds. (1949). *Bede's Ecclesiastical History of the English People* (Oxford)

Collins, R. (1983). 'Theodebert I, "Rex Magnus Francorum"', in P. Wormald (ed.), *Ideal and Reality in Frankish and Anglo-Saxon Society* (Oxford), 7–33

——— (1998). *Charlemagne* (Basingstoke)

Collinson, S. (1994). *Europe and International Migration*, 2nd ed. (London)

Constantinescu, M. et al., eds. (1975) *Relations Between the Autochthonous Population and the Migratory Populations on the Territory of Romania* (Bucharest)

Cooper, J., ed. (1993). *The Battle of Maldon: Fiction and Fact* (London)

Coupland, S. (1995). 'The Vikings in Francia and Anglo-Saxon England', in McKitterick (1995), 190–201

Courtois, C. (1955). *Les Vandales et l'Afrique* (Paris)

Crawford. B. (1987). *Scandinavian Scotland* (Leicester)

Cribb, R. J. (1991). *Nomads in Archaeology* (London)

Croke, B. (1977). 'Evidence for the Hun Invasion of Thrace in AD 422', *Greek, Roman and Byzantine Studies* 18, 347–67

Cross, S. H. and Sherbowitz-Wetzor, O. P., eds. (1953). *The Russian Primary Chronicle: Laurentian Text* (Cambridge, MA)

Cunliffe, B. (1997). *The Ancient Celts* (Oxford)

Cunliffe, B. and Rowley, T., eds. (1976). *Oppida: The Beginnings of Urbanization in Barbarian Europe: Papers Presented to a Conference at Oxford, October 1975* (Oxford)

Curta, F. (1999). 'Hiding behind a Piece of Tapestry: Jordanes and the Slavic Venethi', *Jahrbücher für Geschichte Osteuropas* 47, 1–18

——— (2001). *The Making of the Slavs: History and Archaeology of the Lower Danube Region, c.500–700* (Cambridge)

———, ed. (2005). *East Central and Eastern Europe in the Early Middle Ages* (Ann Arbor)

——— (2006). *Southeastern Europe in the Middle Ages 500–1250* (Cambridge)

Daim, F. (2003). 'Avars and Avar Archaeology: An Introduction', in Goetz et al. (2003), 463–570

Dark, K. R. (2002). *Britain and the End of the Roman Empire* (Stroud)

Davies, W. (1978). *An Early Welsh Microcosm: Studies in the Llandaff Charters* (London)

——— (1990). *Patterns of Power in Early Wales* (Oxford)

De Boor, C. and Wirth, P., eds. (1972). *The History of Theophylact Simocatta* (Stuttgart)

Defarri, R. J. (1964). *Orosius: Seven Books of History Against the Pagans* (Washington).

Delehaye, H. (1912). 'Saints de Thrace et de Mesie', *Analecta Bollandia* 31, 161–300

Demandt, A. (1984). *Der Fall Roms: die Auflösung des römischen Reiches im Urteil der Nachwelt* (Munich)

Demougeot, E. (1979). *La Formation de l'Europe et les invasions barbares: II. Dès l'avènement de Dioclétien (284) à l'occupation germanique de l'Empire romain d'Occident (début du VIe siècle)* (Paris)

Denison, D. (1993). *English Historical Syntax* (Harlow)

Dennis, G. T., ed. (1981). *Das Strategikon des Maurikios* (Vienna)

———— (1984). *Maurice's Strategikon: Handbook of Byzantine Military Strategy* (Pennsylvania)

Dindorf, L., ed. (1832). *Chronicon Paschale* (Bonn)

Dittrich, Z. R. (1962). *Christianity in Great Moravia* (Groningen)

Dolukhanov, P. M. (1996). *The Early Slavs: Eastern Europe from the Initial Settlement to the Kievan Rus* (London)

Donat, P. (1983). 'Die Entwicklung der Wirtschaftlichen und Gesellschaftlichen Verhältnisse bei den slawischen Stämme zwischen Oder und Elbe nach archäologischen Quellen', in *Settimane* (1983), 127–45

Donat, P. and Fischer, R. E. (1994). 'Die Anfänge slawische Siedlung westlich der Oder', *Slavia Antiqua* 45, 7–30

Drijvers, J. W. and Hunt, D., eds. (1999). *The Late Roman World and its Historian: Interpreting Ammianus Marcellinus* (London)

Drinkwater, J. F. (1989). 'Patronage in Roman Gaul and the Problem of the Bagaudae', in *Patronage in Ancient Society*, ed. A. Wallace-Hadrill (London), 189–203

———— (1992). 'The Bacaudae of Fifth-Century Gaul', in Drinkwater and Elton (1992), 208–17.

———— (1998). 'The usurpers Constantine III (407–411) and Jovinus (411–413)', *Britannia* 29, 269–98

———— (2007). *The Alamanni and Rome 213–496* (Oxford)

Drinkwater, J. F. and Elton, H., eds. (1992) *Fifth-Century Gaul: A Crisis of Identity?* (Cambridge)

Duczko, W. (2004). *Viking Rus: Studies on the Presence of Scandinavians in Eastern Europe* (Leiden)

Dulinicz, M. (1994). 'The Problem of Dating of the Strongholds of the Tornow Type and Tornow-Klenica Group', *Archeologia Polski* 39, 31–49 (English summary)

———— (1997). 'The First Dendrochronological Dating of the Strongholds in Northern Mazovia', in Urbanczyk (1997a), 137–42

Dumville, D. N. (1977). 'Kingship, Genealogies and Regnal Lists', in Sawyer and Wood (1977), 72–104

Dunbabin, J. (2000). *France in the Making, 843–1180*, 2nd ed. (Oxford)

Dvornik, F. (1949). *The Making of Central and Eastern Europe* (London)

———— (1956). *The Slavs: Their Early History and Civilization* (Boston)

———— (1969). *Les légendes de Constantin et de Méthode vues de Byzance*, 2nd ed. (Hattiesburg, MS)

Earle, T. K., ed. (1984). *On the Evolution of Complex Societies: Essays in Honor of Harry Hoijer, 1982* (Malibu)

————, ed. (1991). *Chiefdoms: Power, Economy and Ideology* (Cambridge)

Eggers, H. J. (1951). *Der römische Import im freien Germanien*, Atlas der Urgeschichte 1 (Berlin)

Eggers, M. (1995). *Das 'Grossmährische Reich': Realität oder Fiktion?: eine Neuinterpretation der Quellen zur Geschichte des mittleren Donauraumes im 9. Jahrhundert* (Stuttgart)

Elton, H. (1996). *Frontiers of the Roman Empire* (London)

Esmonde-Cleary, S. (1989). *The Ending of Roman Britain* (London)

Evelyn White, H. (1961). *The Works of Ausonius* (London)

Faith, R. (1997). *The English Peasantry and the Growth of Lordship* (Leicester)

Farrell, R. T., ed. (1978). *Bede and Anglo-Saxon England* (Oxford)

————, ed. (1982). *The Vikings* (London)

Faulkner, N. (2000). *The Decline of Roman Britain* (Stroud)

Favrod, J. (1997). *Histoire politique du royaume burgonde (443–534)* (Lausanne)

Fellows-Jenson, G. (2001). 'In the Steps of the Vikings', in Graham-Campbell et al. (2001), 279–88

Fielding, A. (1993a). 'Mass Migration and Economic Restructuring', in King (1993), 7–18

———— (1993b). 'Migrations, Institutions and Politics: The Evolution of Emigration Policies', in King (1993), 40–62

Fisher, P., trans., ed. Davidson, H. E. (1996). *Saxo Grammaticus: The History of the Danes: Books I–IX* (Woodbridge)

Font, M. (2005). 'Missions, Conversions, and Power Legitimization in East Central Europe', in Curta (2005), 283–296

Forte, A. et al. (2005). *Viking Empires* (Cambridge)

Foulke, W. D. (1974). *History of the Lombards* (Philadelphia)

Fouracre, P. (2000). *The Age of Charles Martel* (Harlow)

————, ed. (2005). *The New Cambridge Medieval History: vol. 1, c.500–c.700* (Cambridge)

Franklin, S. and Shepard, J. (1996). *The Emergence of Rus 750–1200* (Harlow)

Freeman, E. A. (1888). *Four Oxford Lectures 1887* (London)

Frendo, J. D. (1975). *Agathias History* (Berlin)

Fried, M. H. (1967). *The Evolution of Political Society: An Essay in Political Anthropology* (New York)

Friedrich, G. (1907). *Codex Diplomaticus et Epistolaris Regni Bohemiae*, vol. 1 (Prague)

Friedrichs, J. (2004). *European Approaches to International Relations Theory: A House With Many Mansions* (London)

Frolova, N. A. (1983). *The Coinage of the Kingdom of Bosporos* AD 242–341/2 (Oxford)

Fulford, M. (1985). 'Roman Material in Barbarian Society c.200 BC–AD 400', in Champion and Megaw (1985), 91–108

Garipzanov, I. et al., eds. (2008). *Franks, Northmen, and Slavs: Identities and State Formation in Early Medieval Europe* (Turnhout)

Garzya, A. (1989). *Opere di Sinesio di Cirene: epistole, operette, inni* (Turin)

Geary, P. (1983). *Ethnic Identity as a Situational Construct in the Early Middle Ages* (Vienna)

———— (1988). *Before France and Germany: The Creation and Transformation of the Merovingian World* (New York)

———— (2002). *The Myth of Nations: The Medieval Origins of Europe* (Princeton)

Gebuhr, M. (1974). 'Zur Definition alterkaiserzeitlicher Fürstengräber vom Lubsow-Typ', *Prähistorische Zeitschrift* 49, 82–128

Gellner, E. (1983). *Nations and Nationalism* (Ithaca)

Geuenich, D., ed. (1998). *Die Franken und die Alemannen bis zur 'Schlacht bei Zulpich'*

Gillet, A., ed. (2002). *On Barbarian Identity: Critical Approaches to Ethnicity in the Early Middle Ages* (Turnhout)

Godja, M. (1988). *The Development of the Settlement Pattern in the Basin of the Lower Vltava (Central Bohemia) 200–1200* (Oxford)

———— (1991). *The Ancient Slavs: Settlement and Society* (Edinburgh)

Godlowski, K. (1970). *The Chronology of the Late Roman and Early Migration Periods in Central Europe* (Cracow)

———— (1980). 'Das Aufhören der Germanischen Kulturen an der Mittleren Donau und das Problem des Vordringens der Slawen', in Wolfram & Daim (1980), 225–32

———— (1983). 'Zur Frage der Slawensitze vor der grossen Slawenwanderung im 6. Jahrhundert', in Settimane (1983), 257–302

Godman, P. and Collins, R., eds. (1990). *Charlemagne's Heir* (Oxford)

Goehrke, C. (1992). *Frühzeit des Ostslaventums* (Darmstadt)

Goetz, H.-W. et al., eds. (2003). *Regna and Gentes: The Relationship between Late Antique and Early Medieval Peoples and Kingdoms in the Transformation of the Roman World* (Leiden)

Goffart, W. (1980). *Barbarians and Romans AD 418–584: The Techniques of Accommodation* (Princeton)

———— (1981). 'Rome, Constantinople, and the Barbarians in Late Antiquity', *American Historical Review* 76, 275–306

———— (1988). *The Narrators of Barbarian History (AD 550–800): Jordanes, Gregory of Tours, Bede, and Paul the Deacon* (Princeton)

———— (2006). *Barbarian Tides: The Migration Age and the Later Roman Empire* (Philadelphia)

Goodacre S. et al. (2005). 'Genetic Evidence for a Family-Based Scandinavian Settlement of Shetland and Orkney during the Viking Periods', *Heredity* 95, 129–35

Gordon, D. C. (1966). *The Age of Attila* (Ann Arbor)

Gorecki, P. (1992). *Economy, Society and Lordship in Medieval Poland 1100–1250* (New York)

Gottmann, J., ed. (1980). *Centre and Periphery: Spatial Variation in Politics* (London)

Gould, J. D. (1980). 'European Inter-Continental Emigration, The Road Home: Return Migration from the USA', *Journal of European Economic History* 9.1, 41–112

Graham-Campbell, J. et al., eds. (2001). *Vikings and the Danelaw* (Oxford)

Graus, F. J. and Dostál, A. (1966). *Das Grossmährische Reich: Tagung der wissenschaftlichen Konferenz des Archäologischen Instituts der Tschechoslowakischen Akademie der Wissenschaften, Brno-Nitra, 1.–4.x.1963* (Prague)

Graus, F. and Ludat, H., eds. (1967). *Siedlung und Verfassung Böhmens in der Frühzeit* (Wiesbaden)

Green, D. H. (1998). *Language and History in the Early Germanic World* (Cambridge)

Gyuzelev, V. (1979). *The Proto-Bulgarians* (Sofia)

Haarnagel, W. (1979). *Die Grabung Feddersen Wierde : Methode, Hausbau-, Siedlungs- und Wirtschaftsformen sowie Sozialstruktur* (Wiesbaden)

Hachmann, R. (1971). *The Germanic Peoples* (London)

Haldon, J. F. (1990). *Byzantium in the Seventh Century: The Transformation of a Culture* (Cambridge)

Hall, C. (1983). *Periphrastic Do: History and Hypothesis* (Ann Arbor)

Halsall, G. (1992). 'The Origins of the Reihengräberzivilisation: Forty Years On', in Drinkwater and Elton (1992), 196–207

——— (1995a). *Settlement and Social Organization: The Merovingian Region of Metz* (Cambridge).

——— (1995b). *Early Medieval Cemeteries: An Introduction to Burial Archaeology in the Post-Roman West* (Skelmorlie)

——— (1999). 'Movers and Shakers: The Barbarians and the Fall of Rome', *Early Medieval Europe* 8, 131–45

——— (2001). 'Childeric's grave, Clovis' Succession and the Origins of the Merovingian Kingdom', in Mathisen and Shanzer (2001), 116–33

——— (2005). 'The Barbarian Invasions', in Fouracre (2005), 35–55

——— (2007). *Barbarian Migrations and the Roman West 376–568* (Cambridge)

Handley, M. (2001). 'Beyond Hagiography: Epigraphic Commemoration and the Cult of the Saints in Late Antique Trier', in Mathisen and Shanzer (2001), 187–200

——— (2003). *Death, Society and Culture: Inscriptions and Epitaphs in Gaul and Spain, AD 300–750* (Oxford)

Hannestad, K. (1960). 'Les forces militaires d'après la guerre gothique de Procope', *Classica et Mediaevalia* 21, 136–83

Hansen, G. C., ed. (1995). *Sokrates: Kirchengeschichte* (Berlin)

Harhoiu, R. (1977). *The Treasure from Pietroasa, Romania* (Oxford)

Härke, H. (1989). 'Early Saxon Weapon Burials: Frequencies, Distributions and Weapon Combinations', in Chadwick-Hawkes (1989), 49–61.

——— (1990). ' "Warrior Graves?" The Background of the Anglo-Saxon Weapon Burial Rite', *Past and Present* 126, 22–43

——— (1992). *Angelsächsische Waffengräber des 5. bis 7. Jahrhunderts* (Cologne)

——— (1998). 'Archaeologists and Migrations: A Problem of Attitude?', *Current Anthropology* 39.1, 19–45

Hart, C. (1992). *The Danelaw* (London)

Haubrichs, W. (1996). 'Sprache und Sprachzeugnisse der merowingischen Franken', in *Die Franken: Wegbereiter Europas: vor 1500 Jahren, König Chlodwig und seine Erben* (Mainz), 559–73.

——— (2003). 'Remico aus Goddelau: Ostgermanen, Westgermanen und Romanen im Wormser Raum des 5./6. Jahrhunderts', in *Runica – Germanica – Mediaevalia*, ed. W. Heizmann and A. van Nahl (Berlin), 226–42

—————— (forthcoming). 'Ein namhaftes Volk: Burgundische Namen und Sprache des 5. und 6. Jahrhunderts', in *Die Burgunder: Ethnogenese und Assimilation eines Volkes*, ed. V. Gallée (Worms), 135–84

Heather, P. J. (1986). 'The Crossing of the Danube and the Gothic Conversion', *Greek Roman and Byzantine Studies* 27, 289–318

—————— (1988). 'The Anti-Scythian Tirade of Synesius' De Regno', *Phoenix* 42, 152–72

—————— (1989). 'Cassiodorus and the Rise of the Amals: Genealogy and the Goths under Hun Domination', *Journal of Roman Studies* 79, 103–28

—————— (1991). *Goths and Romans 332–489* (Oxford)

—————— (1993). 'The Historical Culture of Ostrogothic Italy', in *Teoderico il grande e i Goti d'Italia, Atti del XIII Congresso internazionale di studi sull'Alto Medioevo* (Spoleto), 317–53

—————— (1995a). 'The Huns and the End of the Roman Empire in Western Europe', *English Historical Review* 110, 4–41

—————— (1995b). 'Theodoric King of the Goths', *Early Medieval Europe* 4.2, 145–73

—————— (1996). *The Goths* (Oxford)

—————— (2001). 'The Late Roman Art of Client Management and the Grand Strategy Debate', in Pohl and Wood (2001), 15–68

—————— (2003). 'Gens and regnum among the Ostrogoths', in Goetz et al. (2003), 85–133

—————— (2005). *The Fall of Rome: A New History* (London)

—————— (2007). 'Goths in the Roman Balkans c.350–500', in Poulter (2007), 163–90

—————— (2008a). Review of John Drinkwater (2007), *Nottingham Medieval Studies* 52, 243–5

—————— (2008b). 'Ethnicity, Group Identity, and Social Status in the Migration Period', in Garipzanov et al. (2008), 17–50

—————— (forthcoming). 'Why Did the Barbarian Cross the Rhine?', *Late Antiquity*

Heather, P. J. and Matthews, J. F. (1991). *The Goths in the Fourth Century: Translated Texts for Historians* (Liverpool)

Heather, P. J. and Moncur, D. (2001). *Politics, Philosophy, and Empire in the Fourth Century: Select Orations of Themistius, Translated Texts for Historians* (Liverpool)

Hedeager, L. (1987). 'Empire, Frontier and the Barbarian Hinterland: Rome and Northern Europe from AD 1–400', in Rowlands et al. (1987), 125–40

—————— (1988). 'The Evolution of Germanic Society 1–400 AD', in Jones R. F. J. et al. (1988), 129–44

—————— (1992), trans. John Hines. *Iron-Age Societies: From Tribe to State in Northern Europe, 500 BC to AD 700* (Oxford)

Hedeager, L. and Kristiansen, K. (1981). 'Bendstrup – A Princely Grave from the Early Roman Iron Age: Its Social and Historical Context', *Kuml* (1981), 150–62

Helgason, A. et al. (2000). 'Estimating Scandinavian and Gaelic ancestry in the male settlers of Iceland', *American Journal of Human Genetics* 67, 697–717

—————— (2001). 'mtDNA and the Islands of the North Atlantic: Estimating the Proportions of Norse and Gaelic Ancestry', *American Journal of Human Genetics* 68, 723–37

—————— (2003). 'A Reassessment of Genetic Diversity in Icelanders: Strong Evidence from Multiple Loci for Relative Homogeneity caused by Genetic Drift', *Annals of Human Genetics* 67, 281–97

Henning, J. (1991). 'Germanen-Slawen-Deutsche: Neue Untersuchungen zum frühgeschichtlichen Siedlungswesen ostlich der Elbe', *Praehistorische Zeitschrift* 66, 119–33

———— (2005). 'Ways of Life in Eastern and Western Europe during the Early Middle Ages: Which Way was Normal?', in Curta (2005), 41–59

Herrmann, J. (1968), *Siedlung, Wirtschaft und gesellschaftliche Verhältnisse der slawischen Stamme zwischen Oder/Neisse und Elbe* (Berlin)

———— (1983), 'Wanderungen und Landnähme im Westslawischen Gebiet', in Settimane (1983), 75–101

———— (1984), *Germanen und Slawen in Mitteleuropa* (Berlin)

Hermann, J. et al., eds. (1985). *Die Slawen in Deutschland: Geschichte und Kultur der slawischen Stämme westlich von Oder und Neisse vom 6. bis 12. Jahrhundert*, 2nd ed. (Berlin)

Higham, N. (1992). *Rome, Britain and the Anglo-Saxons* (London)

———— (1994). *The English Conquest: Gildas and Britain in the Fifth Century* (Manchester)

————, ed. (2007). *The Britons in Anglo-Saxon England* (London)

Hills, C. (2003). *The Origins of the English* (London)

Hines, J. (1984). *The Scandinavian Character of Anglian England in the pre-Viking Period* (Oxford)

————, ed. (1997). *The Anglo-Saxons from the Migration Period to the Eighth Century: An Ethnographic Perspective* (Woodbridge)

Hodder, I. (1982). *Symbols in Action: Ethnoarchaeological Studies of Material Culture* (Cambridge)

———— (1991). *Reading the Past: Current Approaches to Interpretation in Archaeology* (Cambridge)

Hodder, I. and Hutson, S., eds. (2003). *Reading the Past: Current Approaches to Interpretation in Archaeology* (Cambridge)

Hodges, R. and Bowden, W., eds. (1998). *The Sixth Century: Production, Distribution, and Demand* (Leiden)

Hodgkin, T. (1886). *The Letters of Cassiodorus: A Condensed Translation* (London)

Holmes, C., ed. (1996). *Migration in European History* (Cheltenham)

Holt, J. C. (1987). '1086' in Holt, ed., *Doomsday Studies* (Woodbridge), 41–64

Holtzmann, R., ed. (1935). *Thietmari Merseburgensis episcopi chronicon*, Monumenta Germaniae Historica: Scriptores rerum Germanicarum, vol. 9 (Berlin)

Hooke, D. (1997). 'The Anglo-Saxons in England in the Seventh and Eighth Centuries: Aspects of Location in Space', in Hines (1997), 65–99

———— (1998). *The Landscape of Anglo-Saxon England* (London)

Hummer, H. J. (1998). 'The Fluidity of Barbarian Identity: The Ethnogenesis of Alemanni and Suebi AD 200–500', *Early Medieval Europe* 7, 2–26

Ilkjaer, J. (1995). 'Illerup Adal (Danemark). Un lieu de sacrifices du IIIe siècle de n.è. en Scandinavie meridionale', in Vallet and Kazanski (1995), 101–12

Ilkjaer, J. and Lonstrup, J. (1983). 'Der Moorfund im Tal der Illerup-A bei Skanderborg im Ostjutland', *Germania* 61, 95–126

Ionita, I. (1975). 'The Social-Economic Structure of Society During the Goths' Migration in the Carpatho-Danubean Area', in Constantinescu (1975), 77–89

James, E. F. (1988). *The Franks* (Oxford)
———— (1989). 'The Origins of Barbarian Kingdoms: The Continental Evidence', in Bassett (1989), 40–52
James, S. (1999). *The Atlantic Celts: Ancient People or Modern Invention?* (London)
Jarnut, J. (2003). 'Gens, Rex, and Regna of the Lombards', in Goetz et al. (2003), 409–28
Jerome, H. (1926). *Migration and Business Cycles* (New York)
Jirecek, H. (1867). *Codex Juris Bohemici*, vol. 1 (Prague)
Johnston, D. E., ed. (1977). *The Saxon Shore* (London)
Jones, A. H. M. (1964). *The Later Roman Empire: A Social Economic and Administrative Survey*, 3 vols. (Oxford)
Jones, M. E. (1996). *The End of Roman Britain* (Ithica).
Jones R. F. J. et al., eds. (1988). *First Millennium Papers: Western Europe in the First Millennium* (Oxford)

Kaiser, D. H. (1980). *The Growth of the Law in Medieval Russia* (Princeton)
————, ed. (1992). *The Laws of Rus': tenth to fifteenth centuries* (Salt Lake City)
Kaiser, D. H. and Marker, G., eds. (1994). *Reinterpreting Russian History: Readings 860s–1860s* (Oxford)
Kaiser, R. (1997). *Die Franken: Roms Erben und Wegbereiter Europas?* (Idstein)
Kantor, M. (1983). *Medieval Slavic Lives of Saints and Princes* (Ann Arbor)
———— (1990). *The Origins of Christianity in Bohemia* (Evanston)
Kapelle, W. E. (1979). *The Norman Conquest of the North: The Region and Its Transformation 1000–1135* (Chapel Hill, NC)
Kazanski, M. (1991). *Les Goths (Ier–VIIe siècle après J.-C.)* (Paris)
———— (1999). *Les Slaves: les origines, Ier–VIIe siècle après J.-C.* (Paris)
Kelly, C. (2008). *Attila the Hun: Barbarian Terror and the Fall of Rome* (London)
Kelly, G. (2008). *Ammianus Marcellinus: The Allusive Historian* (Cambridge)
Kennedy, H. (2004). *The Court of the Caliphs: The Rise and Fall of Islam's Greatest Dynasty* (London)
———— (2007). *The Great Arab Conquests: How the Spread of Islam Changed the World We Live In* (London)
Keydell, R., ed. (1967). *Agathias Historiae, Corpus Fontium Historiae Byzantinae* (Berlin)
Keynes, S. (1986). 'A Tale of Two Kings: Alfred the Great and Aethelred the Unready', *Transactions of the Royal Historical Society* 36, 195–217
———— (1997). 'The Vikings in England, c.790–1016', in Sawyer (1997a), 48–81
Keynes, S. and Lapidge, M., eds. (1983). *Alfred the Great: Asser's Life of King Alfred, and Other Contemporary Sources* (London)
Khazanov, A. M. (trans. Julia Crookenden, with a foreword by Ernest Gellner) (1984). *Nomads and the Outside World* (Cambridge)
King, R., ed. (1993). *Mass Migration in Europe: The Legacy and the Future* (London)
King, R. and Oberg, S. (1993). 'Europe and the Future of Mass Migration', in King (1993), 1–4

Kivisto, P. (1989). *The Ethnic Enigma: The Salience of Ethnicity for European-Origin Groups* (Philadelphia)

Klose, J. (1934). *Roms Klientel-Randstaaten am Rhein und an der Donau: Beiträge zu ihrer Geschichte und rechtlichen Stellung im 1. und 2. Jhdt. n. Chr.* (Breslau)

Kmiecinski, J. (1968). *Odry: cmentarzysko kurhanowe z okresu rzymskiego w powiecie chojnickim* (Lódz)

Knabe, C. and Herrmann, P., eds. (1931–57). *Saxonis Gesta danorum* (Copenhagen)

Knoll, P. W. and Schaer, F. (2003). *Gesta principum Polonorum* ['The deeds of the princes of the Poles'] (Budapest)

Kobylinksi, Z. (1997). 'Settlement Structures in Central Europe at the Beginning of the Middle Ages', in Urbanczyk (1997a), 97–114

Koch, R. and Koch, U. (1996). 'Die frankische Expansion ins Main und Neckargebiet', in *Die Franken: Wegbereiter Europas: vor 1500 Jahren, König Chlodwig und seine Erben* (Mainz), 270–84

Kokowski, A. (1995). *Schätze der Ostgoten* (Stuttgart)

Kolendo, J. (1997). 'Central Europe and the Mediterranean World in the 1st–5th Centuries AD', in Urbanczyk (1997a), 5–22

Kossinna, G. (1928). *Ursprung und Verbreitung der Germanen in vor- und frühgeschichtlicher Zeit* (Leipzig)

Kostrzewski, J. (1969). 'Über den gegenwärtigen Stand der Erforschung der Ethnogenese der Slaven in archäologischer Sicht', in Zagiba (1969), 11–25

Krader, L. (1963). *Social Organization of the Mongol-Turkic Pastoral Nomads* (The Hague)

Kristiansen, K. and Paludan-Muller, C., eds. (1978). *New Directions in Scandinavian Archaeology* (Copenhagen)

Krüger, B. et al., eds. (1976–83). *Die Germanen: Geschichte und Kultur der germanischen Stämme in Mitteleuropa*, 2 vols. (Berlin)

Krusch, B. and Levison, W., eds. (1951). *Gregory of Tours: Historiae*, Monumenta Germaniae Historica: Scriptores rerum Merovingicarum 1.1 (Berlin)

Kuhn, W. (1963). 'Die Siedlerzählen der deutschen Ostsiedlung', in *Studium Sociale. Karl Valentin Muller dargebracht* (Cologne), 131–54

———— (1973). *Vergleichende Untersuchungen zur mittelalterlichen Ostsiedlung* (Cologne)

Kuhrt, D., ed. (1984). *The Politics of Return: International Return Migration in Europe* (New York)

Kulikowski, M. (2000a). 'Barbarians in Gaul, Usurpers in Britain', *Britannia* 31, 325–45

———— (2000b). 'The Notitia Dignitatum as a Historical Source', *Historia* 49, 358–77

———— (2002). 'Nation versus Army: A Necessary Contrast?', in Gillet (2002), 69–84

———— (2007). *Rome's Gothic Wars* (Cambridge)

Kunzl, E., ed. (1993). *Die Alamannenbeute aus dem Rhein bei Neupotz*, 4 vols. (Mainz)

Kurnatowska, Z. (1995). 'Frühstadische Entwicklung an den Zentrum der Piasten in Grosspolen', in Brachmann (1995), 133–48

———— (1997a). 'Territorial Structures in West Poland Prior to the Founding of the State Organization of Miesco I', in Urbanczyk (1997a), 125–36

———— (1997b). 'Die Christianisierung Polens im Lichte der Materiallen Quellen', in Urbancyzk (1997a), 101–121

Kurze, F., ed. (1895). *Annales Regni Francorum 741–829*, Monumenta Germaniae Historica: Scriptores rerum Germanicarum, vol. 6 (Hanover)

Lappenberg, J. M. and Schmeidler, B., eds. (1909). *Helmoldi presbyteri Bozoviensis Cronica Slavorum*, 2nd ed. (Hanover)

Leahy, K. and Paterson, C. (2001). 'New Light on the Viking presence in Lincolnshire: the Artefactual Evidence', in Graham-Campbell et al. (2001), 181–202

Leach, E. R. (1954). *Political Systems of Highland Burma: A Study of Kachin Social Structure* (London)

Lemerle, P. (1979–81). *Les plus anciens recueils des miracles de Saint Démétrius* (Paris)

Lenski, N. (1995). 'The Gothic Civil War and the Date of the Gothic Conversion', *Greek Roman and Byzantine Studies* 36, 51–87

———— (2002). *Failure of Empire: Valens and the Roman State in the Fourth Century AD* (Berkeley)

Levaskova, V. P. (1994). 'Agriculture in Rus', in Kaiser and Marker (1994), 39–44

Lewitt, T. (1991). *Agricultural Production in the Roman Economy AD 200–400* (Oxford)

Leyser, K. (1982). *Medieval Germany and its Neighbours, 900–1250* (London)

———— (1989). *Rule and Conflict in an Early Medieval Society: Ottonian Saxony* (Oxford)

Liapushkin, I. I. (1958). *Gorodishche Novotroitskoe* (Moscow).

Liebermann, F., ed. (1903–16). *Die Gesetze der Angelsachsen* (Halle)

Liebeschuetz. J. H. W. G. (1990). *Barbarians and Bishops: Army, Church, and State in the Age of Arcadius and Chrysostom* (Oxford)

———— (1992). 'Alaric's Goths: Nation or Army?', in Drinkwater and Elton (1992), 75–83

Likhachev, D. S. (1970). 'The Legend of the Calling-in of the Varangians', in K. Hannestad et al. (eds.), *Varangian Problems*, 178–85

Lindner, R. (1981). 'Nomadism, Huns, and Horses', *Past & Present* 92, 1–19

Lohmann, H.-E. and Hirsch, P., eds. (1935). *Widukindi monachi Corbeiensis Rerum Gestarum Saxonicarum*, Monumenta Germaniae Historica: Scriptores rerum Germanicarum, 5th ed. (Hanover)

Loseby, S. T. (1992). 'Marseille: a Late Antique Success Story?', *Journal of Roman Studies* 82, 165–85

———— (1998). 'Marseille and the Pirenne Thesis I: Gregory of Tours, the Merovingian kings and 'un grand port', in Hodges and Bowden (1998), 203–29

———— (2000). 'Urban Failures in late Antique Gaul', in Slater (2000), 72–95

Lowmianski, H. (1960). 'Economic Problems of the Early Feudal Polish State', *Acta Poloniae Historica* 3, 7–32

Loyn, H. (1995). *The Vikings in Britain* (Oxford)

Lubke, C. (1984–88). *Regesten zur Geschichte der Slaven an Elbe und Oder (vom Jahr 900 an)* (Berlin)

———— (1994). 'Zwischen Triglav und Christus: Die Anfänge der Christianisierung des Havellandes', *Wichmann-Jahrbuch des Diozesangeschichtsverein Berlin*, n.f. 3, Jahrgang 34–5, 15–35

———— (1997). 'Forms of Political Organization of the Polabian Slavs (until 10th c.)', in Urbanczyk (1997a), 115–124

Lund, N. (1984). *Two Voyagers at the Court of King Alfred: Othere and Wulfstan* (York)
———— (1986). 'The Armies of Swein Forkbeard and Cnut: leding or lith?', *Anglo-Saxon England* 15, 105–18
———— (1993). 'Danish Military Organization', in Cooper (1993), 109–26
———— (1994). 'Cnut's Danish Kingdom', in Rumble (1994), 27–42
———— (1995). 'Scandinavia', in McKitterick (1995), 202–27.
———— (1997). 'The Danish Empire and the End of the Viking Age', in Sawyer (1997a), 156–81
Lund Hansen, U. (1987). *Römischer Import im Norden: Warenaustausch zwischen dem Römischen Reich und dem freien Germanien während der Kaiserzeit unter besonderer Berücksichtigung Nordeuropas* (Copenhagen)

McCormick, M. (2001). *Origins of the European Economy: Communications and Commerce, AD 300–900* (Cambridge)
MacGeorge, P. (2002). *Late Roman Warlords* (Oxford)
McKitterick, R. (1977). *The Frankish Church and the Carolingian Reforms, 789–895* (London)
———— (1983). *The Frankish Kingdoms under the Carolingians, 751–987* (London)
———— (1989). *The Carolingians and the Written Word* (Cambridge)
————, ed. (1994). *Carolingian Culture: Emulation and Innovation* (Cambridge)
————, ed. (1995). *The New Cambridge Medieval History. Vol. II, c.700–c.900* (Cambridge)
Maenchen-Helfen, O. J. (1945). 'Huns and Hsiung-nu', *Byzantion* 17, 222–43
———— (1973). *The World of the Huns* (Berkeley)
Magie, D. (1932). *The Scriptores Historiae Augustae* (London)
Maleczynski, K., ed. (1952). *Anonima tzw. Galla Kronika* (Cracow)
Mango, C. and Scott, R. (1997). *Chronographia: The Chronicle of Theophanes Confessor* (Oxford)
Manteuffel, T., translated with an introduction by A. Gorski (1982). *The Formation of the Polish State: The Period of Ducal Rule, 963–1194* (Detroit)
Manteuffel, T. and Gieysztor, A., eds. (1968). *L'Europe aux IX.–XI. siècles aux origines des états nationaux: Actes du Colloque international sur les origines des états européens aux IX.–XI. siècles, tenu à Varsovie et Poznan du 7 au 13 septembre 1965* (Warsaw)
Markus, R. A. (1997). *Gregory the Great and his World* (Cambridge)
Mathisen, R. W. and Shanzer, D., eds. (2001). *Society and Culture in Late Roman Gaul: Revisiting the Sources* (Aldershot)
Matthews, J. F. (1970). 'Olympiodorus of Thebes and the History of the West (AD 407–425)', *Journal of Roman Studies* 60, 79–97
———— (1975). *Western Aristocracies and the Imperial Court AD 364–425* (Oxford)
———— (1989). *The Roman Empire of Ammianus* (London)
Mayr Harting, H. (1972). *The Coming of Christianity to Anglo-Saxon England* (London)
———— (1994). *Two Conversions to Christianity: The Bulgarians and the Anglo-Saxons* (Reading)
Melnikova, E. A. (1996). *The Eastern World of the Vikings* (Gothenburg)
Mierow, C. C. (1915). *Jordanes: Getica* (New York)

Minns, E. H. (1913). *Scythians and Greeks: A Survey of Ancient History and Archaeology of the North Coast of the Euxine From the Danube to the Caucasus* (Cambridge)

Minor, C. (1996). 'Bacaudae – A Reconsideration', *Traditio* 51, 297–307

Miquel, A. (1966). 'Ibn Jacub', *Annales* 23, 1048–63

Moderan, Y. (2002). 'L'établissement des Vandals en Afrique', *Antiquité Tardif* 10, 87–122

Moisl, H. (1981). 'Anglo-Saxon Royal Genealogies and Germanic Oral Tradition', *Journal of Medieval History* 7, 215–48

Mommsen, T., ed. (1882). *Jordanes Romana et Getica*, Monumenta Germaniae Historica: Auctores Antiquissimi 5.1 (Berlin)

———, ed. (1892). *Chronica Minora 1*, Monumenta Germaniae Historica: Auctores Antiquissimi 9 (Berlin)

———, ed. (1894a). *Chronica Minora 2*, Monumenta Germaniae Historica: Auctores Antiquissimi 11 (Berlin)

———, ed. (1894b). *Cassiodori Senatoris Variae*, Monumenta Germaniae Historica: Auctores Antiquissimi 12 (Berlin)

Mommsen, T. and Kreuger, P., eds. (1905). *Codex Theodosianus* (Berlin)

Moorhead, J. (1992). *Victor of Vita History of the Persecution in Africa* (Liverpool)

Moravcsik, G. and Jenkins, R. J. H., eds. (1967). *Constantine Porphyrogenitus: De administrando imperio* (Washington)

Mueller, K., ed. (1851–70). *Fragmenta Historicorum Graecorum*, vols. 4–5 (Paris)

Muller-Wille, M. (1999). *Opferkulte der Germanen und Slawen* (Darmstadt)

Myhre, B. (1978). 'Agrarian Development, Settlement History and Social Organization in Southwest Norway in the Iron Age', in Kristiansen and Paludan-Muller (1978), 224–35, 253–65

Nelson, J. L. (1991). *The Annals of St-Bertin* (Manchester)

——— (1997). 'The Frankish Empire', in Sawyer (1997a), 19–47

Nichols, J. (1998). 'The Eurasian Spread Zone and the Indo-European Dispersal', in Blench and Spriggs (1998), 220–66

Niebuhr, B. G., ed. (1839–41). *Theophanes Chronographia* (Bonn)

Nixon, C. E. V. and Rodgers, B. S. (1994). *In Praise of Later Roman Emperors: The Panegyrici Latini* (Berkeley)

Noble, T. F. X. and Head, T., eds. (1995). *Soldiers of Christ: Saints and Saints' Lives from Late Antiquity and the Early Middle Ages* (Pennsylvania)

Noll, R. and Vetter, E., eds. (1963). *Eugippius Vita Sancti Severini, Schriften und Quellen der Alten Welt* 11 (Berlin)

Noonan, T. S. (1997). 'Scandinavians in European Russia', in Sawyer (1997a), 134–55

——— (1998). *The Islamic World, Russia and the Vikings, 750–900: The Numismatic Evidence* (Aldershot)

O Corrain, D. (1997). 'Ireland, Wales, Man, and the Hebrides', in Sawyer (1997a), 83–109.

——— (1998). 'The Vikings in Scotland and Ireland in the Ninth Century', *Peritia* 12, 296–339

Ørsnes, M. (1963). 'The Weapon Find in Ejsbøl Mose at Haderlev: Preliminary Report', *Acta Archaeologica* 34, 232–48

—————— (1968). *Der Moorfund von Ejsbøl bei Hadersleben. Deutungsprobleme der grossen nordgermanischen Waffenopferfunde: Abhandlung der Akademie der Wissenschaft in Göttingen* (Göttingen)

Painter, K. (1994). 'Booty from a Roman Villa found in the Rhine', *Minerva* 5, 22–7

Palade, V. (1966). 'Atelierele pentru lucrat pieptini din os din secolul at IV-leas e.n. de la Birlad-Valea Seaca', *Arheologia Moldovei* 4, 261–77

Parczewski, M. (1993). *Die Anfänge der frühslawischen Kultur in Polen* (Vienna)

—————— (1997). 'Beginnings of the Slavs' Culture', in Urbanczyk (1997a), 79–90

Parmentier, L. and Hansen, G. C., eds. (1998). *Theodoret: Kirchengeschichte*, 3rd ed. (Berlin)

Paschoud, F. (1971–89). *Zosimus: Historia Nova*, 6 vols. (Paris)

Pearson, M. P. (1989). 'Beyond the Pale: Barbarian Social Dynamics in Western Europe', in Barrett (1989), 198–226

Peregrinatio Gothica 1. (1986). *Archaeologia Baltica* VII (Lodz)

Peregrinatio Gothica 2. (1989). *Archaeologia Baltica* VIII (Lodz)

Perin, P. (1980). *La datation des tombes mérovingiennes* (Geneva)

—————— (1987). *Les Francs à la Conquête de la Gaule* (Paris)

—————— (1996). 'Die Archäologische Zeugnisse der frankischen Expansion in Gallien', in *Die Franken: Wegbereiter Europas: vor 1500 Jahren, König Chlodwig und seine Erben* (Mainz), 227–32.

Perin, P. and Feffer, L. C. (1987). *Les Francs à l'Origine de la France*, 2 vols. (Paris)

Perin, P. and Kazanski, M. (1996). 'Das Grab Childerichs I', in *Die Franken: Wegbereiter Europas: vor 1500 Jahren, König Chlodwig und seine Erben* (Mainz), 173–82

Pertz, G. H. and Kurze, F., eds. (1891). *Annales Fuldenses, sive, Annales regni Francorum orientalis ab Einhardo, Ruodolfo, Meginhardo Fuldensibus, Seligenstadi, Fuldae, Mogontiaoi conscripti cum continuationibus Ratisbonensi et Altahensibus* (Hanover)

Petrov, N. I. (2005). 'Ladoga, Ryurik's Stronghold, and Novgorod: Fortifications and Power in Early Medieval Russia', in Curta (2005), 121–37

Petschenig, M., ed. (1881). *Victoris Episcopi Vitensis Historia persecutionis Africanae provinciae* (Vienna)

Pharr, C. (1952). *The Theodosian Code and Novels, and the Sirmondian Constitutions* (New York)

Phillips, J. R. S. (1988). *The Medieval Expansion of Europe* (Oxford)

—————— (1994). 'The Medieval Background', in Canny (1994), 9–25

Pichlmayr, F., ed. (1911). *Sexti Aurelii Victoris liber de Caesaribus* (Leipzig)

Pinder, M., ed. (1897). *Ioannis Zonarae Epitomae historiarum libri XVIII* (Bonn)

Pirling, R. (1966). *Das römisch-fränkische Gräberfeld von Krefeld-Gellep* (Berlin)

Pirling, R. and Siepen, M. (2003). *Das römisch-fränkische Gräberfeld von Krefeld-Gellep 1989–2000* (Stuttgart)

Platnauer, M. (1922). *The Works of Claudian* (London)

PLRE: The Prosopography of the later Roman Empire, ed. A. H. M. Jones et al., 3 vols. (Cambridge, 1971–92)

Pohl, W. (1980). 'Die Gepiden und die Gentes an der mittleren Donau nach dem Zerfall des Attilareiches', in Wolfram and Daim (1980), 239–305

———— (1988). *Die Awaren: ein Steppenvolk im Mitteleuropa, 567–822 n. Chr.* (Munich)

———— (2000). *Die Germanen* (Munich)

———— (2003). 'A Non-Roman Empire in Central Europe: The Avars', in Goetz et al. (2003), 571–595.

Pohl, W. and Erhart, P. (2005). *Die Langobarden : Herrschaft und Identität* (Vienna)

Pohl, W. and Wood, I. N., eds. (2001) *The Transformation of Frontiers from Late Antiquity to the Carolingians, Proceedings of the Second Plenary Conference, European Science Foundation Transformation of the Roman World Project* (Leiden)

Poulik, J. et al., eds. (1986). *Grossmähren und die Anfänge der tschechoslowakischen Staatlichkeit* (Prague)

Poulter, A., ed. (2007). *The Transition to Late Antiquity on the Danube and Beyond* (Oxford)

Powlesland, D. (1997). 'Early Anglo-Saxon Settlements: Structures, Forms, and Layout', in Hines (1997), 101–24

Preucel, R. and Hodder, I., eds. (1996). *Contemporary Archaeology in Theory* (Oxford)

Preussler, W. (1956). 'Keltischer Einfluss im Englischen', *Revue des Langues Vivantes* 22, 322–50

Prinz, F. (1984). *Böhmen im mittelalterlichen Europa* (Munich).

Proussa, P. (1990). 'A Contact-universals Origin for Periphrastic Do, with Special Consideration of OE-Celtic Contact', in Adamson (1990), 407–34

Raev, B. A. (1986). *Roman Imports in the Lower Don Basin* (Oxford)

Rafnsson, S. (1997). 'The Atlantic Islands', in Sawyer (1997a), 110–33

Randsborg, K. (1980). *The Viking Age in Denmark: The Formation of a State* (London)

Rau, G. (1972). 'Körpergräber mit Glasbeigaben des 4. nachchristlichen Jahrhunderts im Oder-Wechsel-Raum', *Acta praehistorica et archaeologica* 3, 109–214

Reichmann, C. (1996). 'Frühe Franken in Germanien', in *Die Franken: Wegbereiter Europas: vor 1500 Jahren, König Chlodwig und seine Erben* (Mainz), 55–65

Renfrew, C. (1987). *Archaeology and Language: The Puzzle of Indo-European Origins* (London)

Renfrew, C. and Bahn, P. (1991). *Archaeology: Theories, Methods and Practice* (London)

Renfrew, C. and Cherry, J. F., eds. (1986). *Peer Polity Interaction and Socio-Political Change* (Cambridge)

Reuter, T. (1985). 'Plunder and Tribute in the Carolingian Empire', *Transactions of the Royal Historical Society*, 75–94

———— (1990). 'The End of Carolingian Military Expansion', in Godman and Collins (1990), 391–407

———— (1991). *Germany in the Early Middle Ages c.800–1056* (London)

———— (1992). *The Annals of Fulda* (Manchester)

————, ed. (1999). *The New Cambridge Medieval History: vol. 3, 900–1024* (Cambridge)

Ridley R. T. (1982). *Zosimus: New History* (Canberra)

Ritchie, A. (1993). *Viking Scotland* (Bath)

Robinson, C. H. (1921). *Anskar: The Apostle of the North, 801–865* (London)

Roesdahl, E., trans. S. Margeson and K. Williams (1982). *Viking Age Denmark* (London)

————, trans. S. Margeson and K. Williams (1991). *The Vikings* (London)

Rolfe, J. C., ed. (1935–39). *Ammianus Marcellinus* (London)

Rostovzeff, M. I. (1922). *Iranians and Greeks in South Russia* (Oxford)

Rowlands, M. et al., eds. (1987). *Centre and Periphery in the Ancient World* (Cambridge)

Rudkin, D. (1986). *The Saxon Shore* (London)

Rumble, A. R., ed. (1994). *The Reign of Cnut: King of England, Denmark and Norway* (London)

Rystad, G. (1996). 'Immigration Policy and the Future of International Migration', *International Migration Review* 26.4, 555–86

Sabbah, G. (1978). *La méthode d'Ammien Marcellin: recherches sur la construction du discours historique dans les Res Gestae* (Paris)

Salt, J. and Clout, H., eds. (1976). *Migration in Post-War Europe: Geographical Essays* (Oxford)

Salway, P. (2001). *A History of Roman Britain* (Oxford)

Sanchez-Albornoz, N. (1994). 'The First Transatlantic Transfer: Spanish Migration to the New World', in Canny (1994), 26–36

Santini, C., ed. (1979). *Eutropius Breviarium ab urbe condita* (Stuttgart)

Sartre, M. (1982). *Trois études sur l'Arabie romaine et byzantine: Revue d'Études Latines* (Brussels)

Sasse, B. (1982). *Die Sozialgeschichte Böhmens in der Frühzeit: Historisch-archäologische Untersuchungen zum 9.–12. Jahrhundert* (Berlin)

Sawyer, B. (1991). 'Viking-Age Rune-Stones as a Crisis Symptom', *Norwegian Archaeological Review* 24, 97–112

Sawyer, B. and Sawyer, P. H. (1993). *Medieval Scandinavia: From Conversion to Reformation c.800–1500* (Minneapolis)

Sawyer, P. H. (1962). *The Age of the Vikings* (London)

———— (1978). *From Roman Britain to Norman England* (London)

———— (1982), 'The Causes of the Viking Age', in Farrell (1982), 1–7

———— (1993). 'The Scandinavian Background', in Cooper (1993), 33–42

———— (1994). 'Cnut's Scandinavian Empire', in Rumble (1994), 10–22

————, ed. (1997a). *The Oxford Illustrated History of the Vikings* (Oxford)

———— (1997b). 'The Age of the Vikings, and Before', in P. Sawyer (1997a), 1–18

Sawyer, P. H. and Wood, I. N., eds. (1977). *Early Medieval Kingship* (Leeds)

Schenkl, H. et al., eds. (1965–74). *Themistii Orationes* (Leipzig)

Schmeidler, B., ed. (1917). *Adam von Bremen: Gesta Hammaburgensis ecclesiae pontificum*, Monumenta Germaniae Historica: Scriptores rerum Germanicarum (Hanover)

Schmidt, L. (1933). *Geschichte der deutschen Stämme bis zum Ausgang der Völkerwanderung: Die Ostgermanen*, 2nd ed. (Munich)

Scholz, B. W. (1972). *Carolingian Chronicles* (Ann Arbor)

Schröder, F. J. (1977). *Völker und Herrscher des östlichen Europa im Weltbild Widukinds von Korvei und Thietmars von Merseburg* (Munster)

Scragg, D. (1991). *The Battle of Maldon, AD 991* (Oxford)

Searle, E. (1988). *Predatory Kinship and the Creation of Norman Power 840–1066* (Berkeley).

Seeck, O., ed. (1962). *Notitia dignitatum* (Frankfurt am Main)

Service, E. R. (1975). *Origins of the State and Civilization: The Process of Cultural Evolution* (New York)

Settimane [Centro italiano di studi sull'alto Medioevo: Settimane di studio] (1983). *Gli Slavi occidentali e meridionali nell'alto medioevo, Atti del xxx Congresso internazionale di studi sull'alto Medioevo* (Spoleto)

Shanzer, D. and Wood, I. N., eds. and trans. (2002). *Avitus of Vienne: Letters and Selected Prose* (Liverpool)

Shchukin, M. (1975). 'Das Problem der Cernjachow-Kultur in der sowjetischen archäologischen Literatur', *Zeitschrift für Archäologie* 9, 25–41

——— (1977). 'Current aspects of the Gothic Problem and the Cherniakhovo Culture', *Arkheologichesky sbornik* 18, 79–92 (English summary)

——— (1990). *Rome and the Barbarians in Central and Eastern Europe: 1st Century BC–1st Century AD* (Oxford)

——— (2005). *The Gothic Way: Goths, Rome, and the Culture of the Chernjakhov/ Sintana de Mures* (St Petersburg)

Shennan, S. (1989). *Archaeological Approaches to Cultural Identity* (London)

Shepard, J. (2005). 'Conversions and Regimes Compared: The Rus' and the Poles, ca. 1000', in Curta (2005), 252–82

Sherwin-White, A. N. (1973). *The Roman Citizenship*, 2nd ed. (Oxford)

Siegmund, F. (1998). 'Social Structure and Relations', in Wood (1998), 177–98

Sinor, D. (1977). *Inner Asia and its Contacts with Medieval Europe* (London)

——— (1990). *The Cambridge History of Early Inner Asia* (Cambridge)

Skalnik, P., ed. (1989). *Outwitting the State* (New Jersey)

Slater, T., ed. (2000). *Towns in Decline AD 100–1600* (Aldershot).

Smith, A. D. (1986). *The Ethnic Origin of Nations* (Oxford)

Smith, J. M. (1992). *Province and Empire: Brittany and the Carolingians* (Cambridge)

Smyth, A. P. (1977). *Scandinavian Kings in the British Isles, 850–880* (Oxford)

——— (1979). *Scandinavian York and Dublin: The History and Archaeology of Two Related Viking Kingdoms* (New Jersey)

——— (1995). *King Alfred the Great* (Oxford)

Speidel, M. P. (1977). 'The Roman Army in Arabia', *Aufstieg und Niedergang der Antiken Welt* II. 8

Stallknecht, B. (1969). *Untersuchungen zur römischen Aussenpolitik in der Spätantike* (Bonn)

Stenton. F. (1971). *Anglo-Saxon England* (Oxford)

Steuer, H. (1982). *Frühgeschichtliche Sozialstrukturen Mitteleuropa* (Göttingen)

——— (1998). 'Theorien zur Herkunft und Entstehung der Alemannen: Archaeologische Forschungsansatze', in Geuenich (1998), 270–334

Stockwell, S., ed. (2008). *The British Empire: Themes and Perspectives* (Oxford)

Syme, R. (1968). *Ammianus and the Historia Augusta* (Oxford)

——— (1971a). *Emperors and Biography: Studies in the Historia Augusta* (Oxford)

——— (1971b). *The Historia Augusta: A Call for Clarity* (Bonn)

Szydlowski, J. (1980). 'Zur Anwesenheit der Westslawen an der mittleren Donau im ausgehenden 5. und 6. Jahrhundert', in Wolfram and Daim (1980), 233–7

Tejral, J. et al., eds. (1999). *L'Occident romain et l'Europe centrale au début de l'époque des Grandes Migrations* (Brno)

Theuws, F. and Hiddinck, H. (1996). 'Der Kontakt zu Rom', in *Die Franken:*

Wegbereiter Europas: vor 1500 Jahren, König Chlodwig und seine Erben (Mainz), 64–80

Thomas, M. G. et al. (2006). 'Evidence for an Apartheid-like Social Structure in Early Anglo-Saxon England', *Proceedings of the Royal Society 273*, 2651–7

Thompson, E. A. (1965). *The Early Germans* (Oxford)

———— (1966). *The Visigoths in the Time of Ulfila* (Oxford)

———— (1995). *The Huns* (Oxford)

Thorpe, L. (1974). *Gregory of Tours: The History of the Franks* (London)

Todd, M. (1975). *The Northern Barbarians 100 BC–AD 300* (London)

———— (1992). *The Early Germans* (Oxford)

Trillmich, W. et al., eds. (1978). *Quellen des 9. und 11. Jahrhunderts zur Geschichte der hambürgischen Kirche und des Reiches*, 5th ed. (Darmstadt)

Tschan, F. J. (1959). *Adam of Bremen: History of the Archbishops of Hamburg-Bremen* (New York)

———— (1966). *Helmold, Priest of Bosau: The Chronicle of the Slavs* (New York)

Turek, R. (1974). *Böhmen im Morgengrauen der Geschichte von den Anfängen der slawischen Besiedlung bis zum Eintritt in die europäische Kulturgemeinschaft (6. bis Ende des 10 Jahrhunderts)* (Wiesbaden)

Ucko, P. 1995). *Theory in Archaeology: A World Perspective* (London)

Urbanczyk, P., ed. (1997a). *Origins of Central Europe* (Warsaw)

———— (1997b). 'Changes of Power Structure During the 1st Millennium AD in the Northern Part of Central Poland', in Urbanczyk (1997a), 39–44

————, ed. (1997c). *Early Christianity in Central and East Europe* (Warsaw)

————, ed. (2001). *Europe around the Year 1000* (Warsaw)

————, ed. (2004). *Polish Lands at the Turn of the First and the Second Millennia* (Warsaw)

———— (2005). 'Early State Formation in East Central Europe', in Curta (2005), 139–51

Vallet, F and Kazanski, M., eds. (1995). *La noblesse romaine et les chefs barbares du IIIe au VIIe siècle* (Paris)

Van Dam, R. (1985). *Leadership and Community in Late Antique Gaul* (Berkeley)

Van Es, W. A. (1967). *Wijster: A Native Village Beyond the Imperial Frontier 150–425 AD* (Groningen)

Van Ossel, P. (1992). *Etablissements ruraux de l'antiquité tardive dans le nord de la Gaule* (Paris)

Van Ossel, P. and Ouzoulias, P. (2000). 'Rural Settlement Economy in Northern Gaul in the Late Empire: An Overview and Assessment', *Journal of Roman Archaeology 13*, 133–60

Verlinden, C. (1955). *L'Esclavage dans l'Europe Médiévale* (Bruges)

Vertovec, S. and Cohen, R. (1999). *Migration, Diasporas, and Transnationalism* (Cheltenham)

Vince, A. (2001). 'Lincoln in the Viking Age', in Graham-Campbell et al. (2001), 157–80

Voets, S. et al., eds. (1995). *The Demographic Consequences of Migration* (The Hague)

Vogel, F., ed. (1885). *Ennodius Opera*, Monumenta Germaniae Historica: Auctores Antiquissimi 7 (Berlin)
von Schnurbein, S. (1995). *Vom Einfluss Roms auf die Germanen* (Opladen)

Waitz, G., ed. (1883). *Annales Bertiniani* (Hanover)
Wallace-Hadrill, J. M. (1960). *The Fourth Book of the Chronicle of Fredegar* (London)
Ward Perkins, B. (2000). 'Why did the Anglo-Saxons not become more British?', *English Historical Review* 115, 513–33
—————— (2005). *The Fall of Rome and the End of Civilization* (Oxford)
Warner, D. A. (2001). *Ottonian Germany: The Chronicon of Thietmar of Merseburg* (Manchester)
Weale, M. E. et al. (2002). 'Y Chromosome Evidence for Anglo-Saxon Mass Migration' *Molecular Biology & Evolution* 19.7, 1008–21
Wegener, W. (1959). *Böhmen/Mähren und das Reich im Hochmittelälter* (Cologne)
Welch, M. (1992). *English Heritage Book of Anglo-Saxon England* (London)
Wells, P. S. (1999). *The Barbarians Speak: How the Conquered Peoples Shaped Roman Europe* (Princeton)
Wenskus, R. (1961). *Stammesbildung und Verfassung: Das Werden der frühmittelalterlichen gentes* (Cologne)
Werner, J. (1935). *Münzdatierte austrasische Grabfunde* (Berlin)
—————— (1950). 'Zur Entstehung der Reihengräberzivilisation', *Archaeologica Geographica* I, 23–32
Weski, T. (1982). *Waffen in germanischen Gräbern der älteren römischen Kaiserzeit südlich der Ostsee* (Oxford)
Whitby, L. M. (1988). *The Emperor Maurice and his Historian: Theophylact Simocatta on Persian and Balkan Warfare* (Oxford)
Whitby, L. M. and Whitby, J. M. (1986). *The History of Theophylact Simocatta* (Oxford)
—————— (1989). *The Chronicon Paschale* (Liverpool)
Whitelock, D., ed. (1955). *English Historical Documents, vol. 1, c.500–1042* (London)
Whitelock, D. et al. (1961). *The Anglo-Saxon Chronicle: A Revised Translation* (London)
Whittaker, C. R. (1994). *Frontiers of the Roman Empire: A Social and Economic Study* (Baltimore).
Whittow, M. (1996). *The Making of Orthodox Byzantium, 600–1025* (London)
Wickham, C. (1992). 'Problems of Comparing Rural Societies in Early Medieval Western Europe', *Transactions of the Royal Historical Society*, 6th series 2, 221–46
—————— (2005). *Framing the Early Middle Ages: Europe and the Mediterranean, 400–800* (Oxford)
Wieczorek, A. (1996). 'Die Ausbreitung der frankischen Herrschaft in den Rheinland vor und seit Chlodwig I', in *Die Franken: Wegbereiter Europas: vor 1500 Jahren, König Chlodwig und seine Erben* (Mainz), 241–61
Wiet, G. (1957). *Les Atours Précieux* (Cairo)
Williams, A. (1991). *The English and the Norman Conquest* (Woodbridge)
Wilson, J. (2008). 'Agency, Narrative, and Resistance', in Stockwell (2008), 245–68
Winterbottom, M., ed. and trans. (1978). *Gildas: The Ruin of Britain, and other works* (London)
Wolfram, H. (1979). *Conversio Bagoariorum et Carantanorum: das Weissbuch der*

Salzburger Kirche über die erfolgreiche Mission in Karantanien und Pannonien (Vienna)

——— (1985). *Treasures on the Danube: Barbarian Invaders and their Roman Inheritance* (Vienna)

———, trans. T. J. Dunlap (1988). *History of the Goths* (Berkeley)

——— (1994). 'Origo et Religio: Ethnic Traditions and Literature in Early Medieval Texts', *Early Medieval Europe* 3, 19–38

——— (1995). *Salzburg, Bayern, Österreich: die Conversio Bagoariorum et Carantanorum und die Quellen ihrer Zeit* (Vienna)

———, trans. T. J. Dunlap (1997). *The Roman Empire and its Germanic Peoples* (Berkeley)

Wolfram, H. and Daim, F., eds. (1980). *Die Völker an der mittleren und unteren Donau im fünften und sechsten Jahrhundert, Denkschriften der Österreichischen Akademie der Wissenschaften*, phil.-hist. Kl. 145 (Vienna)

Wolfram, H. and Pohl, W., eds. (1990). *Typen der Ethnogenese unter besonderer Berücksichtigung der Bayern, Denkschriften der Österreichischen Akademie der Wissenschaften*, phil.-hist. Kl. 193 (Vienna)

Wood, I. N. (1985). 'Gregory of Tours and Clovis', *Revue Belge de Philologie et d'Histoire* 63, 249–72

——— (1990). 'Ethnicity and the Ethnogenesis of the Burgundians' in Wolfram and Pohl (1990), 53–69

——— (1994). *The Merovingian Kingdoms* (London)

———, ed. (1998). *Franks and Alamanni in the Merovingian Period: An Ethnographic Perspective* (Woodbridge)

——— (2001). *The Missionary Life: Saints and the Evangelization of Europe, 400–1050* (Harlow)

Woolf, A. (2003). 'The Britons: from Romans to Barbarians', in Goetz et al. (2003), 345–80

——— (2007). 'Apartheid and Economics in Anglo-Saxon England', in Higham (2007), 115–29

Wormald, C. P. (1978). 'Bede, Beowulf and the Conversion of the Anglo-Saxon Aristocracy', in Farrell (1978), 32–95

——— (1982). 'Viking Studies: Whence and Whither', in Farrell (1982), 128–53

——— (1986). 'Celtic and Anglo-Saxon Kingship', *Studies in Medieval Culture* 20, 151–83

Zagiba, F., ed. (1969). *Das heidnische und christliche Slaventum* (Wiesbaden)

Zollner, E. (1970). *Geschichte der Franken bis zur Mitte des sechsten Jahrhunderts* (Munich)

INDEX